BEHAVIOR MATTERS FOR CATS AND DOGS

Also by Frania Shelley-Grielen:

Cats and Dogs, Living with and Looking at Companion Animals from their Point of View

Confessions of a Bed and Breakfast Diva, Hospitality Lessons from the Other Side of the Desk

Behavior Matters for Cats and Dogs

Frania Shelley-Grielen

Cover Design by Frania Shelley-Grielen

Cover Illustration by Vincent Apollo

ISBN 9798218428532 (Paperback)

and welfare with cats and dogs, and this book provides up-to-date insights on many current topics of concern. I have to admit that I did not know the meaning of "pica" for cats until reading about it in this book (an atypical desire for eating substances not normally eaten). I am not very experienced with caring for cats, but having brought a new rescue dog ("Simba") into our family over the past year, I delved straight into sections such as the ones on dealing with separation anxiety and resource guarding. I have a keen interest in bioacoustics and animal communication in general, and it was also really interesting to see a section on "what is your dog trying to say." Finally, I was pleased to read the section dealing with shock collars for dogs, and I wholeheartedly agree that they are at odds with an empathetic approach to building a positive relationship with dogs and should never be used. Much of my own research is with livestock such as goats and cattle, and I wish a similar approach of banning the use of shock collars could be taken. Overall, I believe that we should always try our very best to find ways to work with and understand our animals in positive ways, whether they are cats, dogs, or livestock, as they are all similarly sentient."

- **Dr. Alan McElligott**, Centre for Animal Health and Welfare, Jockey Club College of Veterinary Medicine and Life Sciences, City University of Hong Kong.

For all the cats and all the dogs: your behavior matters,
ours too

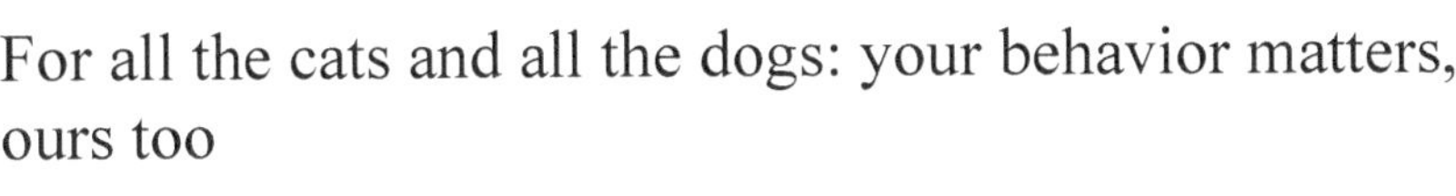

*"Pets enrich our lives and those of our children. We admire the tiger
not only for its fearful symmetry but as a symbol of freedom itself, so
we offer it rather more freedom than we would think fit for the
chicken. It is impossible, however, to avoid the issue that both the
chicken and the tiger are living on our terms."* – **John Webster**

TABLE OF CONTENTS

INTRODUCTION

*Behaviour n. the physical activity of an organism, including overt bodily movements and internal glandular and other physiological processes, constituting the sum total of the organism's physical *responses to it's environment. The term also denotes the specific physical responses of an organism to particular *stimuli or classes of stimuli. US* behavior *- Colman (2006)*

We would use the word behaviour for both these extremes, and for many other activities in between in complexity. It will include all types of activities in which animals engage, such as locomotion, grooming, reproduction, caring for young, communication, etc. Behaviour may involve one individual reacting to a stimulus or a physiological change, but may also involve two individuals, each responding to the activities of the other. And why stop there? We would also call it behaviour when animals in a herd or an aggregation coordinate their activities or compete for resources with one another. No wonder ethology is such a complex science, when the phenomena we study is so disparate. -Jensen (2009)

In essence, humans see the world as being 'in front', and we move 'into' it…In essence, birds probably see the world as 'around them' and they move 'through' it.- Martin (2011)

Why read a book about behavior matters for cats and dogs? For the dog that cannot be left alone without panicking, turned positively phobic, with full on, anxiety fueled, incessant barking? So much barking, that the neighbors leave well-meaning notes under your door or call the management company? Or the cat that overly stresses to the point of wetting on your bed or otherwise eliminating on the clothes on the floor? Or the dog that fearfully or ferociously lunges at breakneck speed at every other dog on the street but wants to smother any strange human with kisses? Or the cat that totally, no holds barred, beyond hates the very cat/kitten/puppy "new friend" you brought home to them so they would not feel lonely? Or the dog that guards everything, from people, to food bowls, to used tissues, no matter what, anything, just as long as someone else is looking at it? Or the cat that meets every misplaced, to their mind, caress with a scratch or bite in return? That? Or for not just the "how to fix," but the "how come's," and why they matter?

My first book on behavior was written for those who asked, wondered, and wanted to know more about what the science might say about their pets' behavior from the cat and dog point of view along with how to manage and modify those behaviors in meaningful response. This second book goes further in depth, covers more science on matters of behavior; training approaches and protocols, resource guarding, separation anxiety, mounting, deeper dives in multi cat households, bioacoustics, pica, scratching, canine and feline aggression, pet care, loss, research applications, interventions and more. Starting with the necessary sections on the basics of natural history, sociality and communication for each species, the book is formatted to be the most useful for readers – separate chapters, that can stand alone, on individual behavior issues.

Much of the time, where animal behavior does not meet human expectations, at issue is our own human predisposition to assume that what is meaningful to our species, is the standard from which all things need to be measured. What we term "aggression" and "unacceptable elimination" can be straightforward transmissions of information or orchestrated persuasive sallies or responses in cat or dog societies but not necessarily welcome or understood in our homes. Such misunderstandings can be dangerous for the cat voiding stress related pheromones, asking for space with hissing or the dog using growling as a warning. Nuanced animal communications or actions taken as distance increasing behaviors meant to avoid further confrontation are often misinterpreted as adversarial to our way of thinking. Human propensity to mistakenly view such behaviors as "challenges" can deprive animals of the possibility of avoiding conflict and escalate aggression in their efforts to defend themselves.

Angry retorts and advances in reply to the dog growling or the cat hissing is exactly not what the cat or dog is asking for when those signals or messages to stop forward motion were communicated. Crossed wires on both ends. None of this ends well; behavior issues, namely "aggression" and "unacceptable elimination," are top reasons for surrender at animal shelters.

Our human manners of posturing and tacit conventions for conflict avoidance are unique to our species and idiosyncratic to individual cultures, but avoiding fight, keeps all animals safe to see another day. Nature may have given a necessary collective language to all living beings when it comes to alerting to possible danger. Take guttural growls and yowls, larger and rigid postures, rapid forward movements; all unmistakable threats and calls for retreats to safety and not necessarily a call to arms. Each and every species has a suite of ritualized display behaviors that signal the need or desire to increase or reduce distance. The universal idea seems to be that a good bluff is the best offense and the best defense. No one really wants to go there, injured animals are compromised in their efforts to hunt, forage, mate, find shelter or carry on very well the business of living. For humans along with the rest of the animal kingdom, deference and space is what keeps the kingdom a peaceable one. Behavior read or unread, sets the rules of engagement for all species. Apotreptic or epitreptic, in response to, or to influence another. Equally significant for humans and just as much, for cats and dogs. Still, what is significant to us as humans means one thing to us and another to dogs and cats.

Context can be everything for behavior matters. That basic bodily function, elimination, can be quick to be seen in Western cultures as carrying ulterior motives in placement, beyond a biological necessity. Intended signaling or communication around such deposits are thought to purposefully offend. Such thinking, with a necessary basis in sanitary concerns, can associate transgression to bodily functions relating to urine and feces, as if cats and dogs have something insulting or angry to say when relieving themselves in our homes.

What is being communicated in such leavings for cats and dogs, is not of the sort we might imagine, it is the more useful kind. Certain components of odor and pheromones ("chemosignals") we miss, carry untold data; who I am, what I ate, my availability to mate, what belongs to me, how stressed or not I am. So many more messages we have yet to uncover, and at times we cannot even picture, lacking the essential Vomeronasal or Jacobson's organ, that cats, dogs and other animals but not humans have, to break down pheromone streams. This possibly adds new and untold meaning to the other reasons that might exist for the cat not using the litter box, aside from it being covered, overloaded and in need of cleaning. While history and learning can never be discounted in anything an animal does, agendas are mostly human concerns. Non-human animals are perhaps more immediate in their responses to environment.

Ludwig Wittgenstein's famous line: "If a lion could talk, we wouldn't be able to understand it," can be taken to mean that lion speak might be total gibberish to us. But not total. Some, of what the lion, sentient, and mammalian, communicated as important, must overlap as equally important for sentient human mammals sharing the same ecosystem. Still, what can we know is of value to lions, beyond those interfaces we meet on? How many such mysteries are there that we cannot even relate to in the unknown worlds of meaning in what they dream of in private, smell, taste, hear or see with sensory powers only possessed in the such separate world of lions?

Our studies often seek to ascertain how much non-human animals can learn that we can teach them and less about their own worlds of being or umwelts. Learn what we say, in the language we use, and our own rules of acceptable behavior to adhere to. We create language boards and devices so they might communicate to us about the things we think matter. What might they say with their own individual, salient meanings, if they picked the buttons to push? Absent our biases, interpretations, manipulations, and interventions, what have we really learned that they might say independently? Could we even recognize it if we heard it?

We can never get in anyone's head, human, or non-human. We can study the mechanisms of sensation and perception for different animals to better understand the hardware, but knowing how to run the software is an almost impossible task as a human. The world we move forward into and occupy, the very essence of its surrounds and resources, shifts and varies according to the limitations and expanses of senses and needs of disparate species. How do we, can we, comprehend the very meaning of something, when we cannot perceive it? See hues, shades, and colors, lost to us in the full spectrum of light? Find order, messages, and missives in the cacophony of sounds we cannot discern in frequencies unheard? Find the vast universes of scent in a room or on the street or in the wild? Realize a natural world lived without borders and its demands, dangers, and satisfactions of foraging, hunting, roosting, hiding, denning, or mating?

Senses take us into the world outside us, defining its limits and reaches, but it has always been our own we imagine that trump others. Our most undeveloped sense, smell, is the dog's greatest and the cat's second after hearing. Olfaction is the canine's primary way to process information, hearing for cats, sight is ours. How are we all even watching the same movie?

It is difficult to progress in our understanding of disparate species when most of what we are looking at is more about us rather than them. Western cultures have strong fundamental beliefs as to the place of animals in society. Beyond property, which they are considered legally, animals are deprived of agency, not just by operation of law but a tacit acceptance that they somehow do not deserve it, needing to be and existing for human conveniences. Training animals and modifying behaviors have strong roots in authoritarian methods. Such legacy also corresponds with a history of opposing thought. Not as enduring or commercially viable, considering animal intelligence, sociality and even soul on par with humans, dates back to Aristotle's writing on animals in ancient Greece, along with human animal interactions beyond the five freedoms as they relate to welfare:

> Recognizing farmed animals to be purposeful beings, his framework goes well beyond negative theories of welfare as freedom from harms. Aristotle also shows that stockpersons have a key role in identifying and promoting positive welfare states." (Grumett, 2019)

Xenophon, a contemporary of Aristotle, writing on horses, where coercion is routine in human interactions, also takes a higher approach, the first treatise on training and riding horses stressing gentle handling:

> "The one great precept and practice in using a horse is this, - never deal with him when you are in a fit of passion. A fit of passion is a thing that has no foresight in it, and so we often have to rue the day when we gave way to it. Consequently,

> when your horse shies at an object and is unwilling to go up to it, he should be shown that there is nothing fearful in it, least of all to a courageous horse like him; but if this fails, touch the object yourself that seems so dreadful to him, and lead him up to it with gentleness. Compulsion and blows only inspire the more fear; for when horses are at all hurt at such a time, they think that what they shied at it is the cause of the hurt." (Xenophon, 355 BC)

The use of excessive force is becoming more scrutinized in equestrian sport but present-day horse training and handling continues to be overwhelmingly punishment and force based.

Punishment and force, as cruel, inhumane and trust destroying as they are, persist and endure because they have immediate effect. Might may not make right but it will make you do something. And it continues to have its advocates, for the results and for the ego. Even from those we might believe due to education and training to be more impartial. It is easy to find science writers and veterinarians who are still publishing and advising spraying water or sounding air horns on cats to control behavior. (Horwitz & Landsberg, nd), (Fauzia, M. (2023). Scruffing too, multiple professional organizations have released positions against the practice as inhumane but the quick and easy result of an immobilized cat often takes precedence over welfare concerns and/or learning alternate approaches as seen and routinely demonstrated by multiple professionals and influencers on viral videos and more.

Without punishment for the quick fix, finesse in handling, training and behavior modification take skill and time. Carving space out for either, can be a detraction for not just the owners, care takers and guardians. Even more so, time counts for those in the business. For those of us paid to make a difference in behavior matters. I have heard often in the back rooms of the groomers, shelters, doggy daycares, at the lunch and cocktail hours at veterinary behavior conferences, the same phrase repeated, over, and over: "I don't have time for behavior."

Whichever professional will not take the time for behavior for a pet, another gets to take up that slack, often at a shelter or a veterinarian's office that pet does not get to walk out of. Such time is not only begrudged for animal behavior.

Clients can be in the same sights for service providers not taking time for their own very human behavior. There is a growing trend on social media, on the Twitter (now X) and Facebook feeds, to gleefully post disdain, shame, and ridicule clients for their ignorance or high emotions concerning the pets they are devoted to, all from the professionals they have paid to see. Such public derisions are not openly shared or seen where clinicians work directly with human patients, yet seem acceptable to providers working with human clients caring enough to bring animal patients for paid services. These behaviors of scorn are themselves shameful and even more so coming from professionals being consulted for help.

That clients who come looking for assistance to veterinarians, trainers, behaviorists, or groomers are the ones to be lauded for seeking out the help of those they perceive as experts, not ridiculed. We ask always to rule out the medical before questioning the behavioral or the training. The questions asked are to be better informed to benefit animals, the sick patient brought in hopes of healing, the matted to be shorn and groomed, the stressed, anxious, and reactive to be helped. The scorn we harbor belongs to those persons the professionals never get to meet or see. The ones who never pay us to care because they don't either. We are better to practice all that +R, low-stress, fear-free, force-free and relationship-centered practices with each other first. Treat clients who do care with kindness. It is the ones who never become clients or look for the right answers, the stories of the ones that we will never get to hear, that are to be chastened. A chance we will never get and neither will their animals.

The internet is one hell of a drug and the animals in our lives can often be found in the center of it. Viral likes and sixty second videos are more about vanity and validation and little about welfare or how the animal themselves perceive their world, and what is being done to them in it. Everybody, even machines, can now publish anything online, or at least most things. The focus and consideration, we can, and do spend, to so much content, on so many little screens is limited by form, substance, and withering attention spans.

There is true and virtual avalanche of information out there about companion animals, perhaps too much of it, self-interpretations, misguided explanations, with little basis in fact or science. A lot of it, not so good, and some of it, downright dangerous to put into practice.

There are justifications rife with confirmation bias or engineered solely to create a need to sell you something. "Saw it online" is not enough of a credential to rely on, a caveat that sounds even louder with AI glitches. The web isn't going anywhere, bigger and bolder every day. Determining trustworthy sources; asking who stands to benefit, who is doing the writing or posting and why, is a learned and valuable ability to acquire not just for trusting how the material impacts animal welfare but how we are made to think about it.

Research confirms there is a reluctance to spend money on professional advice on animal behavior offered in person, less hesitancy to buy a book and finish it, with people most likely to search for advice that comes with no cost, online or in person. There are thousands of posts, hundreds of pages and groups that tell of separate communities, threads, all manners of information exchanges, of hours upon hours spent on Dr. Google and Reddit, all of which tells a story of how much we would like to know about animals, if it is delivered in a way that is palatable to us, without hoops to jump through, of dollars spent, or differences in opinion.

A note on style. I use the word "pet" throughout the book. I do know that these are our companion animals, more correctly termed in the literature, and in our lives. "Pet" is shorter, more readily familiar and accepted, the outdated Victorian overtones of cossetting may not follow as strongly as we might imagine, and the conventions of living language and animal loving families unify a more current meaning in the word in its everyday connotation.

Other readings on behavior and training delve into fine tuning distinctions and digressions into learning theory. Here, basics as to learning theory is covered as it relates to training and modifying behaviors. Understanding the fundamentals is essential and where we can effectively proceed from, as long the grasp is solid. I have not seen, in years of teaching students how to work professionally with pets, or clients to modify behaviors or train companion animals, a lack of the basic premises of learning theory. We seem to get this. Associative learning and operant conditioning, are not necessarily hard to assimilate or apply concepts. Understanding another species' perceptual world along with timing is harder to grasp. We can more easily comprehend what markers or secondary reinforcers are, than be able to execute them without pausing, to stop and wait for some other unknown event to signal when we should execute them.

Timing confounds with all its applications in manipulations of distance, duration, and intensity. And is further complicated by the human time clocks we place on non-human animal processes. Changes in behavior take the time the species and individuals need for them to take hold, often far longer than we would like, or allow for. Timing also affects our difficulties with rates of reinforcement as well, along with a parsimony with rewards (cultural concerns with who should be doing the listening and why get into the mix). We pause and pause too long and often, where and when we need to continue in the mechanics of association. We wait when we need to stay in sync with the process.

Body language of different species can be hard for us to learn and further complicated by individual differences needing consideration. Skill in behavior modification depends heavily on sensitivity to environmental shifts, history, and individual thresholds, which can be even more difficult for us to recognize and to measure. Pushing close to or past thresholds can undo and further damage progress and increase trauma. We can, given due careful attention, modify behavior but never "cure" it. No matter how much we would like it in the now, change is a process with its own schedule, aided yes, by the efforts and consideration we supply, to keep, and not a singular event.

We amplify lives worth living for ourselves and for those animals around us by making behavior matter, human and non-human.

<u>References</u>

- Colman, A.M. (2006). *A Dictionary of Psychology*. New York, Oxford University Press

- Fauzia, M. (2023) *Inverse*. Why does my cat ignore it's scratching post? A vet offers and ideal solution. Retrieved September 2, 2023 from https://www.inverse.com/science/train-my-cat-use-scratching-post-ideal-solution

- Grumett, D. (2019) Aristotle's Ethics and Farm Animal Welfare. *Journal Agricultural and Environmental Ethics* **32**, 321–333. https://doi.org/10.1007/s10806-019-09776-1

- Horwitz, D. & Landsberg, G. (n.d.) *VCA Hospitals*. Preventing and Punishing Problem Behavior in Cats. Retrieved September 22, 2020 from https://vcahospitals.com/know-your-pet/preventing-and-punishing-undesirable-behavior-in-cats

- Jensen, P. (2009). *The Ethology of Domestic Animals, 2nd Edition. An Introductory Text.* Oxfordshire, CABI

- Martin, G. (2011). Understanding bird collisions with man-made objects: A sensory ecology approach. *Ibis*. 153. 239 - 254. 10.1111/j.1474-919X.2011.01117.x.

- Sandøe, P., Palmer, C., Corr, S., Springer, S., Lund, T. B. (2023) Do people really care less about their cats than about their dogs? A comparative study in three European countries. *Frontiers in Veterinary Science* 10, 2297-1769, 10.3389/fvets.2023.1237547

- Shore, E. R., Burdsal, C., & Douglas, D. K. (2008). Pet owners' views of pet behavior problems and willingness to consult experts for assistance. *Journal of Applied Animal Welfare Science*, *11*(1), 63-73.

- Xenophon, (1894) *The Art of Horsemanship* (Morgan, M.H., Translator). (Original work published 355 BC) London, J.A. Allen and Company Limited

CATS

When it comes to domesticated animals, horses, dogs, and cats changed the world for us. Farming, exploration, travel, war and more all look immensely different with animal support. Western cultures are a mostly dog centric society, with cats having more of a divided appeal for modern day humans, aside from the utilitarian. Not to be outranked but perhaps underappreciated, we value what cats do for us, what they did for us when we started to store grain and foodstuffs, keeping warehouses, wharves and ships secure for stored safekeeping. Bodega, barn yard cats and others are still on the job. And so are our house cats.

Agrarian society depended in no small way on cats to continue from season to season. Cats changed the world for us too. Humans were no match for enterprising rodents competing for those stored resources for voyages, overwintering and next year's plantings. Admiration for who might best them came at a distance. Stealthy and independent hunters of vermin might lack the appeal of the dog's easy affection for people. Research on popular sentiment on dislike of cats shows 17.4% of Americans agreeing on such disdain, where only 2.6% dislike dogs. We imagine dogs to be more human like somehow and cats to be more alien. Dogs are likened as man's best friend and cats the familiar of witches and lonely older women. Neither checks out. Both species coevolutionary journeys have changed them and how they interact with us. Less can be said for the humans. And where else can you find a show about modifying environment and behavior to help humans and their cats with positive methods called "My Cat from Hell" while a show with forceful and negative training methods is called "The Dog Whisperer"?

Fair to point out, is that our own attention to the dog has influenced much of this affiliation and affection. The consistency of close and positive association with our dogs in interactions; conversation, training, multiple daily walks, games, romps, teamed efforts, add up in shoring connections with canine companions. Where and when we do focus similarly on our cats, rapport and relationship, can emerge more easily. Cats are social animals who respond to affiliates and seek connection. Cats can be trained also, most any animal can. Learning theory applies equally across species along with requisite differences in application for individual and species-specific protocols. Equally, fair to point out 'though, for many a cat, place matters most. Devotion to home and territory count. They can be more devoted to our homes and its comforts than to us. When we move house, some cats might be just as happy if they got to stay and some other nice people moved in to take our place.

With, and in addition to, the dog lovers among us, are numbers of devoted cat lovers. Enthralled with the exotic, the other, the fearful symmetry, a devotion based in awe. How to coexist in such incongruence? In a world where there are cat people and dog people, why does the love of one need to cancel out the love of the other? When, why and how does liking cats equal the negatives? Even the veterinarians can take issue with cats. A vet practice owner highlights the bias reflecting on efforts to hire another vet:

> "I was surprised to learn most of them regard cats as pariahs of the animal kingdom. It seemed that the majority of them who were interested in companion animal practice wanted to work with dogs. Apparently, I missed the veterinary school class in which cats are branded as vicious and mean."

More widely accepted research now shows cats as social, sensitive, sharing in maternal caregiving, friendly in turn, viewing their own human caretakers as familial and affiliates but often popularly and unfairly branded as fractious, solitary and disposable. That is unless you know them.

Cats were the first family pet I have the strongest memories of as a little girl growing up in New York City. Before cats, are vague memories of transient pets- counting: a bird not meant for babies' grabbing that did not survive it, hamsters that escaped from cages not adequate to contain them and not ever to be found again, those too small turtles; the ones that are barely an inch or two, sold in certain parts of town and pet shops that no one should be keeping anyway. And later a dog that did stay in the family for years, and who I cannot recall leaving and why. But the cats were always around.

My father was the cat lover and tabby cats were his favorite. That love for cats was a good thing for his children to share. We were probably not the best for the cats however. I now know, children or even the adults, in the family did not pet or hold the cats or kittens properly. Whiskers got cut short for no good reason and one cat's "accidental" fall out a six-story window (the cat somehow survived) was no accident.

My parents, typical of their generation and beyond, had a basic understanding of how to care for cats with definite feelings on what was appropriate for animals in the home. Neutering was not one of them. Keeping them inside was, for as long as we lived in the city. When we moved outside of town to a house in the suburbs, those cats became indoor-outdoor cats.

Unneutered with outdoor access means more cats and kittens. Kittens which were not adopted out, that were instead, set "free." Such a practice is a terrible, terrible thing to do. As children, we knew that then and this remains. We would protest to my father, plead for a different ending, to no avail, there was no different ending. My father's response was this was "giving the kittens a chance." That this chance was to slow starvation, predation and sickness were forbidden topics. No matter whether we liked it or not, we were too young and too powerless to change this.

Two of those kittens somehow had a different fate. Why those two cats got to stay around our house and were not dumped, I could not say. They had grey coats, one long haired and one shorter, both with piercing green eyes. I called them Julius and Cesar and I was smitten with them. They were not allowed in the house but I would spend as much time as I could around them and so for them, became a somewhat trusted presence. A high school friend of mine mentioned needing cats for a mouse problem at her home. I knew that Julius and Cesar would be the perfect candidates and they would find a home in the bargain.

Julius and Cesar were still feral at that point. Most humans, aside from feeding, had little social relevance as far as they were concerned.

But they were young, selectively approachable and I was agile enough to be able to catch them. I would bring each one in turn inside the house to my room and despite their initial struggles would hold them close and stroke their heads. Such a method, and one I would not recommend today, worked. So called now, and having a moment among rescuers as "forced love socialization," is basically flooding. The animal ends up submitting to interactions due to "learned helplessness," they are left no choice.

Flooding overwhelms and incapacitates, leaving the animal helpless in defense. Initial reactions of compliance do not displace the fear and trauma of the process and the animal is left to still work through negative associations of loss of choice and control. Subjecting the animal to the method has the often-desired quicker response, force works that way. Long term, the effects can be supplanted by fear and displacement behaviors. Providing resources, removing force and gradual desensitization, counter-conditioning and socialization can help the animal going forward. For Julius and Cesar, it happened, they tamed to the touch, to being inside, got rehomed, and went to live long lives as house cats after that.

Years have passed, and with them, cat after cat has passed through my life, each as they go leaving large and gaping holes in the fabric of days, of when to feed, who to come home to, bowls left empty, toys untouched, vacant spots uncurled upon on chairs, bed, windowsill and in heart. That rending decision of when to call time, has never gotten less wrenching, easier, no matter how long, no matter how much I have learned. No metric, no scale, measures what is in their own hearts and heads. There is never, ever, enough time with them. None can take the place of another or console in what we might have done differently for each. But we try.

Not long ago, I adopted street kittens, two littermates, a brother and sister and a third, older hard luck female kitten. We advocate for bringing in bonded siblings, keeping families together, and in doing so, taking the pressure off our older original cats with two kittens as opposed to just one. The thinking is, kittens have each other to match development, play styles and activity budgets. Good reasoning, as long as attention paid to the original cat is not overly impacted.

Along with the necessary considerations, can come the unintended consequences of certain details. Namely, bringing in intact juveniles who want to be together all the time as they mature and the resulting complications, as one goes into the early heat street cats are prone to while waiting for neutering. Sex hormones are integral to physical development and as such are beneficial to health, they also are integral to reproduction. That's the fine line and subject of other points not discussed here.

Responsible pet owners spay and neuter, and they are subject to the simultaneous growth of the animals and provider's schedules, early heat or not. We had to wait months for an appointment. "Watch her," was the caveat I heard when I asked for an earlier date. And these are not the only schedules that count. Estrus goes in cycles, ten days or so on and then off again. It is a lot of work for the body to be on such a heightened state all the time. There is no choice in either feelings felt or behavior for the cat in season, biology drives the process. The fevered announcements of reproductive status on the ready does not present as pleasurable for female cats, there is a lot of deeply plaintive posturing and calling. And no possible relief available for my little rescue. Brother and sister were separated despite the deep desperation in the manifestations of the need of that little, still a kitten, body. Best on offer instead, in response to cycles rising, were human comfort, play and valerian root, all not enough. For that female kitten, mournful meows and calls went unanswered, no doubt deeply unhappy, but in the end, no literal baby momma.

Cat distribution system at work. Plagued by the vagaries and whims of adopters, the pandemic's severe kitten explosion exacerbated by the withdrawal of most, if not all, available sterilization, each part of what led me to capitulating on the request to take on yet another cat to join in who got to become part of the family. Further fueled by the guilty realization, that as adorable as the charismatic baby brother and sister were, this other thin, and ungainly oriental tuxedo cross would be no one's first or last pick. She was an older street feral kitten, traumatized by the harsh realities of urban street life; food scarcity, severe illness, and adverse human impacts. The last survivor of her litter decimated by feline panleukopenia, with resulting mild cerebellar hypoplasia (CH), she had endured and displayed; the trauma of capture and over restraint, unwanted and at times harsh handling, lack of socialization to humans, extended and overly protracted cage life in less-than-ideal or adequate space for equally less, than ideal or adequate extended recovery post neuter.

The legacy of such history is indelible. Her wobbliness is most evident in certain failed jumps, splayed digits anchoring her holding in place, teetering in tight circling, and learning delays. Her time processing redirection or responding to alternate stimuli seems to have its own circling. A lot, and a lot, and then more, of repetition is required. Impacts to development or damage to the cerebellum from CH, can affect more than balance, learning new skills and timing is also affected. All of these are challenges for a wobbly cat to assimilating, integrating into a new environment, relating to the original resident cat's communications, where timing in all things cat, counts.

My thoughts were the interactions, contact and social support of those younger kittens to be a significant support and aid to pad past traumas, achieve stability, and comfort, and this has been borne out. But overcoming timing obstacles, processing time, learning differences, losing her strongly entrenched fear of human contact morphing to insecure attachment, accepting the original cat, have been mountains upon mountains for her to scale.

It's an applied lab around here these days. Moments I have stopped in the middle of a room and looked at my approach, or an encounter where the distance increasing behaviors from one cat to another do not seem to be working to my mind fast enough, even if for the cats, they are. Time remains the hardest for us to work with animals in every modality, intervention, and application. Throughout these assimilations, I have questioned my timing on stages of introductions, on understanding their timing, rethought placement of those vital additions of vertical space. For these and more "am I doing it right" queries that I wanted to respond intuitively to, I have reminded myself repeatedly, that there is a science here that I have trained in and studied that can help. What does the research say? What am I seeing? What have I learned? What would I tell a client? What would I tell them to do right now? What would I write about in answer to what I am seeing, to what intervention is best? I need to do the applying now. And to tailor it in response to the responses of the individual. Distinct and different, as is each animal. Cats as alike in certain natures, as self-possessed in mien and inward musings, are each, their own entity, to attend. And as I listen closely, I can see in the careful, considered, and applied, put to work, how that amalgamation of insights, interventions, and science can aid the process when we let it.

<u>References</u>

- https://www.dvm360.com/view/why-some-veterinarians-dont-cats-and-how-change, retrieved November 5, 2022

CAT BEHAVIOR – DOMESTICATION, SOCIALITY, AND COMMUNICATION

The very first step in the scientific method is observation. Next up comes a question, and then a possible explanation or hypotheses related to it to be tested, considered, evaluated. And the process repeats. Modern science owes this method, not to the scientists, but to 16th and 17th century philosophers, Francis Bacon and Rene Descartes' beliefs that research needs to come from close examination of necessary facts presented and not predetermined conjecture and abstract ideas of how the physical world works. The world must be seen and verified to be believed and not just imagined. To look objectively at surroundings, environment, or another's reality, we also need to consider our own bias, to account for or take ourselves out of the picture, as difficult as this might be to do. Such a look, in all its dispassionate recitation of research, facts and observations, is where understanding cat behavior best starts.

We can begin with close and careful considerations in looking at the cat's natural history, their rich repertoire of communication, how this often-misunderstood social species came into the lives of humans, and why they have stayed around.

Cats are relatively newcomers as companion animals go, they come much later into the homes of humans than does the dog, whose close association with humans may have started as long as 15,000 years ago. There is much difficulty, in saying with certainty, when domestication of cats began; a jawbone found in ancient in Cyprus in 6000 BC, Egyptian drawings in 1600 BC. It is widely accepted that the African wildcat (*Felis libyca*), more territorial and docile than other wildcat species such as the European wildcat, is the ancestral species of the domestic cat. Also known is that at least one period of significant domestication originated in the fertile crescent of West Asia. Harder to pinpoint, are dates certain in time or how such processes took place.

Human domestication of cats is typically tied in history to moves from hunter gatherer societies to agrarian ones. Theories as to the cat domesticating themselves in a commensal system, where humans turned farmers, store grain, and benefit from feline rat catching skills, are the most familiar. Cultivating and stockpiling foods like vegetables and grains benefit multiple species aside from ours. Feline obligate (requiring the unique amino acids and proteins found in meat to survive) carnivores were attracted with the smorgasbord of prey that comes on offer with barnyards and farmlands, the birds, rabbits, voles, mice, rats, and other foragers, drawn to flowering orchards and fields with plentiful harvests and piles of stored excess. History similarly tells us that such arrangements were not species specific, ancient Greeks and Romans used ferrets and or polecats for pest control. But it is the cats that has stayed with us through time. Just as with the origin of the domestic dog, regarded as an opportunistic scavenger feeding on human refuse, those cats that were less wary of humans benefited. And the humans benefited in turn.

In staying in closer proximity to us, cats became more socialized to our presence and us to theirs. Other arguments have been made that inter-species affiliations and the cat's charisma and selective social nature began the process. Human fascination with the cat or with the animals surrounding us, is not just of modern times. Past or present, there have always been certain people drawn to animals and captivated easily by juveniles, taking them into their homes. No doubt a combination of both may be true. In either or both scenarios, bolder and tamer individual cats were more likely to tolerate or even welcome human presence and solicit contact. As these individuals multiplied so did their corresponding traits.

Domestication may have begun both as an accidental and a coincidental process, fueled by favorable circumstances for both human and non-human animals. With those animals choosing to stay around human settlements, progression in such processes can be said to have led to purposely keeping animals around that were easy to capture, handle and feed without deliberate tampering of breeding cycles or temperaments. That comes many years later in the second wave of domestication.

Francis Galton, reflecting in 1865 on which characteristics lent most easily to animal domestication, or in pondering why there was more successful outcomes for those certain animals, such as the horse versus the zebra, came up with six essentials for keeping animals under human control and to our advantage:

> "1, they should be hardy;
> 2, they should have an inborn liking for man;
> 3, they should be comfort-loving;
> 4, they should be found useful to the savages;
> 5, they should breed freely;
> 6, they should be gregarious." (Galton, 1865)

Galton links gregarious to the necessary "easy to tend," expanding on this criterion, as it relates to husbandry and taming of wild animals, he outlines how this one obligatory characteristic might apply to all but the cat:

> *Easy to tend.*-They must be tended easily. When animals reared in the house are suffered to run about in the companionship of others like themselves, they naturally revert to much of their original wildness. It is therefore essential to domestication that they should possess some quality by which large numbers of them may be controlled by a few herdsmen. The instinct of gregariousness is such a quality. The herdsman of a vast troop of oxen grazing in a forest, if he sees one of them, knows pretty surely that they are all in reach. If they are frightened and gallop off, they do not scatter, but are manageable as a single body. When animals are not gregarious, they are to the herdsman like a falling necklace of beads whose string is broken, or as a handful of water escaping between the fingers.

> The cat is the only non-gregarious domestic animal; It is retained by its extraordinary adhesion to the comforts of the house in which it is reared." (Galton, 1865)

Galton has a point when it comes to cats. Territorial nature or location preferences, it is the cat's adhesion to their homes with us that can be said to have kept cats arguably more attached to house (and its recourses, mice included) than owner and staunchly protective of home base and wary where intruding cats and others are concerned.

Galton is certainly and most definitely not alone in signaling out the cat as not belonging to comfortable and or easy groupings. The cat has been targeted throughout history and into today for what some people, writers, and religions, have great difficulty with, female sexuality. James Serpell eloquently notes:

> "A powerful element of misogyny also seems to have underpinned this animosity toward cats. Medieval and early modern Christianity was dominated by an overwhelming male priesthood with distinctly ambivalent attitudes towards women…Medieval clerics also accepted Aristotle's evaluation of the female cat as a particularly lecherous creature that solicits sexual attentions indiscriminately from any available male. Thus, a strong metaphorical connection was established between cats and the more threatening aspects of female sexuality.

> …Buffon also loudly reasserted medieval ideas concerning the female cat's insatiable craving for sex: 'she invites it, calls for it, announces her desires with piercing cries, or rather, the excess of her needs…and when the male runs away from her, she pursues him, bites him and forces him, as it were to satisfy her.' In nineteenth-century zoological literature, according to Ritvo, cats were the most frequently and energetically vilified of all domestic animals. Whereas the dog was admired for its loyalty and obedience, the cat was despised and distrusted for its lack of deference and its failure to acknowledge human dominion. Cats were also negatively portrayed as 'the chosen allies of womankind '" (Serpell, 2014).

Manifestations of sexuality, are not viewed as contemptible for male animals in human society or history; prowess, might, and fearsomeness, are more the associations given. The stallion keeping harems in check, rutting stag, butting ram, or dripping, musthing elephant has never been so disdained for the innate biological hardwiring, displays, and necessity of breeding behaviors. Such intense disdainful remarking's are reserved for and selectively written about and observed for cats and females.

As the domestication process advances, humans progress to selecting for breeding, and behaviors more favorable to our own conveniences. Incidental to these selections are changes seen in body and leg size, coat colors (whiter and piebald), ears that flop, tails that curl, brains, and heads that are smaller and more frequent cycles of reproduction. For the cat, the shorter period of domestication and frequent occasions of out-breeding means less selective vanity breeding for what "we" want from a cat. Or at least it was.

Breeding to enhance functional utility in dogs, which carries its own compromises, has progressed to breeding for looks with its often, adverse side-effects. Selecting for traits in cats for vanity and convenience can also be problematic. Up to recently, the cat has been spared the bold eugenics plied with dogs, where every aspect from temperament to size to the tip of the nose through the end of the tail has been tweaked. We are catching up though; "designer" cats; touted for tractability such as the Rag Doll, for looks we admire that compromise welfare, the hairless Sphinx that requires clothing to stay warm, the cartilage deficient Scottish fold, the flat face and too narrow stenotic nares of the Persian, and the loss of function and agility seen in the shortened forelimbs of the Kangaroo and dwarf stature of the Munchkin.

Flat faced dogs and cats may appeal to some for baby faced attractiveness but these animals have compressed airways, affecting breathing ability as well as compromised grooming ability for cats. Kangaroo and Munchkin cats are often valued for not being able to jump on furniture, Sphinx cats not to shed, Scottish folds for floppy ears and Ragdolls for docility. Comprising temperament and movement impacts facility to escape danger. Losing coat cover affects the protections lost to skin and body in regulating temperature and skin condition. Changes in body shape and type can adversely alter physical health where cartilage and bone density is compromised, as well as communication strategies used with tail, eye, ear, etc. movement.

Popular culture often ridicules cats for what we believe they might represent. We are prone to call cats "aloof," "sneaky," "sly" or the more recently popular, "grumpy." Of course, these are human attributes that are being wrongly passed off as belonging to cats. The cat simply does not have the intricate facial muscle mobility seen in humans, horses, or dogs. In those other animals, facial muscles are connected to each other and skin by a layer of connective tissue (known as "SMAS") not present in cats. Highly sensitive and essential, cat whiskers or "vibrissae" underlying structures have occupied necessary facial architecture supplanting SMAS (Burrows, Kaminski, et al. 2021). This can be a good thing in certain circumstances, tuning into the cat's facial expressions, is mostly not what we see people looking for when selecting cats in shelters. One study showed adopters focused more on whether cats made contact as opposed to facial appearance (Caeiro, Burrows, et al. 2017).

That stoic stare of the cat is more likely attributed to what we are subjectively inferring from the cat's facial features instead of what they might be thinking. More mobile and telling for cats and faces are to be found in elements of varying degrees of: rotating and or flattening ears, eyes and pupils narrowed or wide, focused on or away from a subject, muzzle and whisker positioning and movements and facial muscle tension.

Looks matter to us. Humans are a highly visual species. Our primary sense helps us in reading facial expressions in each other to determine much about the other person we are looking at. Humans have tremendously mobile and expressive faces thanks to having the most (52) separate facial muscles of any animal. Covered with fur, lacking underlying connective tissue for facial mobility, feline faces cannot compare with humans for manifesting emotion we can easily determine. But cats process information through primary senses not our own. Scent, pheromones, and sound are how they "see" the world.

Knowing this, attention to those body parts that take in chemical signals ("chemosignals") and sound information, become more telling. Other body parts and their functions, are more flexible and developed in cats, than humans, such as feline ears and noses. And some additional required body parts, not present in people, are necessary for added sensory processing, such as the vomeronasal organ for detecting and "tasting" chemosignals such as pheromones, rich in information and influence.

Cats use odors in multiple ways, to delineate territory, provide cautions, reestablish social bonds, broadcast available reproductive status, and in hypotheses arrived at in my work with stress related litter box avoidance, to communicate distress through urine and feces deposits in our homes, either through alarm pheromones or cortisol. Glands along the cat's muzzle, between the eyes, ends of the lips and chin, base of the ears and tail, all serve to deposit pheromones or chemicals when rubbed on surfaces that are believed to act as hormones outside the body and influence responses from other cats or provide the cat chemosignals about their environment. We can see this when observing momentary rubbing action by cats, whether on surfaces, us, or each other, starting with the cheek/muzzle area and moving along to the base of the tail. Chemosignals in pheromones secreted by glands can result in changes to the receiver's behavior (certain animals in season can become immobile when exposed to male pheromones) and or physical functioning (similarly breeding season can be induced in certain animals). These processes are not well understood but progress in research continues.

Cat hearing is more sensitive than human or dog hearing. Feline ears are built to maximize receiving sounds; incredibly mobile and able to rotate independently 180 degrees at speeds at least ten times faster than a guard dog. The cat has 32 muscles to control the movement of the outer ear compared to six muscles controlling the human ear. These highly mobile and expressive ears are good indicators of both cat perception and communication. Cats will rotate an ear towards a sound to better hear it. Emotion and intent is also signaled with the ears. Cat ears that start to flatten or are totally flattened against the head ("airplane ears"), are typically signs of stress and serve to protect the inner ear as well. Ears bear watching, all behavior exists on a spectrum, defensive can progress to offensive, ear position in cats with flattened ears can indicate defense, where the ears start to turn on themselves and rotate against the side of the head, a more aggressive encounter may be expected.

Cats rely greatly on their second primary sense, hearing. When it comes range in hearing, the cat is superior to not just people but to dogs as well. Our human hearing operates comparably well on the low end of the scale, cats share perceiving those softer sounds found there, almost equally with us. Worlds of sound closed off to humans but where non-human animals move easily in, are the ultrasonic.

When it comes to higher pitch and into the ultrasonic range, cats excel in hearing even above dogs. Rodents, among other animals, communicate in both audible and ultrasonic ranges. Cats hear them.

The cat's superior ability as an independent hunter is owed in part to these unique physical attributes. Cats can hear (up to 65 kHz, we can hear up to 20 kHz), smell (cats have 60-80 million olfactory cells, humans have 5-20 million) and see (six times better at night) more than we can. Such enhanced hunting prowess extends to proficiency in capturing all manner of prey. This universal attraction to all manner of creatures - flying, crawling, or walking, makes the cats a fearsome predator for multiple species. And while the cat can target many victims, those that crawl rather than fly are favored, with rodents and lizards figuring heavily in successful hunts. Feline infinite focus, patience and determination is the consistent force and factor in a cat's success when the "catch" rate is close to 20% of all attempts. Such proficiency is learned, not innate. House cats that have not learned such skills are more likely to "play" with their catches when they encounter them, often fatally. Mother cats and related females bring home live prey to kittens to teach the "killing bite" delivered with deliberation and precision to the nape of their target's neck. The art of the kill is a learned accomplishment.

While time has been a main factor in limiting human interference with cat reproduction, a stronger factor in allowing cats to remain closer to their heritage and wild ancestors is just how well they do at providing useful feline functions in our lives as they remain independent in their own. Hunting requires knowing the terrain, agency, and spatial reach. Cats carve out territory in both their home, natal and roaming areas. Free living females typically roam up to an acre while males can roam territories up to four acres and each return daily to home ranges in barns and houses.

The cat's skill as the consummate huntsman did not go unnoticed and as such was quickly utilized as a benefit to humans. Where dogs were valued as pack animals and hunting partners, cats were the first species primarily domesticated for the purpose of controlling pests. In addition to their additional appeal as religious icons and occasional companions, cats rid homes, storage places and ships, of any number of other animals, we might consider vermin, competing for and spoiling food. Drawn as they may have been, to the bounty of food available and people as caretakers, as fiercely territorial, as devoted to where we lived, we did not always keep them at home. Translocation of the domestic cat throughout the world was enabled by those persons keen for the predation services protecting human food stores from rodents and other bandits on transports making long voyages to far off, often cat free lands. However, the proliferation of cats once relocated to new environments would not always be valued in return of their initial or ongoing beneficial services.

Like many social species, cats share territory depending on the availability of resources. Sufficient food, shelter, safe habitat, environment, and mates are factors in peaceable kingdoms. Carrying capacity is a key characteristic of cat society and how closed the community is for survival. Closely knit, with defined places and little tolerance for strangers, cat society depends on highly ritualized communication strategies to maintain the status quo. In the cat colony, intruders are mostly not allowed and deference keeps actual physical fighting mainly to competitions amongst males during mating. Assimilation into existing colonies is best accomplished gradually over time and distance scaled to cat criterion, with specific and careful display behaviors to indicate intent and acceptance of existing dynamics while waiting for an opening once a toe hold can be granted. This can be most easily observed in gaze direction and the slow creep forward from the periphery of spatial boundaries accompanied with lowered look and non-threatening postures over lessening distance over time.

How cats determine when to advance, are granted license, and how far, may be more subtle and less apparent to us than more overt distance increasing behaviors but the cats are clearly picking up on the signals. None of this is rushed. Every animal has a flight distance, an envelope of space they feel secure in. Respectful distance to this space is crucial to cat integration protocols. The cat settled on the colony perimeter intent on moving forward, measures and is measured in every step. The same tenacity to careful contemplation of potential prey is needed to work progress in the hunt forward.

Selectively social and contemplative, may not be our first thoughts when thinking about cats, traits of independence and reclusiveness are still associated with them. Such a characteristic can be a detraction for pet owners or appealing, dependent on how much maintenance we are looking for where our pets are concerned. And then there is the individual. Even as individual differences, personality and chemistry have always existed, personality types, mainly not studied in biology, zoology or even animal behavior are more recognized. That we continue to group cats still as a mostly solitary species and this still lingers in the common parlance, even after decades of research revealing contrasting hallmarks of cat sociality, speaks more to culture than science.

While hunting for many cat species is a solitary affair, cat social life includes other cats, including:

> "consistent membership over long periods of time, exhibit individual recognition, engage in a variety of social interactions, and have a complex social organization" (Crowell-Davis, 2007)

Mainly a matriarchal society, with parenting duties shared by related and other females, cats give birth to multiple offspring with corresponding high mortality rates. Average litter size is between four to six kittens, 25% of which will survive. Males, once mature, typically leave natal areas and if they remain, will avoid mating with kin. Essential building blocks of these societies are the cat's "Preferred social partners" and "lineages" (Turner (2014) –which count as important resources for shared parenting, affiliation, protection and possibly more. And then there are the persons cats live with.

Consistent, positive relationships with caregivers have been cited as necessary to cat wellness. This is now understood as a given to secure good welfare. Often underestimated is the worth the cats themselves give to such individual human interactions. Valued affiliates, connections, attention, and affection are as important resources for cats as are food, shelter, and environmental enrichments:

> "In some cases where owner interaction is highly valued by cats, it may be necessary to consider the role of human company as a resource and to alter the level of owner involvement in providing access to other vital resources such as food or access to outside. Cases where this may be an issue include those where cats are fed regimented means and cannot gain access to food in any other way or where cats are actively

let in and out of the house by the owner at human determined times." (Heath, 2016)

"A definitive answer on how cats perceive us thus remains elusive, for the time being, the most likely explanation for their behavior towards us is that they think of us as part mother substitute, part superior cat" (Bradshaw, 2013)

Feline affiliations are initiated in part and maintained in myriad ways, including greeting behaviors counting nose touch, muzzle rubbing or "bunting," tail-up and flank rubs. Social bonds are sustained and reinforced with mutual grooming ("allogrooming"), allorubbing, play and contact lying (where part or a plane of the body is touching the other). These behaviors can and do carry over to the cat's individual relationships with people. Close cat human relationships can be characterized by similar instances of proximity, sleeping together or next to caregivers, contact lying, playing, bunting, vocalizations and twining underfoot for contact or when cans finally get opened for meal time.

Cats are now more widely recognized as social animals who communicate quite effectively with each other. Feline body language, including highly ritualized display behaviors, chemosignals, and vocalizations are used to communicate and alert others to presence, availability, mood, intent, and request for more or less distance, as the case may be. Amongst cats, posturing, avoidance, and deference are the main stratagems successfully employed to avoid conflict. When conflict does occur, there is usually a resource issue at root that cannot be more tactically resolved. Physical violence is more typically circumvented through ritualized threats and warnings contained in chemosignals and distance increasing behaviors such as body posturing that includes arched back and hair erect to increase size, vocalizing from hissing to yowls and growls and visual communiqués including stare downs and gaze alternations.

* * *

Cat Communication

"Intraspecies communication takes three major forms: vocal expression, body postures, and visual or olfactory marks. For most

animals, body language is the primary messenger, not vocalizations. Chemical communications are often underappreciated by humans. Interspecies communication is more complicated than intraspecies communication because animals of different species generally are not considered to have the innate ability to understand the communication of other." –Beaver (2003)

"We have only a limited understanding of the nuances of animals' lives—of what is important to them, of how they perceive the world around them—but we make assumptions based on our very limited knowledge." – Slobodchikoff (2013)

A working definition of communication is an exchange of information or signals from sender to receiver. More recently, communication theory has started to favor that such information is sent not to inform the receiver but rather to influence them.

Humans communicate through writing, verbally and nonverbally, and mostly nonverbally. Foundational work on communication done by Albert Mehrabian in 1971 found human responses based 7% on words, 38% on tone of voice and 55% for body language (93% nonverbal). Our recent overreliance on writing via email and texting over physical contact is now supplemented by a raft of emojis to enhance and clarify meaning lest we be too easily misunderstood. We can and do miss much of what is being conveyed without hearing tone of voice or seeing each other. But body language to communicate is universal amongst species. Movement, position, and behaviors can be deliberate in signaling emotion and intent. So is context and culture of that language. For humans, culture is characterized by group identifiers that include region, language, religion, politics, gender, and economics. Animals have similar identifiers depending on region, species specific body language, sensory capacities, and gender.

Then, how does non-human communication compare to human communication? It does and it does not. We are a different species. Communication itself is not species specific but how it is processed and employed as distinct species and individuals is. How we process and experience our world is dependent on sensory capacity and ability, e.g., echolocation, range of sound, light, odor even speed of cognition. To communicate, the receiver must be able to interpret or understand the signals and information transmitted by the sender. For the cat, those signals begin with one of the primary senses we are least equipped and find hardest to relate to, compared to them – the sense of smell. Odors left and received, signal and mark among other things, presence, place, sociality, emotion, and availability.

Other communications beyond the purely functional may exist in chemical signals that cats are versed and aware of and we have yet to uncover. We know that scent and pheromones, are deposited by cats through rubbing or bunting behaviors on favored individuals and objects. Some researchers suggest scratching is used for scent communication but this has not been borne out by observations of other cats when investigating areas where those marking have been left behind. The cats simply do not display interest in scratch marks. Other leavings do get attention. Scent in excrement is widely accepted to mark territory and advertise sexual status. Male (and some females) spray urine against surfaces and unlike dogs that will mark over another dog's urine, cats mark under or above other markings. Leaving feces unburied and in conspicuous locations or "middening" is less common and thought to relate strictly to marking territory boundaries.

Sound, along with odor, alerts the cat to the presence and movement of prey and is used to communicate with others. The cat has a wide range of vocalizations typically characterized by open or closed mouth sounds.

There are two major types of feline vocalizations or calls. Single type calls like a growl or a hiss or mixed calls combining varied vocalizations. If you live with a cat and have spent any time trying to figure out what it is that they want or just what it is they are trying to tell you, you have probably been helped along by a purr here and a trill there. And if you didn't pick up on it right away, you were no doubt reminded. A hungry cat or a cat who wants attention has a definite way of letting you know it and our responses reinforce it.

The most familiar of feline vocalizations is the meow, not often used by cats between themselves or after kittenhood, and which depending on the tone used in the meow may indicate request or alarm. Cat cries can indicate confusion, requests, mating calls, anger, and sounds in the ultra-sonic range which functions are more difficult to determine for humans. Cats trill or murmur to acknowledge and for social contact. Purring signals contentment or self-soothing. Cats will also purr to solicit contact and along with trills, these sounds are included in distance reducing behaviors.

Hissing, another familiar vocalization, is used as a preliminary warning and or distance increasing behavior, while this is a highly effective signal, it is not overtly aggressive, serving more as a stop sign than a green light to conflict. Hissing and even growling can be seen in play interactions when the contact has become unwelcome for one of the cats. Dialing back the unwelcome behavior allows for play to resume. As with other "apotreptic" (hoping to dissuade or deter) or distance increasing behaviors, if distance is not obtained or the provoking trigger not stopped, vocalizations can progress to spitting, yet a stronger admonition, coupled with aggression and becoming more severe with snarls and yowling.

While dogs, whose primary sense is smell, have keyed more into our reliance on the visual, honing shared eye contact with us, our feline companions have come to rely on our ability to interpret the meaning in a range of auditory signals or words for humans and purrs and meows for cats. We tend to be deficient in deciphering such "cat speak" seen in tail up ("greetings friend/happy to see you"), whiskers back ("definitely concerned here"), airplane ears ("not happy/angry") or eye blinks ("let's be friends/it's all good"), for starters, it seems cats communicating with humans have developed what appears to be a particularly effective repertoire to help us help them.

Meowing, is found in only about five of the forty cat species that exist. The African wild cat - thought to be the ancestor of the house cat, is one of the cats that do meow. But what those meows sound like make a difference. When a 2004 study in the *Journal of Comparative Psychology,* tested the reactions of human listeners comparing the meows of the African wild cat to those of the domestic cats the respondents were in favor of domestic cats over the wild cat.

It seems, cats often "talk" because and for us to listen. And cats know who is doing the listening. A 2009 study in *Current Biology,* found that cats purr differently in proximity to people when they are seeking food. Mother cats and kittens also purr when nursing. A purr for "need food now," doesn't just sound different, it is different. Even without the technical equipment to break down the sound components, individual owners can gauge the message contained within "solicitation" purrs which are perceived as "more urgent and less pleasant."

The earlier 2004 article notes that meows may be the most widespread cat-to-human verbal communication and can last for a fraction of a second or several seconds. Meows can begin or end with a trill or a growl and typically rise and fall in tone. These calls are not so common cat-to-cat, kittens both undomesticated, and not, will vocalize but usually stop when they reach adulthood. That cats will continue to meow to us as adults, especially around mealtimes, is not so surprising when you realize we continue to provide cats with their main source of food, just as their first and primary caregiver, their mother did when they were babies. Researchers have suggested cats view human caretakers as if we are standing in for maternal cat caregivers, such a view keeps the cat in an extended juvenile role.

A different study looking at the effects of "gentling" or stroking, on shelter cats and found that responses were most significant when petting was done without vocalizing by the person doing the petting. Another indication that those "conversations" are unique to known affiliates only. And, we can add that the cats know who their person is, if not who the boss of them is.

A separate study found that certain owned cats reacted with ear, tail or other body movements, pupil dilation, etc., on hearing prerecorded human voices. The reactions were dependent on just who was doing the talking. When a stranger called the cat's name, there was a reaction but when the owner called the name, the response was markedly greater. The cats were also observed being less responsive to hearing their owner's voice directed to another human and significantly more responsive to their owner's voice directed to them. The cats reacted to those familiar voices even when the person they belonged to was not in the room. A stranger's voice directed to another human and to the cats garnered no response from the cats. Why pay attention to a non-threatening stranger? In science speak, the study suggests that cats rely and know the voice of their owner and that such one-to-one associations are likely central for cats and humans to form strong relationships.

A later study, comparing if related and unrelated human listeners were able to judge if meows occurred in scenarios the cats would consider negative or positive, was done in 2014. For this particular study, the researchers tested recordings of meows related to feeding times versus those recorded while waiting to visit a veterinarian. The food related meows were characterized by rising tonal patterns while the vet related meows had falling tonal patterns. Cat owner or not, the cats were clear in being "heard." The human listeners were able to correctly identify the context 65% of the time, for those listeners who had experience with cats the percentage rose to 70%.

Generally, with cats and most other animals, including humans, vocalizations that are associated with fear or affiliation tend to be higher and more tonal where vocalizations associated with aggression tend to be lower in frequency and not tonal. Think growl versus happy to see you barks where dogs are concerned.

How we think about vocalizations we are hearing from cats or other animals, starts with how those signals are perceived in the first place. As to how the human brain responds to positive and negative vocalizations, a 2008 study examined MRI results when humans are exposed to positive and negative vocalizations of rhesus monkeys and cats compared to similar contexts for human vocalizations. Human animal or nonhuman animal, the communications are effective. When asked to identify the animal vocalizations, the humans responded correctly 67% of the time for positive contexts and 71% for negative, compared to 78% for positive human contexts and 70% for negative.

Cognition and emotion take different paths. The researchers found differences in the human brain response to negative and positive vocalizations in areas of the brain that are associated with auditory processes as opposed to emotional and decision-making processes. Sounds made in adverse circumstances tend to be of longer durations. With these, there was a greater response in the secondary auditory cortex to negative vocalizations. For those negative utterances, there were greater reactions observed in the region of the brain (orbitofrontal cortex or OFC) associated with emotional response. In the OFC region of the brain, there was no difference in whether the response seen, was to human or animal vocalizations. This suggests that even when we may not be consciously aware of what the message may be from a sender, we are aware of it on another level of processing and no doubt respond to it in some way. Perhaps why we're buying all those different cat food flavors each time we shop.

A study launched in 2016, aimed to record and analyze hundreds of cats and their vocalizations to humans in their home environments in different regions of Sweden for prosodic patterns or those patterns of rhythm and pattern found in language. While, there is no disagreement that cats have learned to produce different meows directed at humans for different purposes, there is not an agreement that all cats know what they are saying. Cats might just be meowing for the response getting sake of it. But are they? In this study, the researchers point out, cat expert, John Bradshaw's belief that meows are: "an arbitrary, learned, attention-seeking sound rather than some universal cat– human 'language.'" In other words, each individual cat develops certain meows to each individual human for shared specifics and not as meows with generic or all encompassing, meanings.

The study authors contend in response:

> "If each cat and owner develop their own arbitrary vocal communication codes, other humans would be less able to identify meows uttered by unfamiliar cats." (Schötz, Eklund, et al. 2016).

Good point and why the authors are also hoping to determine if most cats are using similar vocalization patterns and if experienced human listeners are responding to them.

The authors have additionally proposed testing how responsive "baby talk" is with cats. Such research is questionable on the face of it. This sort of speech pattern tends to utilize a higher pitch which is more acceptable and effective with dogs than it is with cats. Notably, when baby talk is directed towards dogs in an excitatory manner, the response is contagion. Mostly, where one would not want to go with cats. While both cats and dogs have exquisite sensory processes where acoustics are concerned, the cats' is superior. The cat's highly sensitive hearing and their primary sense, is finely tuned to discord and pitch and as both prey and predator, higher pitch can be alarming. It has been found it best to appeal to the cat in a softer voice. Professionals working directly with cats recommend using your "elevator voice" around cats. Cat or dog, there's a reason the melodic tones of classical music sooth and benefit both, amongst other animals, as opposed to the more discordant types of music like heavy metal or rap.

Additional research is needed to identify if adult cats do vocalize to each other in cat colonies with human caretakers, especially those where social relationships exists including shared parental care, common among related felines. And how much meowing is going on, compared to other vocalizations in those cat-human households where cats are allowed regular outside hunting forays.

Visual aspects of cat communication can be subtle for the unobservant and can happen in front of us without being noticed. Such notice does not escape the cats.

Like most non-primates, cats avoid direct eye contact with each other as it can be considered adversarial. Direction of gaze is also significant. A look away from another cat can signal a green light for the other cat's passage, think of passing in front of each other on the way to the water bowl or cat box. Without recognizing the slight shift or greater, 45-degree head turn, one may think nothing has happened but it has and the cats know it. Seeing and being seen is significant for cats. Whether or not visual envelopes of space are afforded and offer additional protections when sight lines are blocked are suggested by the cat's inclinations to seek out resting spaces with raised sides. Even if those raised sides are part of a cardboard box in the middle of a busy room. Ceasing gaze is similarly communicative. Tempo counts. Blinking slowly is an affiliative or calming gesture that cats use with each other. Cat friendly advice now includes a slow blink directed at your cat to let the cat know of your own friendly intentions. A receptive cat and your proper cat fashion is confirmed, if the blink is returned.

There is a whole spectrum of pupil size and eye width along with a whole spectrum of behavior to correlate with the cat eye. Alert, scared, or trying to see, the state of the dilated pupil depends on environment. Widened eyes, slit eyes, and corresponding pupil size need to be viewed in context of what else is being observed in the whole body of the cat and the immediate surroundings including other participants. Pupil size additionally responds to light and can be indicative of neurological or other conditions in conjunction with affective states.

Direction and duration of gaze figures prominently. For the cat as hunter or prey, surveillance and vigilance are locked together. Releasing gaze from another cat, signals affiliation in blinking and alternating the direction of gaze allows for an antagonist to move away. Barriers to sight lines, are security sought in the cats' seeking dens and resting spaces with raised sides. Relying on the safety of sight lines, when vulnerable, can be seen in preferred elimination with uncovered litter boxes and why removing box coverings can be effective with unacceptable elimination in the home.

The muzzle region of the cat's face is highly expressive for felines with extremely sensitive and highly mobile underlying structures connecting and padding whiskers that function to position objects and detect movement directly in front of the cat. Even with tremendous night and good distance vision, cats are somewhat far sighted and find objects as close as up to ten inches in front of them, harder to see without those proprioceptive and responsive whiskers to locate them. Muzzle tension is significant and practical. When whiskers are placed either pulled back onto the face or pushed forward, they can also be a sign of arousal or be functional (pulled back and protected/protective). Mouth movements with lip licking out of the context of food or being hungry (this can be seen with dogs as well), panting or a swallow that is pronounced or exaggerated, can indicate stress. Much of what we look at in studies indicates stressed states. It is important to also take the measure of animals in relaxed and content states to have a baseline to compare from, to look at a range of emotional states in individuals to have reference points and remember them. Careful observations of animals encountered in our homes and surrounding environments can provide this.

The development of grimace scales in animals to measure pain along with advancements in technology to measure facial architecture by locating "landmarks" and action units of movement have advanced standard recognition of painful cats and heightened awareness of the most expressive aspects of the cat's face. No surprise, ears, eyes, whiskers, and body positioning figure prominently. Leading researchers to create the validation and standardization of a "Feline Grimace Scale" in which:

> "Five action units (AU) were identified: ear position, orbital tightening, muzzle tension, whiskers change and head position." (Evangelista, Watanabe, et al. 2019)

Body movements in the cat can be even more expressive than facial ones, dependent on the scenario. Distance reducing behaviors calling for space or changes in the immediate environment can indicate initial stress, may be reactive, as seen by muscles rippling along the back, turning away, slinking low and slow, backing away and progress to threat displays such as cuffing, batting, or patting with claws retracted or extended. Cuffing, batting, or patting as inhibited movements can also be seen in play.

Another sign of stress, often overlooked by humans, is displacement grooming; usually only on the shoulder and not the rest of the body and typically for a short time, and in the performance, may allow for a moment of feline collection or self-soothing. A more obvious sign of displeasure or complying with distance increasing requests for the cat, is leaving if space allows. Intra-species communication between cats is highly ritualized and dependent on body language, vocalizations, and chemical signals. Individual relationships have their own sets of criteria.

Cat greeting behavior between affiliates typically consists of nose-to-nose almost touching. Cats may sniff at hindquarters; this is more of information or pheromone gathering (flehmen or the "scent tasting" grimace is often seen after this behavior) as opposed to greeting. Bunting, or rubbing briefly along the side of the muzzles can also be seen amongst associates. "Social Roll" or dropping to the ground and rolling on to the side, where the head is often turned upwards and gaze directed to a respondent, is a gesture of friendship, invite to play or contact. Eye blinking signals and maintains positive connections between individuals. Cats will blink, social roll, bunt and rub against their people in moments of affection and attention seeking as well. Tails may be twined and brief rubbing flank to flank in affiliates. Flank rubbing and twining are most often seen with humans, as if anticipatory excitement is not to be contained when meals are prepared.

Such behaviors are fleeting in duration and suggest possible reasons why extended stroking by people is not favored by most cats, especially when concentrated along the back – an area cats mostly do not allogroom or rub against with each other. Brushing, done appropriately, may be considered differently where timing is concerned. Cats will leisurely groom each other and this behavior has been posited to either cement social bonds or to deflect aggression depending on individual relationships. Maintaining or developing bonding communication can also be observed with "contact lying," where cats position their bodies so that one plane, aspect, or part of the body is touching and in contact with the other cat. Nice to remember when your cat sleeps alongside you.

Antagonistic behaviors rely heavily on locked eye contact. The peaceable kingdom stayed that way probably because no one was staring at anyone else. Blinking, that signal of neutrality and affiliation, is notable in amicably breaking the fixed stare down. Concurrent or in in escalation of defense or offense are, flattened ears, whiskers back, piloerection (hair standing on end), and body tension. These can elevate to pressured vocalizations and swats. Piloerection and sideways stances are additional mechanisms designed to make the cat look bigger to the observer and deter them from advances.

Conflicts are avoided through stand-offs with one cat deferring to the other to avoid more severe interactions. Movements in response to another's distance increasing behaviors or to pass a would-be objector are often slow and measured. The languorous indolence of the cat has functionality at work. Hunting is best accomplished with measured determination and patience. That slinking, low and slow pacing one can observe between not so neutral associates and unknown conspecifics keeps passage peaceable. Fast movements can invite chase. One of the caveats, humans hear around strange animals is never to run away, backing s l o w l y away is recommended. Cats have this down.

Tail movement in most animals has more than one function, for instance keeping pests away, and serving signaling capacity. Tail movement is understudied and there is still much about it to be learned. We do know that, in cats, a rapidly swishing tail moving from side-to-side signals consternation, while a tail flick or swish may signal some internal conflict. A raised vertical tail is usually an affiliative sign, especially if the tip is bent towards the affiliate. A raised tail with hair standing on end is not to be confused with affiliation. Here the tail, along with the rest of the body, is made to seem larger as defense not invitation.

To complete any discussion of feline body language, a review of detailed attention to focus on every aspect of the whole cat when observing them is essential. Our current understanding of cat communication and corresponding emotional state relies heavily on a study published in 1997 of work done at a British cattery. Researchers, Kessler and Turner, looked at the behaviors of boarding cats left by owners and strays. The scientists devised a method to chart the mood of a cat by scoring for appearances from relaxed to terrified. Cats were evaluated for body appearance and posture, legs, tail, breathing, head position, eyes, ears, whiskers, and sounds. Every feature of the cat went into determining mood. These findings, developed into the "Cat Stress Score" or "CSS," are rated on a scale and continue to assist in reading cats today:

Relaxed cats: may be active or inactive, lying down or sitting. The active cat has a straight back. Breathing is normal. The legs may be laid out if lying down or stretched out when standing. Tail position is loose and either down, up or wrapped. The cat's head is over the body and may move slightly. The eyes may be closed or open and the pupils will have normal sized slits. Ears and whiskers are normal to forward. The cats may be silent or purring.

Alert cats: may be inactive, attentive, or exploring. The body may be sitting or lying down with a straight back. The breathing is normal. The legs are bent in repose and stretched out if standing. The tail may rest on the body or curl backwards. Tail movement may be seen with tension while standing or flicking or twitching. The cat's head is over the body and may move slightly. The eyes are open and the pupils will have normal sized slits. Ears are normal or held either to the back or front of the head, whiskers are normal to forward. The cats may be silent or meowing.

Tense cats: may be inactive, attentive, exploring or trying to escape their surroundings. The body may be sitting or lying down, if standing or moving the cat will hold the back of the body lower than the front or "slink." The breathing is normal. The legs are bent in repose with the front legs stretched out if standing. The tails is on or close to the body or curved onto the front of the body if standing. The tip of the tail may be twitching. The cat's head is over the body or held tightly to the body, movement is minimal. The eyes are tightly closed or open wide and the pupils will either be normal or beginning to dilate. Ears are held either to the back or front of the head, whiskers are normal to forward. The cats may be silent or meowing with the meowing becoming more intense.

Anxious cats: may be attentive or trying to escape their surroundings. The body may be sitting or lying down, if standing or moving, the cat will hold the back of the body lower than the front or "slink." The breathing is normal or rapid. The legs are held under the body in repose and bent if standing. The tail is on or close to the body or curved onto the front of the body if standing. The tip of the tail may be twitching. The cat's head is held in line with the body, movement is minimal or absent. The eyes are open wide and the pupils have dilated. Ears are beginning to flatten; whiskers are normal, forward, or back. The cats may be silent or meowing intensely, growling or yowling.

Fearful cats: may be attentive, crawling or not moving. The body may be shaking. If the cat is standing, the body is lowered. The cat may also be hunched over all paws or on the abdomen. Breathing is rapid. The legs bent and are close to the ground if standing. The tail is next to the body or curved next to the body if standing. The cat's head is held in line with the body and movement is absent. The eyes are open wide and the pupils are completely dilated. Ears are totally flattened ("airplane ears'); whiskers are held back. The cats may be silent or meowing intensely, growling or yowling.

Terrified cats: may be attentive or not moving. The body may be shaking. The cat may be hunched over all paws; the hair is on end on the back and the tail. Breathing is rapid. The legs are rigid and extended or bent (this cat is trying to "look" larger, think Halloween cats). The tail is next to the body. The cat's head is lowered. The eyes are open wide and the pupils are completely dilated. Ears are totally flattened and facing backwards; whiskers are held back. The cats may be silent or meowing intensely, growling, yowling or hissing.

Cat body language is complex and best understood in the context of the surrounds of the environment it is occurring in. What is happening to the cat or has happened or is going to happen in that moment. Behavior happens on a spectrum, it is not fixed. History of the individual and personality also factor in. As with any animal, all aspects and modalities need to be accounted for to accurately assess the communication intended. Tail flicking by itself cannot indicate a mood or intention but a resident vigilant cat with tail flicking, whiskers pulled slightly back, eyes narrowed, ears starting to flatten and a tense body positioned next to a window looking onto a rival cat tells much and speaks volumes. The entire cat must be observed.

For practical purposes, the careful, experienced listener and observer already knows that cats communicate both with each other and humans. The resourceful cat with a deep and rich inventory of sound, scent to pick up on as well as leave behind, and body language, including significant eye, ear, tail, and whisker positioning has already figured out that as visual as we are, we are not visual enough to "read" cat, and has met us on common ground - vocalizations. Use your soft "elevator voice: and remember, the next time your cat says something to you, give them the favor of a response; start with a soft verbal acknowledgement including their name, look around to figure just what it is they are asking you to "listen" to.

The time to consider every aspect of how cats communicate, their individual differences, and their behavior opens a feline world to us when we pay attention to it.

<u>References</u>

- Beaver, B.V. (2003). *Feline behavior: a guide for veterinarians second edition*. St. Louis: Saunders

- Belin, P., Shirley, F., Ian, C., Nicastro, N., Hauser, M. & Armony, J. L. (2008). Human cerebral response to animal affective vocalizations. *Proceedings. Biological sciences/ The Royal Society*. 275. 473-81.

- Bradshaw, J. (2013). *Cat Sense*. New York. Basic Books.

- Burrows, A.M., Kaminski, J, Waller, B.M., Omstead, K.M., Rogers-Vizena, C., Mendelson, B. (2021). Dog faces exhibit anatomical differences in comparison to other domestic animals. *The Anatomical Record*. 304: 231– 241. https://doi.org/10.1002/ar.24507

- Caeiro, C.C., Burrows, A.M., Waller, B.M. (2017). Development and application of CatFACS: Are human cat adopters influenced by cat facial expressions? *Applied Animal Behaviour Science*. 189: 66-78, ISSN 0168-1591, https://doi.org/10.1016/j.applanim.2017.01.005.

- Crowell-Davis, S. (2007). Intercat aggression. *Compendium Continuing Education for Veterinarians* 2007; 29: 541-546.

- de Mouzon, C., Gonthier, M. & Leboucher, G. (2022) Discrimination of cat-directed speech from human-directed speech in a population of indoor companion cats (*Felis catus*). *Animal Cognition*. https://doi.org/10.1007/s10071-022-01674-w

- Evangelista, M.C., Watanabe, R., Leung, V.S.Y., Monteiro, B.P., O'Toole, E., Pang, D.S.J., Steagall, P. V. (2019). Facial expressions of pain in cats: the development and validation of a Feline Grimace Scale. *Scientific Rep*orts **9**, 19128. https://doi.org/10.1038/s41598-019-55693-8

- Galton, F. (1865). The first steps toward the domestication of animals. In *Transactions of the Ethnological Society of London* (pp. 122-138). Royal Anthropological Institute of Great Britain and Ireland

- Heath, S. (2016) Intercat Conflict. In: Rodan, I, Heath, S. (eds.) *Feline Behavioral Health and Welfare*, (pp. 357-373) Dordrecht, The Netherlands: Springer

- Kessler, M.R., Turner, D.C. (1999) Socialization and stress in cats (*Felis silvestris catus*) housed singly and in groups in animal shelters. *Animal Welfare*, 8, 15–26.

- Liu, S., Paterson, M.B., Camarri, S., Murray, L., & Phillips, C.J. (2020). The effects of the frequency and method of gentling on the behavior of cats in shelters. *Journal of Veterinary Behavior-clinical Applications and Research, 39*, 47-56.

- McComb, K., A.M. Taylor, Wilson, C., Charlton, B.D. (2009). The cry embedded within the purr. *Current Biology*. 19. R507-R508.

- Mehrabian, A. (1971). *Silent Messages* (1st ed.). Belmont, CA: Wadsworth. ISBN 0-534-00910-7.

- Nicastro, N. (2004). Perceptual and Acoustic Evidence for Species-Level Differences in Meow Vocalizations by Domestic Cats (Felis catus) and African Wild Cats (*Felis silvestris lybica*). *Journal of Comparative Psychology*. 118, 287-296.

- Schötz, S., & van de Weijer, J. (2014). A Study of Human Perception of Intonation in Domestic Cat Meows. In M. Heldner (Ed.), *Proceedings from Fonetik 2014*. 89-94. Department of Linguistics, Stockholm University.

- Schötz, S., Eklund, R. & van de Weijer, J. (2016). Melody in Human–Cat Communication (Meowsic): Origins, Past, Present and Future, *Proceedings from Fonetik 2016. Speech, Music and Hearing, KTH*, Stockholm.

-Serpell, J. (2014). Domestication and the history of the cat. In D.C. Turner & P. Bateson (Eds), *The Domestic Cat the biology of its behaviour*, (3rd ed., pp. 84-100). Cambridge, UK: Cambridge.

-Slobodchikoff, C. (2012). *Chasing Doctor Dolittle.* New York: St. Martin's Press

-Turner, Dennis. (2013). Social organisation and behavioural ecology of free-ranging domestic cats, (pp. 64-70). In: Turner, D.C. and Bateson, P. (eds) *The Domestic Cat*, 3rd edition, Cambridge: Cambridge University Press. 10.1017/CBO9781139177177.008.

MAKING YOUR NEW CAT FEEL AT HOME

The first cat that came to live with me when living on my own, was not one I chose. A friend showed up one day at my door with a ginger tabby who had lost his owner in sudden and terrible circumstances. That cat had horror stories to tell. What did I know then to make such a cat, any cat, feel at home? Especially for a traumatized cat who could not speak in ways I could understand at the time? Canned cat food, a scratching post, a few cat toys, and a covered litter box seemed enough. Friends, pet shops and magazines confirmed this. Wrong. It wasn't. I just didn't know that. Fair or not, cats in turn, supplied tolerance, appreciation, affection, selective contact, found their cat spots on the bed, chairs, or sofa. Cats made every place a home. But it was only in the living with them when things were not going so well, that I began to learn more about cats beyond what I could intuit. I read and asked more about how to keep a cat healthy and happy mostly when they seemed lacking in either. Well-being and welfare seemed the same, even as I would learn later and the cats always knew, they were very different.

Most of us who have had a cat in our lives have been there; bringing that adorable new cat, whether rescued or fostered, home. Expecting the best of all the ways it will go, the amazing relationship we will get to have, how much the cat will love us on sight or with unending gratitude for the magnitude of the foster, and love us even more in their forever home, the difference such devotion and love will make in our lives. And then the reality check. What happened to acknowledging what we have done for them? The cat does not seem to be sharing the dream or even want to be in this amazing new place.

What else makes home for a cat, especially when the new cat feels as if this new place is anything but? What then? What happened to all those expectations and why don't they get it?

How we define our present-day relationships with the cats we live with can run a range from an "open relationship," "remote association," "casual relationship," "co-dependent," or "friendship," each depending on how the human interacts with their cat according to a 2021 study looking at cat human bonds. Researchers surveyed cat owners, the majority of which respondents were mainly women located in the United Kingdom. Answers to the survey questions allowed for categorization of the types of interspecies relationships those guardians most often encountered. Of particular and interesting note is the following qualification on how we/owners/guardians look at cat companions (emphasis added):

> "The data from the principal component analysis can be viewed in terms of both what the cat is perceived to do and how this impacts the owners' feelings about the cat. Thus, from the owner's perspective, the components could be construed to represent their emotional investment in the cat, the perceived loyalty/faithfulness of their cat to them, or its emotional closeness and enjoyment of mutual physical interaction. **However, in terms of the cat's behaviour, it seems that these perceptions might arise from the degree to which the cat's responses are compatible with the owner's needs**, (perhaps akin to the social sensitivity that is widely recognized as important in our relationship with dogs), the gregariousness of the cat (a commonly recognized personality trait of many species), the ability to cope with the owner's absence (a trait which might relate more to frustration tolerance rather than attachment and tolerance of physical contact (a highly variable feature between and within cats." – (Ines, Ricci-Bonot, et al. 2021)

Another study from 2020, looking at reasons given for pet relinquishment to shelters in Denmark, noted that it was owner reasons, not behavior problems that prompted surrendering pets. All this to say, that while we may often be seeking what our pets can best do for us, when we bring them home, how might we provide the best homes for them to do it in? What ideas on outcomes might be reframed? And how best to understand that your new cat can be happier in new surroundings with your help. Here, a case history to help you help them:

A rescue group in the New York City area reached out to me, for advice in integrating a new cat into a household. Expectations and timing needed adjustment on both sides of the dynamic. The adopter and cat had been together for several weeks, with the owner becoming very concerned that the cat was not acclimating to the surroundings fast enough and that the present state of affairs was more permanent than temporary. A shared email from the new owner eloquently voiced frustration and apprehension for cat and guardian both:

"It has been a few weeks since I have had (name withheld) and although he comes out at night to eat and drink, he simply will not come out and say hello. I have tried his favorite treats but he is very stubborn in his ways. I'm not sure if he will bond with me as he still acts like the first day I got him. He allows me to pet him and seems to love it very much but even that won't get him out from hiding. He has stayed in my bedroom since and will only come out while I am asleep (sometimes waking me up when using the litter box). He is eating, drinking, and using the litter box on a regular basis.

I have been very patient but afraid he will not snap out of his shell. I hate to give him back but feel there is nothing I can do at this point. Please let me know if you have any advice."

When it comes to a new environment, cats, like humans, need time to feel at home. We meet a new cat and are entranced by their beauty, how wonderous they appear. How great we think it will be to have such a new and fascinating companion. Surely, they must know we love cats, that we are animal people. Certainly, they will gratefully accept our advances, our caresses, our words of welcome. Meanwhile, this is no easy feat for this stranger coming into a strange land.

Coming into a new place they have never been before, they enter uncharted territory, a strange and foreign world with different smells, sights, and sounds, including the uncertainty of what to suppose with different human and non-human animals. Know that they are also leaving behind familiar territory and relationships and this loss of what they knew and who they had bonded with before, also has to be processed. Sayings can be based in wisdom and the ones that tells us to "go slow to go fast with cats" and "let them come to you" are profound and worth adhering to here.   Depending on the individual history and the personality of the cat, the acclimation process may vary from hours, days, weeks or even months.

Know that the cat would like nothing more than to know there is safety here, to feel comfortable and at home immediately. If the cat were a human instead of a cat, he might be able to do just that. He might be able to use human reasoning to know that he is in a good home with a friendly, respectful human who will not harm him, who will give him time to settle in, who knows the right way to pet a cat, likes to play, who will provide comfy beds and towers, fill the best puzzle feeders with dry food and treats and the food bowl with good wet food, keep the water fresh and keep the litter box cat standard clean.

But the cat is not a human so he cannot know this until he experiences it. This cat can only trust his environment, what he sees, hears, smells, and feels. And he can only trust the stability and safety of the environment with the passage of time. Cats wear a different watch than we do to decide when there is time enough.   What is sufficient passage of cat time is determined on their individual cat count not ours, to feel safe and then comfortable, and then truly at home.

If the cat has been around dangerous humans (ones who harmed or neglected him) or non-humans, the cat needs to take the time to make sure the absence of those dangerous animals is permanent and not temporary. If the cat could take your assurance that humans and other species to be feared, do not live here surely the cat would. Or to know that the new humans to be encountered are different. Why waste the energy on vigilance? But, in the wild and in the world of humans, watching and waiting is the only way the cat can know that this new world is a safe one.

But you as a human know these things that the cat does not know. With your help the process can be an expedited one:

- Set up your new cat initially in a closed off separate room, such as your bedroom or an office. Choose a room you will also be spending part of your day in. Avoid garages, basements and even the bathroom. These spaces and rooms are the least welcoming in our homes for a cat and do little in comforting or acclimating cats to new environments and establishing how to best live with us. If you do not have a separate room, concentrate on a corner of a room to add an ex-pen, or bathroom where you can set up cat furniture (cat bed, igloos, carboard box, toys, scratchers, etc.), litter box (placed in the farthest spot from food and cat furniture), etc.

- The cat's natural instinct will be to find security. This may mean hiding first to get the lay of the land from a "safe" perspective. Which is why a cat in a new home is often found seeking refuge under the bed, behind the sofa or toilet. Offer an alternative safe space; a cardboard box turned on the side with some fleece for a secure haven, a cat bed or cat igloo or basket. Keep the separate room set up for the first week or two. After a week to ten days, leave the door to this room ajar, so kitty can return from exploring to his now known safe space.

- A new place can mean a loss of appetite. Most cats will not openly eat or drink or even use a litter box for the first day or two, preferring to do this under cover of darkness or night. This is natural, your cat is stressed and operating on "safe mode" on arrival. Make sure to give the very most enticing food available, this is not the time for special diets, your cat needs to eat daily to maintain health. That smelly canned commercial cat food brand (pate is usually best), that most cats love no matter how junky, is just what you want to give now. Make sure to discard any uneaten or leftover food and offer a fresh portion at breakfast and dinner.

- Fresh water each day placed as far from food and litter is essential as is clean litter. A pet fountain is a great choice for cats who will seek out the cleanest water in their natural environments. For those cats that have not been exposed to fountains in the past, there may be a slight adjustment period so leaving a water bowl nearby can help. Fountains do require maintaining and monitoring, so being realistic about them can mean a clean and replenished daily water bowl may be the best choice for some owners.

- Scoop litter boxes out daily. Twice a day is optimal. If the cat is installed in one room, gradually move the food, water, and litter into the other rooms. For instance, place the litter box and the food outside the door for several days and then along the hallway (in the direction of their permanent placement) for several days, and then in their permanent positions.

- Cats and dogs are crepuscular animals, meaning they are naturally most active during twilight or dawn and dusk compared to humans, who are diurnal meaning most active during daylight hours. Domesticated animals being familiar and dependent upon us for food and social interaction become accustomed to the diurnal routines of human beings. Because your cat has not yet gotten fully comfortable around and with you, he may only be active during the cover of night time hours when all is quiet. Give your cat the time to feel safe and he will pretty much adjust his schedule to yours. Remember, we are talking cat and not dog, so some night time exploration might still occur and that kittens may take even longer to settle into routine times (and why interactive play sessions before your bed time can help).

- Be careful not to force interactions on your new cat straight away. Relationships take time to develop and cat human relationships take cat time to develop. Make your presence a soft and welcome one when you are around. Cats prefer our elevator voices over those happy puppy voices we use with dogs. Speak in a gentle tone to your new cat even when you do not immediately see them. For instance, when you enter a room where the cat generally spends most of his time, greet the cat by saying his name and a short friendly sentence or two (less is more at first) to help accustom him to your movements and voice. Your cat is very aware of your presence and announcing yourself and speaking to your cat (even when he is not in eyesight) is the first step towards having a conversation.

- Spend time around the cat at cat level so that the cat can get used to you, your smell, and your voice. Do this by reading out loud (the more mellow the material the better here) or talking softly on the phone while sitting on the floor alongside (not facing/blocking) the spot your cat has decided is safest (aka his hiding place). Do this for at least a half hour every day. Starting this process, you do want to keep the pressure to interact off the cat. Again, remember that saying, let them come to you? It is golden.

- Avoid staring at your cat. Direct eye contact is usually perceived as predatory and aggressive by most animals. Cats will signal friendly social overtures by blinking slowly at each other. Begin with one or two daily slow blinks when your cat is looking at you. You will know if you are doing this right if you if you get a blink back and see more of a relaxed body posture in your cat as well. That blink back may take some time as trust develops but can come over time with practice.

- Remove and do not use any punishment. This includes spraying water, shaking cans filled with rocks, scolding or lectures, at all costs. Punishment only makes matters worse and serves to create fear and distrust. Work instead on what you would like your new cat to do and make sure they have the set up/places to do them.

- Make your home cat friendly. Now that you live with a new cat provide the necessary furnishings in each room: Vertical space to access, cat beds with at least three raised sides (I like the fleece oval shaped beds for comfort, easy cleaning and price and have several in each room and to supplement raised resting spaces) to offer security and warmth in good spots are mandatory. Add boxes or cat igloos to curl up, nap and hide in. Make sure there are sturdy cardboard scratch boards or posts in each room for use. And don't forget the toys, sprinkle liberally, including fur mice and balls to chase.

Remember provide the options for what you would like them to do instead. Don't want cats on the counter top? Being prey and predator, cats need elevated resting areas to feel safe - make sure they have a cat tower or cat shelf in the right spots to access instead. Don't want kitty scratching furniture? Cats need to scratch to flex muscles and keep claws in order - make sure to provide stable cat scratchers in every room. Want cats to use the litter box religiously? Keep it clean, in the right spot and without a top. Don't want cats waking you up at 4 am for breakfast? Ditch the food bowls and feed dry food directly in the right puzzle feeders so cats can exercise choice and control over meal time. Avoid those "slow food" feeders which frustrate rather than satisfy and keep feeders replenished especially at bed time to ward off nighttime/early dawn food requests. Don't want cats playing with everything that can move? Make sure to provide objects they get to play with instead; crinkle balls, any toys that are catnip stuffed, ones that rattle and are shaped like mice are good to start with. Experiment with different toys to find the most appealing ones for your cat and give the cat time to decide before removing them.

Boredom counts in indoor living, optimizing environments helps.

- Establish schedules and keep them. Feeding, litter cleaning, and play time on schedule can set routines cats can trust and offer a sense of control over the events in their lives. Such a sense of control is a secure and necessary assurance in a world where we get to decide on when and where mostly every event in their lives get to happen

- Leave the radio tuned to a classical music station and do leave it on when you are not at home. Remember that cats have sensitive hearing so set the volume lower than you think is comfortable for background music. Classical music is soothing and melodic and the announcers' voices on these stations tend to be soothing and melodic as well. This is a positive association with human voices for your cat to form.

- When it comes to petting, take your cues from the cat and avoid full on body stroking and belly rubs better suited for those dogs that like them. It is always best to initially confine caressing to the cats' head area. Cats greet each other by sniffing nose to nose and then rubbing along each other's muzzle areas. Stroking behind and between a cat's ears, under the chin and the back of the head is usually welcomed. Keep petting sessions short at first and even as your relationship develops. Less is more here. When cats themselves solicit or accept caresses, they are most readily received in limited duration more frequently. Think repeated three or four strokes per area.

- Once you know each other, full on contact may be welcome but look closely for consent as in does the cat come closer, lower their head, lean toward you, bunt or rub in return? If you do stroke along the body, what is the response? A displaced groom (quick groom on the shoulder only), turn away, ripple along the body, may signal discontent and leaving always means they would rather be somewhere else. Always, do what is welcome and go slowly. You are getting to know each other after all.

- Offer playtime. Interactive play sessions with a fishing wand toy where you draw the object across from or away from the cats' line of vision, are the most effective ways to create and cement relationships, especially what is called for in getting to know your new cat. For instance, dangling a feather at the end of a piece of string for your cat to pounce on allows you to interact with your cat and that quality time creates bonding for the both of you. Some cats may not initially be up to playtime, that's OK. Go slower with them, and still bring out the fishing wand toys on schedule to demonstrate play for when they are ready or just to think about pouncing on once he gets to know you.

Please consider where they come from, how new this all is to them, what cats' needs are at home, the importance of your relationship to them, and time, that most magical of all ingredients, to allow the cat to settle, trust, and get to know you and the new optimized environment.

References

- Ines, M., Ricci-Bonot, C., & Mills, D. S. (2021). My cat and me—A study of cat owner perceptions of their bond and relationship. *Animals*, *11*(6), 1601.

- Jensen, J. B. H., Sandøe, P., & Nielsen, S. S. (2020). Owner-Related Reasons Matter more than Behavioural Problems—A Study of Why Owners Relinquished Dogs and Cats to a Danish Animal Shelter from 1996 to 2017. *Animals*, *10*(6), 1064. MDPI AG. Retrieved from http://dx.doi.org/10.3390/ani10061064

ENRICHMENT STRATEGIES FOR A CAT FRIENDLY HOME

"For an environmental intervention to be considered enriching, the changes it produces (whether behavioural, psychological, physical or physiological) must be linked to an improvement in the welfare of the animal. Currently, scientific studies investigating potential EE strategies often assume a link between a change in behaviour and improved welfare, which often may not be the case." - Ellis S.L. (2009).

As an American living in New York City, I can be horrified at moments of the very idea of outdoor cats. Ever mindful of the debate over the benefits to the cat of being able to roam free, are not even to be entertained, not with streets crowded with pedestrians, dogs, cars, and rats and mice controlled with poison. For other parts of this country with more bucolic and pastoral landscapes and in other parts of the world, notably Europe and especially the United Kingdom, the conversation may be a different one. Because my work focuses on behavior and welfare with cats in the home, I will try to mostly leave off how to best keep the indoor-outdoor cat a happy one. Such cats are good at doing that themselves and that topic is beset with controversy over cats outdoors and their impacts.

It would not be fair to cats although, not to say, after cautions over predation, food scarcity, exposure, accidents, and disease that can harm them in outdoor encounters, that the issue of how dangerous a predator the cat is for wildlife, birds in particular, needs to be mentioned. Cats cannot be single handedly blamed for decimating bird populations. It is easy and black and white to point to the cat as being the only responsible party rather than taking a harder look at anthropogenic causes. The science shows differently. Loss of habitat, pesticides killing off insect prey and collisions with manmade structures are the leading cause of bird fatalities, far, far edging cats out (Buchan, Franco, et al., 2022). How we even count cats in outdoor environments is suspect and needs confirmation before we can assume impact (King, 2013). Additionally, rodents and reptiles are known to be preferred prey for cats over birds. The very domestication of cats, centers on their innate focus on small mammals to hunt for that live on the ground and under it. The world over, cat keeping for vermin control continues. Another verifiable note on bird deaths with cats is, these are often the result of "compensatory predation," as in older, sick, injured or fallen birds being victims over others (Baker, Molony, et al., 2008).

Cats live alongside us in farms, docks, shipyards, villages, streets and in our homes. Their presence can be tolerated, valued as pest control, despised, or welcomed. Thankfully for cats, there will always be those who regard the cat as more than just utilitarian pest control, whose affection for cats seeks to provide opportunities for cats to have their best cat life. What a feline life worth living looks like is very clear for cats but can vary culturally where people are concerned.

Providing indoor cats and captive animals enrichment in their environments has recently gained much needed attention. The impact and extent of domestication of the cat continues to be debated and the need for an appropriately complex living situation goes along with the discussion. Cats can return to free living states (feral) and large populations of cats exist independently of humans. Optimizing welfare of home living owned cats is also worthy of careful consideration

Enrichment is the popular term we use and one I use here but, I wish we had a better word. "Enrichment" has the connotation of a favor granted, a gift given, an extra added, rather than a needed optimization of an impoverished and unnatural environment. Such small worlds we confine the inside only cat to - the same humans, same other pets, same processed meals fed on our schedules, walls, windows, furniture meant for people and little to satisfy their behavioral or emotional needs. Boredom, frustration, stress and trauma of new and unknown people and environments come with such small territory. Comparing indoor cats to captive animals is no stretch:

"In this sense, we may consider the environment of domestic cats in the home to be similar to zoo animals, with the confined space, proximity of conspecifics and other predator and prey animals, combined with limited resources and opportunities to express species-typical behaviour potentially influencing the cat's perceptions of control and threat, which in turn determine its welfare." (Stella, Buffington, 2014).

Keeping cats at home has an added ethical component where enrichment is concerned. These innovations or interventions to an animal's environment can improve welfare when, and if, such improvements are specific to the natural behavior, desires and needs of the individual and the species. There is no "one size fits all strategy." Because welfare or how the animal experiences their environment, is subjective for everyone, in how they in turn, individually perceive their world, defining just exactly how to arrive at improved welfare can be elusive. Cat expert, Sarah Ellis writes that "goals" instead, are a valuable method to define how to arrive at environmental enrichment ("EE"):

"Using quantifiable goals allows outcome(s) of enrichment strategies to be scientifically evaluated, which is of critical importance when measuring the success of application. The most commonly reported goals of EE are to:

• Increase behavioural diversity;
• Reduce the frequency of abnormal behaviour;
• Increase the range or number of 'normal'
  (i.e., species-typical) behaviour patterns;
• Increase positive utilisation of the environment;
• Increase the ability to cope with challenges in a more 'normal' way." – (Ellis, 2009)

How then arrive at goals, to have and identify criteria to put in place? How to then have the best ways to create an optimized environment for a cat friendly home?

Begin with your personal relationship with your pet. While the perception of the cat as a solitary being may persist, research continues to show that cats are social creatures with individual personalities, affiliates and relationships that allow for extended family groupings raising young and teaching hunting skills. Relationships between cats extend beyond shared parenting. Cats manage social bonds, groom, play, rest and sleep with ("contact lying" so called because parts of the body in repose, touch and stay in contact) and next to affiliates and respond to social, alarm or alert vocalizations of surrounding cats.

In multiple ways, cats demonstrate and benefit from time spent with select other cats as preferred associates and the same holds true for time spent with certain people in their environment. Interaction with humans is a key component of your cat's immediate world and may help to substitute for outside interactions and other cats. Our cats may view a feline affiliate or human caretaker, us, with the same or even greater affection and familiarity as another cat.

The closer your relationship with your cat, the longer and more pronounced your contact. Studies show that when cats initiate the communication, the duration is much longer than when initiated by a human, which suggests that for the cat to invite and welcome contact, they need to feel secure in anticipating a positive and safe interaction. This includes the expectation of feline friendly handling, as in, knowing where to touch and pet a cat (side of the muzzle, under the chin, behind and between the ears in the direction of the fur) and where they would prefer not to be touched (mostly on the body). Some cats are more forward in attention seeking and will approach on their own. In this scenario, respond with one or two short strokes along the muzzle. The cat leaning into the contact invites more. For the cat not offering this response, advance from the side, lower the body, and extend a finger to gauge interest (cats initiate contact nose to nose). Proceed by offering cat specific appropriate petting - focusing on the head, just as another cat would in grooming or playing.

Cats groom each other to maintain social relationships. This practice called "allogrooming" is replicated when we stroke or groom our cats. Just as with petting, grooming is best done so it is a pleasant experience for the cat and not one they must submit to. Let the cat sniff the brush when introduced and pair with praise or a treat. Some cats will rub the side of their face straightaway against the brush. Finding the right brush (a slicker or pin brush are good picks) and confining initial brush strokes to the sides of the muzzle is a good start. Keep grooming sessions short in the beginning and leave off on that proverbial good note. Take the time to work up to full length body stroking, keeping those strokes to one or two only. If the cat reacts by tensing, hissing, or leaving, stop and go back to what was last working.

Not every cat enjoys being groomed, and being proficient in self-grooming, they really do not need our assistance here. The exceptions may be with arthritic or flat faced breeds where feline friendly grooming may be beneficial. Keep petting and grooming sessions at the length of time the cat responds positively. The cats that do enjoy the experience of brushing typically make it known – for those and even others, investigate and install the self-groomer brushes that can be affixed to doorways or room corners.

There are those cats that prefer these interactions to have limited duration but to occur frequently and others who would like longer sessions. Paying attention to how the interaction is going by body language, consent and soliciting contact are your best clues.

Cats will also seek out our company for resting. Either in proximity or where a part of the body is in contact. Allow for this "contact lying" that signals preferred associate where cats seek to rest or sleep touching (or close to another). Make sure when initiated or when your invite is accepted, to think ahead– a cat bed next to or on your desk when working, a pillow next to your spot on the sofa, a hand laid softly or forearm touching the cat next to you in bed.

Avoid direct and fixed gazes with cats. This caveat is pretty much universal for all animals even for people where we regulate and monitor the length and alternation of eye contact. Direct eye contact can be adversarial in much of the animal kingdom and this applies especially to cats as well. Soft eyes and slow blinking are signs of affiliation between cats and this can be communicated between our and other species. A slow blink directed at your cat can be understood and best communicated with softened body language. Thinking positive and or sentimental thoughts, while blinking towards the cat can help to effectively capture the energy you are trying to convey with the signal.

As much as we might like our cats to be interested in us, we can be the ones keeping cats at more of a distance unintentionally and then wonder at their lack of interest in us. Directing attention and speaking to pets is often more of a dog centered activity as opposed to cat centered. We may talk more freely and often to our dogs, offer treats and play more readily, and those walks and outdoor excursions all count towards intensifying our relationships with them. Equal consideration to cats can yield benefits for both parties. All animals respond to interaction and reinforcement. Research with captive servals showed a reduction in stereotypic pacing with training sessions either associated with cognitive stimulation or food rewards or both. Cats can also be readily trained with force free approaches and welcome the exchanges when they are positive ones.

Become versed in feline body language and communication. To know the effects of interventions, acceptance of attention and interactions, and how well improvements are being received, you must know what a cat is "saying." Ears, eyes, muzzle tension and whiskers figure prominently in cat communication along with body posture and vocalizations. Cats and dogs are not alike in expressions of body language, and every animal is an individual. Take the time to review what signals and strategies cats use, it's the first chapter in this book.

Greet your cat, using their name in a soft tone on entering your home or in a room you see them in. Studies show cats meowing more around people than other cats once they are past kittenhood. This can certainly be in response to all the talking we do as humans as opposed to reading cat body language.

Always, always avoid any sort of punishment with cats – Aversives and punishment have limited effects, result in displacement behaviors, and create fear and distrust towards the dispenser. Punishment includes not only what we might categorize as punishing but what the cat might perceive as punishing as well, this includes spraying with water, noise makers, tense body language, scolding, lecturing or physical force of any kind. Rather than seeking to penalize behaviors we consider bad, think what you would like your cat to do instead and make sure they have the set up/places to do them. Keep reading for more on how to do just that.

Interactive play is an incredibly bonding experience between owner and cat. This intervention can be one of the most effective to put into place to address stress, divert tension, and build affiliation. Play is also both rewarding for its own sake and necessary as an activity that just might satisfy a cat's natural predatory instincts. John

Bradshaw and Sarah Ellis, writing about how cat play mirrors hunting behavior note:

> "This overlap between play and hunting opens up the intriguing possibility that owners may be able to satisfy cats' predatory instincts simply by playing with them."

Not surprisingly, kittens will play more than adult cats. Intensity in play behavior can be seen declining after four years of age. Individual cats, however, may play consistently for years. Certain cats may not have had sufficient opportunities to engage in play behavior during development or in their earlier years but the joy of play is hard wired in the joy of the hunt for the cat. Give it time. Solitary object play has its own rewards while the unpredictable and dynamic nature of the interplay between person and cat (or cat and cat) cannot be quantified. Whether adding a new roommate or a new cat, playing with your cat is one of the top three must-have-bang-for-the-buck's to add to the life of the indoor cat both for what it brings to them and to the human cat bond or inter-cat bonding.

For that necessary interactive play, use fishing wand toys, "cat dancers" or pieces of string, etc., paired with a human on the other end to move objects imitating prey movements across or away from the cat's line of vision. Keep movements imitative of what cats' prey on: rodents, reptiles, and the occasional bird. Remember these targets are evading the cat and never attacking.

Offer routine, consistently scheduled, play for at least 3-5 minutes in the morning and in the evening. Add an extra session before bedtime for younger cats or to help with additional stressors or zoomies.

Schedule and routine. The free-living cat benefits from freedom and agency. Cats in our homes are hopefully not constrained by leashes or cages but they are subject to when we decide they should eat and what, where to eliminate and in what, where to sleep, with what to play, when we might interact with them, etc., etc., etc.

As contradictory as it may sound, routine and schedule offer a sense of control over events. Predictability over when things will happen removes uncertainty and the stress of expectations. A study on the predictability of daily routines with captive rhesus macaques compared the occurrence of caretaking routines such as feeding, cleaning and enrichment when provided unpredictably, when signaled by a door bell, at a set time and when on time and paired with a door bell. The greatest reduction in stress, as measured in diminished stereotypical and displacement behaviors, was seen when caretaking routines occurred at set moments or predictable times. Life in captivity was better with schedules. With the researchers noting:

> "Because these results are not necessarily applicable to animals that are given control over their environment or housed in a group setting, the management recommendation that can be made from this study is that temporal predictability of feeding reduces stress and anxiety and is thus beneficial to captive indoor single-housed rhesus macaques." (Gottlieb, Coleman, et al. 2013)

Schedule and routine surfaced as another possible variable to counteract anxiety in cats in a separate study focusing on stressors with domestic cats. Researchers at Ohio State University Veterinary Medical Center found an increase in sickness behaviors such as not or lessened eating or drinking, vomiting, loose stools, lack of grooming or interacting, etc. when:

> "cats were exposed to multiple unpredictable stressors which included exposure to multiple unfamiliar caretakers, an inconsistent husbandry schedule, and discontinuation of play time, socialization, food treats, and auditory enrichment." (Stella, Croney, et al., 2013)

The great 2020 pandemic is often faulted for creating stress for companion animals with the disruption of owners formerly being out all day for work and then home all day for work. More likely, stress and anxiety resulted as gone were the routines with the 9 to 5 work day. Routines which offered set times, when meals were fed, litter boxes were changed, caretakers came and went, and attention was provided predictably. During work from home, stay at home, when the cat got to eat or play or being attended to was anyone's guess. No wonder frustration can result.

Learning theory informs training protocols with reinforcement schedules to affect outcomes. We shape, lure, or ask for a behavior and reward it when it happens. The sooner the reward, the closer it is associated with the action requested and performed. Whether in training or naturally occurring, behavior that is rewarded repeats. One of the most effective and the most frustrating reinforcement schedules is the intermittent one. Not knowing when a reward is coming can keep performances forthcoming but at the cost of high anxiety. Consistent delivery of rewards and acknowledgements can greatly reduce reactivity. Schedules and routines include life rewards such as meals, walks, playtime and more.   I have seen over and over in the field, that as simple an intervention as this is, it is one of the most effective to put in place and for humans, often the easiest to institute.

Put schedules and routines in place to keep regularly and time them according to what events are most in keeping with natural behaviors for the cat. Look to morning and evening meals and play sessions to be fed as close to dawn and dusk as possible in keeping with when cats are the most naturally active, seeking prey and interaction. Pairing feeding and interactive play with our own morning routines can help to insure their happening on a routine basis for our own human convenience.

Cat toys. Toys for solitary object play are important for cats too. Experiment with a variety of different types of toys to find the most appealing and give the cat time to decide about them. Stuffed catnip scented toys big enough to grasp and rake, appeal to some cats, while crinkle balls, plastic bottle caps, or fur mice that rattle can be a universal hit requiring only a paw to propel. Never underestimate the appeal of packing paper and those carboard boxes for fort making that come in the mail. Do not forget to also leave these cat toys for independent play available and not piled in a corner or stashed away when you are not around.

Catnip and valerian root. What is it with how we feel about catnip for cats? We already know that most cats love catnip even if we are not sure about how good it is for them. Cats just seem to like it a little too, too much and that must make it wrong- right?

Get ready to relax about catnip, veterinary colleges and veterinarians, like the vet school at Texas A&M and VetStreet's Dr. Marty Becker confirm the plant is nontoxic to cats and can actually benefit them. Adding catnip or valerian root, another relative in the catnip family, to cat toys, scratchers or cat furniture on a routine basis can heighten the appeal of those spaces and encourage use when new and if underutilized.

While the leaves and the flowers of the catnip plant may be the most well-known, it is not the only plant that cats are in love with. Silver vine fruit, Valerian Root, and Tatarian honeysuckle wood are also cat attractants, with the last plant, classified as invasive in certain parts of the United States.

To test which of these plants had the biggest impact on cats, a 2017 study looked at 100 cats in home and shelters along with 14 big cats (bob cats and tigers) and their responses to each one. The scientists were looking for the "cat nip response" (yes, there is an official term for that) defined as:

> "sniffing, licking, shaking their head, rubbing (chin/cheek) and rolling on their back, sometimes accompanied by drooling and raking," (Bol, Caspers, et al. 2017)

Of the 100 domestic cats studied, 79 of them responded positively to the silver vine, 68 to catnip, 53 to Tatarian honeysuckle and 47 to valerian root. While the response to silver vine (79%), was close to the response to cat nip (68%), the response of the cats to the silver vine was significantly more intense. So even as both plants had close to the same initial appeal, the sliver vine was more engaging.

Looking at the differences between how older and younger cats responded to cat nip, the researchers found less mild and more intense responses in younger cats compared to the older cats. What about the big cats? Tigers are known to be less caring than other cats about scent enrichments and true to form, only one out of nine tigers responded positively to cat nip and all tigers either ignored or disapproved of the silver vine. Not a good endorsement for either. Five bobcats were tested with four responding positively to silver vine and one to catnip. The bobcat's response was among the most intense seen lasting from several to 15 minutes.

For those of us, myself included, who are interested in the other plants cats love, you can find powdered valerian root in drug stores where it is sold for its tranquilizing properties (catnip is said to have a similar effect for humans). Valerian root powder is less gritty than catnip to rub onto cat furniture and has added appeal in possibly offsetting some stress in multi cat households along with distribution of resources. Tatarian honeysuckle is considered an invasive species in many states in the Northeast, including mine, so it may be difficult and problematic to obtain. The part of the silver vine that the cats are attracted to is the fruit of the plant after the gall midge has matured and moved out of it. Gall midges are only found in certain areas of Asia but powdered silver vine galls can be found online and were used in the study. The wood of silver vine plants is sold online and available to experiment with how attractive these might be for cats. Always exercise sufficient caution in selecting the pieces and supervising the cats with these or any item. The wood may be too thin or brittle and can be a choking hazard (another reason powders are a good idea). As with any product sold in the unregulated pet market, be doubly cautious and supervise.

Provide puzzle food feeders. Feeding a portion or all your cat's dry food diet this way is a huge positive for your cat's activity budget spent on the things cats want to do, offers more valuable choice, and control over environment, stimulates neural activity and is, additionally, intrinsically fun and satisfying.

Studies show that cats having free access to food on an ongoing basis are less likely to exhibit stress related behaviors such as pica or ingesting foreign objects. Avoid slow feeders designed to prevent pets from rapid eating, this is frustration not enrichment.

Reset access and control by offering dry food not in plates, bowls, or timed mechanisms but in feeders that stimulate thinking, puzzle solving and natural behaviors of batting, pawing and scooping food, kibble piece by kibble piece. Large and complex tray feeders that stimulate natural behaviors are the ones to identify for best use. Set rolling feeders on the largest openings first. Initially adding valued treats to dry food and showing your cat how to use the feeders can also heighten appeal.

Treats and dry food can also be scattered about for sniffing out. Wet food puzzle feeders have yet to be perfected for cats. Portion out wet food in flat dishes, spreading the food out across the surface of the plate to allow for savoring and prevent whiskers from brushing against the sides of a bowl and becoming stressed ("whisker fatigue").

Incomplete nourishment impacts the body and the mind. When we deprive adequate sustenance, we limit how well the body can function, how healthy the animal can be and how wellbeing of mind and mood are affected as well. All animals are attracted to not just nutrition in food but palatability as well. Food must taste good to be enjoyed. In a natural environment, cats will choose a perfectly suited diet of needed vitamins, amino acids and fatty acids and eat it. There is pleasure and satisfaction in a good meal. If we are taking the "they will eat when they are hungry" method of feeding, we are setting up yet another aspect of frustration of the life of an indoor cat. You do not have to make your own cat food but you can give the enjoyment of the highest quality cat food available.

When it comes to water, make sure it is fresh, away from food bowls and think about cat fountains and make sure you can maintain them (if not water bowls with fresh water are a better choice). We may think a cat sitting by a faucet is adorable but a free-flowing source of water is far more satisfying for the cat than having to beg for you to turn it on. In a natural environment, cats avidly prefer running water – drinking away from prey they are consuming or storing. This serves as a built-in instinct not to contaminate water sources.

Remember, cats being cats need to scratch. Scratching feels good, allows for a needed long stretch to flex muscles and strips older keratin layers to groom claws. Confine scratching to acceptable areas by providing stable scratch boards for your pet in areas they will likely use them in. Sisal or corrugated cardboard cat scratchers are intensely reinforcing. Studies show sisal is preferred by more cats but experience may show cats can switch off between both. Test out both fabrications and make sure they are sturdy to offer resistance and not sway or topple over. Avoid hanging scratchers which may look appealing but with all their movement, difficult to manipulate for a good scratch session. Add at least one scratcher to every room and don't forget rubbing in valerian root, dried catnip or applying catnip spray to the scratcher surfaces to increase their cat appeal.

Cats spend a good part of their day time sleeping so make sure to provide soft, comfortable areas such as fleece pads or cat beds with raised sides. Nothing may match a favorite spot on your very own bed. Cats, like dogs, have a highly developed sense of smell and the scent of their favored human guardian is particularly comforting - making your bed and its linens particularly attractive for your cat. In multiple cat families, cats often share favored resting spots, either by taking turns or at the same time, if their relationship is a friendly one.

Your cat's resting spaces should have at least three sides, be located at both ground levels and at higher vantage points to provide your cat comfort and security. The cat's innate nature is to seek higher spaces to securely survey their environment or to be concealed in safe or nesting spots. One of the reasons we often find counter-tops and table tops to be the default when alternate necessary spaces are not provided. Adding vertical resting places expands on the available feline real estate available for cat naps and time shares, more cat specific room, can also lessen stress, especially where multiple cats are at home. Cats are not known for reconciling conflict through negotiation strategies preferring retreat and time to develop tolerance. Time outs leads to solitary down time or at least time out of each other's line of sight and the more places to indulge this, the happier the home.

Provide cat furniture with carefully placed and sited cat baskets or beds, perches, window seats and cat towers. It can seem that except for cardboard boxes (ask any cat, the best cat beds often really do come in the mail), adding new cat spaces can take time to acclimate to. Often because, where we put new cat items can make the difference in being used or ignored. Location matters, so does ingress and egress. Evaluate the surfaces and area for the most cat friendly access and placement, including various access and exit points where more than one cat is using surfaces. The center of a room is typically less than a desirable location, while next to a window or in a corner of a room are better options. Under a chair against a wall might be great for a cat bed but under a chair in the center of a room, not so much. Consider areas where the cat is already using space. What is defining about those spots? What can be replicated in other places?

Characteristics and number of what is being added counts too. The raised sides of cat beds and igloos provide security along with comfort and warmth. Ideal is at least two different resting areas per cat, one at ground level and one raised for optimal contentment. Cat tunnels are another good and versatile add when it comes to placement.

Increasing horizontal and vertical surface area for a cat to utilize amplifies and enriches your pet's living area. Cat towers offer multiple levels for cat interaction but finding floor space for cat furniture in tight quarters may be unrealistic in certain living situations or in a city where space is at a premium. Cat shelves offer a space solution that maximizes additional vertical cat resting areas without using precious floor space. Another often overlooked option, is placing a cat igloo/house, or bed on existing furniture in a room: a dresser, credenza, desk, etc., can also be a great answer in where to find room to add needed feline vertical resting spaces.

Classical music. Studies have found that exposure to classical music is beneficial for many animals, including cats, while exposure to rap, heavy metal and other discordant music is not. Play classical music at a level slightly lower than you would find comfortable keeping in mind the cat's superior sense of hearing. Leaving a classical radio station on, especially when you are not home, in one of your rooms (allows your cat the choice to listen or not), adds in the soothing voices of those velvet toned announcers for positive associations and comfort. And remember to keep cat conversations directed to the cat in a softer voice. Baby talk is often better directed at dogs while our elevator voices are more favored with cats.

References

- Baker, P.J., Molony, S.E., Stone, E., Cuthill, I.C. and Harris, S. (2008), Cats about town: is predation by free-ranging pet cats *Felis catus* likely to affect urban bird populations? *Ibis*, 150: 86-99. https://doi.org/10.1111/j.1474-919X.2008.00836.x

- Bol, S., J. Caspers, L. Buckingham, G.D. Anderson-Shelton, C. Ridgway, C.A. Tony Buffington, S. Schulz, E. M. Bunnik, (2017) Responsiveness of cats (*Felidae*) to silver vine (*Actinidaia polygama*), Tatarian honeysuckle (*Lonicera tatarica*), valerian (*Valeriana officinalis*) and catnip (*Nepeta cataria*). *BMC Veterinary Research*, 13:70/.

- Buchan, C., Franco, A. M. A., Catry, I., Gamero, A., Klvaňová, A., & Gilroy, J. J. (2022). Spatially explicit risk mapping reveals direct anthropogenic impacts on migratory birds. *Global Ecology and Biogeography*, 31, 1707– 1725. https://doi.org/10.1111/geb.13551

- Ellis S.L. (2009). Environmental Enrichment: Practical Strategies for Improving Feline Welfare. *Journal of Feline Medicine and Surgery.*11(11):901-912. doi: 10.1016/j.jfms.2009.09.011

- Gottlieb, D. H., Coleman, K., & McCowan, B. (2013). The Effects of Predictability in Daily Husbandry Routines on Captive Rhesus

Macaques (*Macaca mulatta*). *Applied Animal Behaviour Science*, *143*(2-4), 117–127.
https://doi.org/10.1016/j.applanim.2012.10.010

- King, B. (2013) Do We Really Know That Cats Kill By The Billions? Not So Fast, *NPR,* retrieved April 2, 2024
https://www.npr.org/sections/13.7/2013/02/03/170851048/do-we-really-know-that-cats-kill-by-the-billions-not-so-fast

- Stella, J., Croney, C., & Buffington, T. (2013). Effects of stressors on the behavior and physiology of domestic cats. *Applied Animal Behaviour Science*, *143*(2-4), 157–163.
https://doi.org/10.1016/j.applanim.2012.10.014

- Stella, J.L., Buffington, C.A.T. (2014). Individual and environmental effects on health and welfare. In D.C. Turner & P. Bateson (Eds), *The Domestic Cat the biology of its behaviour*, (3rd ed., pp. 186-200). Cambridge, UK: Cambridge.

WHY CAN'T WE BE FRIENDS? ASSIMILATION STRATEGIES FOR NEW CATS AND KITTENS

A new kitten! Because: "ADORABLE!!!!!" "Kittens are easiest with older cats." A new cat! Because: "Your only cat: "needs a friend," "needs someone to play with," "is lonely when you are not home," "misses the cat you lost," "would stop scratching the couch if you got her another cat," "would stop scratching you," "would _______,"– fill in the blanks. You know you heard them before, maybe even said a few. There are even more well-meaning advices and suggestions.

Next is the cat distribution system. We all know the stories and each of us, doubtless, has at least one of our own to tell. You choose: a rescue cat that breaks your heart, a rescue cat that looks just like the one you lost, any rescue cat, that foster situation that was supposed to be a temporary one failed, another cat showed up on your door step for the umpteenth night in a row, a loved one of your own just passed and their cats need someone to love them, you miss your oldest, best cat every day. For these and so very many more reasons, a new cat or kitten seems like a good idea at the time. A good idea, until the original resident cat finds out about the infiltration and then what?

Has anyone asked the cat's thoughts on the idea? In the cat's natural environment, cats live and socialize with family members or choose select affiliates to socialize with. Cat societies are accessed by outsiders on a strictly by invitation only basis. Dependent on resource availability, communication, preferences and feline conventions, new cats can and do enter existing colonies. As hard as it is from outside of the frame to determine criterion of acceptance, there are factors that easily stand out – body language, time, and distance. The ability to pass time meaningfully and without frustration must be hard wired to the cat as predator and as a member of cat society. Time to steady, seated or settled over haunches on agreed upon perimeters to parse out others' intended meanings and/or convey them, time to reduce or expand distance. Time also measured in stare downs, standoffs and in the deliberately low and slow slink past another cat as conflict is avoided. Waiting time goes with hunting, the game of cat and mouse has no stop watch, it just is, staying engaged in pursuit of goals either out of ready sight or in the future. Other factors are at work in the progression to inclusion. How that is accomplished, what other conditions, needed, carefully calibrated, and well known to cats exist, the particulars of which, humans have yet to uncover.

Multi-cat aggression, is second only to unacceptable elimination or house soiling, in concerns raised by cat owners in connection with both surrendering their animals and consulting animal service professionals. Indications on both fronts, that what we are doing at home is often not in concert with what the cats would do on their own. When we do the picking and choosing for cats, mostly not to their liking or their time tables, research and practice can show that allowing for an appropriate introduction and integration process is most effective when it includes those vital components that reflect essentials seen in optimizing human cat interactions; distribution of resources and feline behavioral needs.

Looking at ways to ameliorate deficiencies and highlight necessary elements in the assimilation journey, along with human liking of simple systems for companion animals - no intervention works if it is not applied - I have found, **four P's:** No **P**unishment, More Cat **P**laces, Interactive **P**lay, and **P**riority of the original Cat, to be instrumental in better outcomes. There are other necessary aspects, wholistic approaches are the foundation of any behavior work. Integration of a new cat into a multi-cat family is best done in increments of increases whatever the variable for established resident cats and newcomers.

Many of us can tell stories of bringing cats home to multi-cat households. And tell stories of those introductions done without the fanfare or bother of protocols, thresholds, engineered steps or measures. It can seem, looking back, that given enough space and time, new cats got along with old cats after a while. An occasional hiss or tangle might happen along the way. Maybe the two were never the best of friends but they appeared to have worked it out. Independently.

Such an approach does follow surely, the main ingredients apparent in cat assimilation – time and distance. Time and distance to desensitize and habituate to the presence of the newcomer. Gradual exposures below threshold, over a staggered and prolonged period, can effect much in desensitization and habituation, including successful assimilations. We certainly see this in free living cat colonies where timed and imperceptible to us, but not cats, step-by-step moves forward to acceptance are meted out according to distinct and measured feline signaling, formulas and individual requirements. Difference being, the vagaries of the home environment can make thresholds tricky. Space and resources can be uneven or constrained and these can be crucial limitations, pushing how much is at an individual's threshold beyond levels of tolerance to stressed and reactive. Accessible exits, sufficient distances, finding other resources, habitats or colonies, are not available temporary or permanent options to indoor cats. Add the concern as to what can be left out; counterconditioning or developing positive associations on which to build relationships in a shared environment with its finite resources, including human attentions and interventions.

Introducing kittens to an older cat is usually an easier sell for the original cat. Nature designs juveniles to be less threatening and elicit caretaking in adults. With intact animals in natural environments and in times of extreme stress, infanticide can occur but this is highly unlikely in the home environment with appropriate resources.

Two kittens are often a better idea than one. The kitten's activity level will far outweigh the older cat and keeping bonded pairs and cat families together advances their welfare and gives them an evenly matched play partner. Caution needs to be taken when bringing home kittens of both sexes that have not yet been neutered. Time passes quickly and you do not want babies having babies. It happens. Schedule those surgeries to get on provider's calendars, even if you are intent on waiting until they are a little older and keep close watch on how the kittens are interacting. Separating male and female kittens from each other will be necessary if the female shows signs of going into heat until after both have been neutered.

In a natural environment, taking stock of surroundings at a distance first is the safer approach and one to follow at home. Seen or unseen, the introduction process begins as soon as the new cat or kitten enters a dwelling. Both cats become aware of each other's presence through scent and sound. While the cats will hear and smell each other, seeing each other initially is often best delayed.

For starters, set up a separate room for the new cat. Limiting the new environment to a single room will enable the cat to navigate this uncharted territory with all the novel infrastructure and attendant sights, smells, and sounds (including you) on a smaller, less stressful scale. Choose the room that affords daily contact with you, such as a bedroom or office, and helps to mainstream the cat into your family life. Avoid attics, garages, and basements and keep bathrooms as a last resort. Provide cat furniture, including a cat bed, raised resting space or perch, scratching pad or post, toys, food, and litter box (keeping the litter box a good distance from the food and bed).

Do not be alarmed if there is a lot of hiding initially. This is normal, the cat needs to get acquainted in this new place with the environment and to trust in its safety, which can only happen with the passage of time.

Resident cats may display little or no interest in a closed off room, aside from the occasional raised head to sniff on a walk past. Unless of course, this is a room they have been using. If your original cat sleeps with you every night, installing the newcomer in your bedroom will be problematic for the resident cat. Consider the best room for all concerned.

To facilitate the adjustment process for each cat, resident, or newcomer, keep in mind that cats are extremely sensitive and reactive to changes in their trusted environment. As we are responsible for and a principal part of this environment, how we interact and influence is key in familiarizing them to both new situations and individuals for more successful assimilations:

Be a calm and soothing presence. Your body language, eye contact, and tone of voice, will set the scene for every encounter. A calm and neutral demeanor is one to put on with any animal. Remember, cats may vocalize more around humans than each other, especially around mealtimes, but those meows are most probably for us and not for them. Silence can be golden with cats.

Announce yourself and acknowledge the cats with the use of their names. Begin with the name of your resident cat when you first enter your home and/or a room they are in. Even if you do not see your original pet, greet them before the new pet on entering each space. Chances are that cat hearing can register your greeting wherever they are at the time. And such a verbalization is a good reminder and affirmation for us to keep primary in our priorities.

Keeping the cat that came first also first in your interactions is imperative. This is one of the four P's cornerstones: **<u>P</u>riority of the original pet**. No matter how adorable, in need or engaging the new cat is. Keep number one, number one. Your affection, attention and engagement are just as much of a resource as raised resting places, feeding stations, litterboxes, etc., to appropriately distribute. There is usually a lot of attention paid to a new cat or kitten when it comes into a home and this is stressful for the resident pet both because of the change in the family and the loss of attention to them. A new cat or kitten is not just an interloper for the resident cat and an insistent and irritating demand for interaction -they can also deprive resident cats of our interest and caring. Children are the not the only family members that can feel neglected when a new baby comes home.

Make sure to also greet the new cat or kitten softly by name each time you come into their space. If they continue to hide as the days go on when you enter the room try sitting down on the floor directly alongside them. This should be sideways to the cat and not facing them. Speak gently (this is a great time to catch up on some reading - out loud) and give the time for cat or kitten to emerge. Ignore them until they approach close enough to be within easy reach of touch and because cats greet nose to nose, muzzle to muzzle, extend a finger to initiate contact and gauge for a welcoming response, as in the cat sniffing or moving their face forward to your finger before extending a hand for petting. Keep the hand below muzzle or at muzzle level and not from over the cat's head so it can easily be tracked by cat eyes.

Confine petting for the time being to the head area only, stroking the top and sides of the head and behind and around the ears. Keep those strokes to one or two initially. For those cats or kittens who take to contact right away, continue. The idea is looking for consent to the interaction by taking your cues from the cat in bunting, social rolling, pressing into contact and relaxed body tension. Offer a slow and steady blink which is an affiliative signal and response in cats. Stiff or rigid posture, hissing, flattened ears, or whiskers, looking away, shrinking and such, all have "No" or "Not Yet" written all over them. Listen. And keep going in trying for more contact at a later time. Expand socialization to human handling by making this consistent, routine, and welcome.

Keep mealtime and playtime on schedule to afford a sense of control over events in a routine that has been interrupted with the presence of this new kitten they did not ask for.

Kittens need to eat three-four times a day depending on age. Your older cat typically has two meals a day. Link kitten feedings with positive associations by offering your original resident cat a delicious treat immediately before or immediately following kitten feeding.

Learn cat body language and know what it is "saying." Communication with most animals, including humans, is mostly nonverbal. Life with cats requires learning cat body language. The entire cat is involved in signaling with the foremost attention usually paid to the movements of the highly mobile ears, whiskers, eyes, tail, and body positioning. Ritualized display signals convey squadrons of information. Stress and tension are typically conveyed in stiff body positioning, piloerection (hair standing on end), flattened ears and whiskers, dilated pupils, tightened muzzles, trashing tails, guttural vocalizations and more. (Please see the first chapter on cats in this book for a more detailed discussion on cat body language.)

Understanding body language cues us into emotional states and responses to environment. It also can indicate consent or concern. Understanding when a behavior is an invitation to move forward and when it is a rebuff is necessary for our understanding of steps onward or backward and how interventions can best facilitate this if necessary. Such understanding also sets a stage to look at context and environment.

Avoid any punishment -no hitting, scolding, recriminations, scat mats, spraying with water bottles or shaking noise-makers of any kind. None of it. Including those things cats may perceive as punishing. Including lectures. Avoid them. They are ineffective at best, no matter who they are directed towards and the emotions associated with them are rarely positive for the person delivering them. Cats pick up on all of this. Additionally, research shows simply removing punishment may be the one intervention that is most effective at reducing aggression in multi cat homes. De Porter, Bledsoe, et al. looked at a pheromone diffuser vs a placebo to treat feline inter-cat aggression. The researchers removed punishment from both study and control group while allowing for this as a confounding variable:

> "Punishment was discouraged (e.g., spraying cats with water) and positive redirection strategies (e.g., food) were suggested. This change in management at home may be sufficient to explain the drop in mean scores between D–7 and D0, irrespective of group. These management strategies, while confounding, were undertaken in consideration for the welfare of the cats and because most interventions that reduce anxiety may be less effective if punishment is imposed. This may have contributed to a relevant portion of the improvement seen in both groups throughout the study." (DePorter, Bledsoe, et al. 2019)

Punishment is cruel, highly stressful, interferes with learning or creating positive associations. Always remember that cats do not respond well to correction or harsh treatment and tend to associate that event with the person and new cat instead. This is the most important of the four P's: **No PUNISHMENT**.

More cats? More cat furniture. Another of the four P's – **More Cat Resting and Raised Places**. Experts recommend, with good reason, multiple litter boxes, food and water bowls, toys and at least one raised resting space and one floor space (igloo/box/bed) per room, per cat. With cats being both prey and predator, retreat to secure vantage points and from danger is more than innate in their behavior. Adding more horizontal resting areas expands cat territory, allows for retreat and that sometimes -necessary hiding/alone time needed in a multi cat household while offering more cat friendly areas for cats to share or time share.

Expand vertical space in surfaces, these can be found in a cat bed or igloo on a dresser, cat trees, or shelves wide enough to turn around on, placed carefully and allowing for access and exit points for multiple cats. Placement is also important for use. Oftentimes cat furniture may go unused because of where it is sited. A middle of a room is an iffy choice. Next to a window or against a wall, or on a corner wall, lets cats look outside from a high vantage point and survey their interior and exterior world safely (more in the chapter on making a cat friendly home).

Floor level resting spaces are also necessary, igloos, cat beds, crinkle tubes and shipping boxes all work nicely. Placement counts here too. Except for shipping boxes, the middle of the room is usually not a preferred spot. And do not forget toys for object play, faux mice, crinkle balls, stuffed animals, etc. Cat time to familiarize with cat furniture and objects is a definite thing. Allow for longer periods than humans might typically expect for cats to accept new well sited additions to their environment. Rubbing cat attractants such as Valerian root on toys and furniture can heighten their appeal and add opportunities for scents they are drawn to in their natural environment.

Add more scratching opportunities for everyone too. Sufficiently stable corrugated cardboard or sisal scratch pads with catnip/valerian root rubbed in offer a soothing and stimulating experience especially beneficial around the presence of another cat. While there is still a lack of consensus on the function of scratching as communication in cat societies, the exercise of the behavior is a necessary one to flex and keep nails and muscles in good condition.

Food and foraging are integral to cat welfare. Wet food should be fed to cats separately and each cat at a distance from the other. Resist the idea that pushing cats together by pushing food bowls together will help in bonding. It can have just the opposite effect and can put pressure on cats to compete for or to scarf food. Cats are not communal eaters or little lions sharing meals, they need space from each other when eating. Even as you progress to same room feeding, space feeding stations for wet food far apart as in the other side of the room or at separate ends of the wall.

Providing the entire portion of each cats' dry food diet ad libitum (freely fed so always available) in a puzzle feeder can detract from vigilance and attention on the other cat and offer engagement and allow for natural behaviors. Add puzzle feeders to both cat's environment. Large and complex tray feeders or rolling feeders allow for natural predation and play behaviors such as batting. Avoid "slow feeders" which can frustrate rather than satisfy. As assimilation continues, selecting appropriate large, flat feeders that are both sizable enough and positioned so both cats can freely access it, can also offer another opportunity for positive associations around each other.

Cats rely heavily on scent to process their environment. Once all cats are in the home, even without the visual, the presence of each cat is already known to the other through their sense of smell. They are also aware of their distance. Scent exchange is a good first step in introducing the residents to the newcomer and vice versa. Have several cat beds and resting spots in each cat's area and swap one out to exchange, this adds the scent without the confrontation.

Try using soft cat toys (and at least one catnip filled toy) and rub the toys behind the ears of the cat, along the muzzle, on flanks and base of tail. Exchange cat scented toys in each area. In addition to exchanging cat scents with the toys, add yours to the process to create a group scent: Soft stroking the new cat along the side of the muzzle and then the original cat, layers three scents to be processed. Another way to add to group scent is to retrieve a small piece of clothing from the bottom of your hamper (skip rubbing this on humans since it is already scented) which can be rubbed along cat areas noted above and one item left in each cat area. Scent exchange should begin after the first two days.

Begin interactive play time with all cats. This is another of the four P's, **Interactive <u>P</u>lay**. Take two minutes to begin, in the morning and in the evening and play with the cats separately using an interactive toy on schedule. Note on time – this can be longer period but not too long to become repetitive or boring for either one of you. Setting it at only several minutes comes from making the prospect less of a chore for the humans. Tying it to our own routines can also make interactive play times easier to accomplish on a routine basis. Each factor is important.

Always start playing with your original cat first. Fishing wand toys where you can engage the cat by drawing a lure across or away from their line of vision fit the bill nicely here. Make sure to have the toy at the end of the string retreat from kitty and not attack kitty. Think of mice, snakes, rats, and birds trying to evade your predator. These are the movements you are trying to imitate. The lures on these wands vary from feathers to objects to cardboard pieces. You may need to try several to see what best attracts the attention of your cat and to work on your approach to engage the cat's interest. Keep it low key, no high-pitched voices here -better suited for dogs not cats. Slower movements with stops and starts rather than flailing about. Watch the cat for their response – which moves get more eye contact and engagement? Those are the ones to build on.

Some cats are not familiar with play or stressed by a new environment. Play with them anyway without pressure to interact. Give them space to register and not feel crowded. Two feet away is a good place to begin with, if the cat shows signs of stress, back up. Start with a minute and build up to two, trying to entice interest in the movement of the wand. Silence with reticent players can be better than sound.

For the first week or so play with your newcomer in the room they are in without asking for any other interaction (unless actively solicited). Try to stage your play sessions with your original cat as close to the door of the newcomer's room as possible. Should your resident cat seem more interested in the door than in playing, move away from the door in the opposite direction to get the attention back on play. After several days, you can move forward gradually as long as the cat remains focused on play. This can also be done by two individuals playing at the same time with each cat.

Individual cats respond differently to play but subdued or ebullient, there is a response and activation of different neural pathways from the stress and vigilance of interloper cats in home territory. Resist "over playing"– keeping these sessions on schedule provides the control of routine to the cat and avoids the unavoidable frustration of intermittent reinforcement of unpredictable play, as in you will not always be able to play on demand and that can be highly stressful. Hide fishing wand toys when not in use – this is important not just for novelty but for safety and scheduled use.

Institute routine and schedules. In a world where we are in control over most, if not all, of the resources for companion animals, afford the control of routine and schedule. Keep mealtimes and interactive play sessions as close to being on schedule as possible – studies show cats benefit from feeding and caretaking at routine times rather than unpredictable ones. Life can happen in between, so even when schedules are upset make sure events happen.

After the first week of the above progress to supervised interaction, play in eyesight of each other. Install removable baby gates in the door of the newcomer— one on top of the other. While the gates are in place practice this routine:

- Use the interactive toy each cat prefers most and play several feet away from the opposing cat location.

  Increase play sessions for up to five minutes.

- Offer high value treats to each cat in view of the other. Small pieces of turkey or chicken usually rocks. For this step, the cats can be within six feet of each other which means equidistant is where you should be standing. Do not push distance.

- Toss treats behind or to the side of cats not in front of them to avoid pressure to advance.

- Feed dinner or breakfast just in sight of each other but not closer than that. Try this, at the end of the week. Place food stations a good ten feet away from the gates. Again, we can put too much pressure on the cats during assimilation and after by placing food too close for comfort. They are not social eaters, even if cats are eating out of hunger or to best each other for food, this does not indicate comfort or formation of positive associations. They need separate food plates and to eat at a comfortable cat distance from each other.

Ideally, you would do this every day but it may not be possible. Even if you miss a supervised interaction do make sure your interactive play session happens.

In scenarios where installing baby gates is not possible, is holding the door separating the cats ajar several inches with your calm nature and supervision possible for a minute or two something you can safely do? Is this something you can repeat several times a day? The idea is for the cats to be able to visually register the presence of the other cat from a safe distance without raising thresholds and in doing so be gradually desensitized. A door opening of about three inches provides registration and visual exposure but prevents physical interaction. The cats should not be close enough to charge or feel the need to respond reactively to this. If that is happening, your tension may be apparent, the cats are too close to the door, the opening is too wide or you are holding it open too long. End the session. Carefully close the door and take the time to wait (at least half an hour) before trying again, so there is a break before and between further trials. Consider what may have been the obstacle and how to overcome that, always safest to go back to what was last working.

Exchange environments. This can happen after the first week or ten days of interactive playing. (Interactive playing continues throughout this process and afterwards.) Allow the newcomer cat to explore the rest of the house or apartment and the resident cat to hang out in the newcomer's room. Stay with the original cat during the first week or two of this process and advance to allowing cats in exchanged environments when you are not there. Do this when you are most comfortable to be out of the room - eating dinner, watching a movie etc. Progress on both phases (being with the cats and without) gradually from an hour, to a few, to a day to overnight.

"Read" where things are before progressing further with your integration. Pay attention to the behavior of your cats. What does the body language say for each cat at each step of the assimilation? What is the body tension telling you? Are you seeing "airplane ears" or neutral ears? Whiskers forward or back? Head hunched over the body or relaxed? Tail twitching? What else?

Know that if things are not proceeding in a positive direction at any step, this means going back to what was last working and starting from there is a good strategy. Should you have more than one original resident cat, use the most resistant cat as your touchstone here for timing. Know also that even as you might not be able to see resistance in a seemingly accepting cat, stress can still be internal. Since you can more easily observe, the most resistant cat, that is the gauge to go from in accepting the newcomer cat as belonging to "her" home/space/apartment. To that end, paying tons of positive attention to this cat by the door, laying treats leading up to the bowl, playing next to the door, any good associations you can create, can help. Make sure to also include the other resident cat(s) in your attentions.

Again, with cats go slow to go fast. Positive or neutral, small exposures in time to each other are always better than longer ones. Taking more time with any part of this process is a good thing not a bad one. If you see stressed, reactive or displacement behaviors from either the new or the original cat, this can indicate steps forward or timing may need to be reevaluated.

When permitting cats access to each other, make sure that the original cat is in the room before the newcomer cat. And do this after feeding times not before. Cats can be competitive over food and are very aware of when mealtimes are. Anticipatory stress can be a positive when winding around your legs for that canned food to be open but add in a strange cat and the recipe changes. Feed first and keep the resident cat in residence in initial shared space encounters. The cats are aware of who came first and who is looking for acceptance as newly introduced.

Cats will defend or stake out territory depending on personality, history, and a need to possess newfound territory. They have ritualized display behaviors they use in new spaces and around new cats. It has been said that cats have no conflict resolution strategies and employ avoidance but watching cats together tells a different story. Seeing cats and kittens at play can show when play behaviors are too exaggerated and border on becoming hurtful the response is a stilling and tensing of the body followed by a vocalization, a sharp meow, or a hiss. Unheeded, there is retreat, the game stops. Overtures can start the game again. Warning signals can similarly stop conflict progression and each transmission of information processed has value in developing effective communication.

Field observations show how frequently distance and time is used by cats as a mediating force in new environments and with new players. There is a certain area of personal space that is determined to be a safe distance to be from a stranger or hostile cat, and a certain amount of time when that distance is fixed. There is a definite rhythm and exchange between cats in how and when they can go forward from that set distance toward the other or stay in place. Kittens get such understandings quickly, there is an immediate response to the rebuff of an older or original cat. They appear very sensitive to avoiding conflict. Watch cats and kittens together and see how their gaze is fixed intently on the other cat to monitor the slightest response in ear, eye, whisker position and body positioning and tension, as they lower in place or slink and creep sinuously, slowly away. They may settle at a determined space apart from the previously objecting cat.

Stationing in place at a certain distance not opposed to by the other cat has benefits. Stationing begins with standing in place, then sitting and can progress to repose with paws folded under. And from there to pass time, nap or even sleep for an extended period in the other's remote presence. For the cats, these behaviors communicate more than simple alleviation of conflict. Whatever feline natured conventions of habituation and integrations are being accomplished, much is yet unknown to humans. Gradually over time they decrease, with consent of the other, the distance to the resident cat. And it is in this fashion, perhaps a bridge of time, constructed with components of some sort, of protocols and context they have determined amongst themselves, is created in inching forward to incorporation.

As kittens settle in, there may be more romping and insistence on engaging the older cat from a single kitten. Older cats may alternate play with hissing and swatting as they establish boundaries for acceptable interaction and that includes acceptable play and time devoted to it. Kittens learn the rules of social interaction from other kittens and from older cats. Typically, kittens have more motivation and desire for more frequent and extended play sessions. The appetite for the level and duration of play for a kitten exceeds that of an adult cat (why again, two are an easier sell initially). The older the resident cat, the less generally tolerant for kitten crazies. Communicating this to kittens can take several forms from disinterest, moving or turning away, vocalizing or contact as in an extended foreleg or swatting, etc.

Hissing is a preliminary vocal warning and although it may sound terrible, it is relatively minor in terms of aggression, more "bark than bite." This also holds true for the swatting, a definite warning display but the claws are retracted. You can see how effective, for both the kitten and the older cat, the use of distance increasing behaviors can be. Marking territory and time or learning social skills, an ebullient approach by the kitten is rebuffed by a hiss. The kitten stops, retreats, settles into a crouch, watches. Learning is happening – a change in behavior based on experience. When the kitten tries to engage again, the hiss may be followed by a growl or a swat. Rarely, if ever is there a third time. In cat society, acceptable distance and time are integral to relationships. Degrees of how to advance in both depending on context are being resolved in these interactions.

Keep an eye on cat/kitten interactions. If the kitten is repeatedly pressing for play where your older cat is adamant against it happening, redirect kitty by offering regular interactive play with a fishing wand toy to distract and engage his attention. Kittens need to discharge energy, learn from play predatory behavior (think of that fishing wand toy object as a stand in for a mouse, snake or bird) and enjoy the sheer fun of play for its own sake. Play will have an added benefit of interesting your original cat as well. Even without engaging, your older resident seeing play, will have mirror neurons stimulated used in play and in that process, offset some stress, anxiety and vigilance.

The dance of acceptance you are helping along still has cat rules of time, distance and how to act as a newcomer here. They already know the rules. Progress in cat assimilation can be measured when these distances, again -determined by the cat, are shortened. Adding opportunities to form positive associations along with raised resting spaces and floor level hiding spaces offers stationing and retreats for both cats. For prey and predator, avoidance, and flight are as valued behaviors as is fight, and less costly.

When allowing the new cat to enter rooms or places, permit the new cat to observe from a few or more feet away, do not crowd and maintain calm energy. Remember each space you find a cat in, is where to softly greet the original cat for recognition and social support. If you are physically close enough and the cats are not vigilant in their monitoring of the other, adding welcomed petting can be a plus. Remember, owner demeanor colors interactions. Anxiety and stress are communicated by chemical signals from all animals including humans.

Maintain exits and escape routes. Once cats are in the same room there is both the opportunity for stationing, friendly contact, or consternation. Make sure all cats have clear paths to exit an encounter. Clear escape routes are key in avoiding actual conflict. Can cats get on to raised resting spaces, around furniture and each other without being backed into a corner?

Keep play part of cat encounters. In practice, I have found that interactive play with the cat that is the most vigilant, defensive, or reactive, while cats are in sight of each other but at a sufficient distance not to provoke, can prolong the length of time new cats can share a room. Perhaps at work is the innate nature of the cat to fix on prey at the expense of competitor cats. Direct engagement with cats interactively is most effective. What has also shown good effect, is the use of motorized cat toys that rotate objects to attract cats. Most interactive toys come with timers to shut off after a fixed amount of time, a helpful feature in turning the focus off. I am also looking at, for distraction purposes only in limited application, bird watching videos. More to come on that. Not to be used are laser pointers which can frustrate without an object to catch or satisfy.

Interactive play along with the right interactive toy shifts the energy and the cognitive processes that fuel aggression and defense while simultaneous exposure to the newcomer cat affords the desensitization and counter conditioning. Adding this step into this protocol in my experience has been significant in lessening the amount of time to assimilation.

Always closely monitor these exercises and use your judgment. You are working toward allowing these cats to be in the same room together, with you in it to supervise. Do not push for more than five minutes initially. This may seem like a lot of work but please remember, again and again –go slow to go fast with cats. Build on multiple, myriad small or limited exposures or interactions to desensitize and counter condition. End each exercise or encounter on a neutral or good note for it to be both a pleasant experience and positive association in relationship building.

Make sure you know what friendly or neutral cat encounters should look like. Sleeping or relaxed lying cats in proximity to each other are ideal. Fixed stationing is often part of determining acceptable distance from or as a newcomer. There is communication going on there. Know also what play looks like, and when to allow for stand-offs, back downs and to impose interventions. Your calm demeanor and slow movements can aid every meeting.

Worth the repetition, emphasis added – do **not** punish, panic, yell, lecture, reprimand, spray water, squirt air, throw items at cats for any behaviors, especially for any behaviors that rebuff another or appear aggressive. It is important for cats to be able to exchange signals and communicate. Distance increasing behaviors are performed for a reason. Monitor context, environment, and exchanges – fixed staring without averted gaze or head turns, and or blocking, and vocalizing are red flags that intervention may be needed.

Play chasing or wrestling is not the same as fighting. Animal play is different than human play but the rules of play typically cross species boundaries – everyone must want to play, listen to each other during the games, keep it fun, keep it relaxed, and engage with inhibited bodies and movements.

Play is announced prior to beginning the fun, but not every species starts the game in the same way. While dogs initiate play with a very defined behavior or ritualized display behavior, the play bow – front legs flattened, rump raised, direct gaze awaiting response, cats do not.

For cats and dogs, chase also works as a play signal (why when we work on recalls we work on backing away from the animal to encourage following). And there are unique defined behaviors to prompt play for felines. Kittens will sometimes use a sidestep or short sideways run that is a play/chase invite. Cats and kittens will also pounce unannounced on an unsuspecting other cat to begin to play. The pounce is an inhibited play pounce and often received as an invite to play. Or not. No one gets offended.

Allowing cats to play is an important part of their development when younger. No human can teach boundaries or acceptable cat interactions as effectively as another cat. Throughout a cat's life, play additionally perfects hunting skills for when needed, relieves stress, expends energy and is intrinsically rewarding for its own sake. Play is fun.

Practicing predatory behavior is part of playtime where hunting and social skills get developed along the way. Cats will chase, bat, wrestle, rabbit kick and take turns in play. Play that is too rough is responded to by stiffening, vocalizing, or refusing to play. Conciliatory attempts to resume play are allowed and can be accepted, but when play continues too aggressively the game may be over. For example, a swishing tale in a cat is usually a sign of stress, if the tail stops swishing and the cat appears calmed, good. Hissing is a warning signal (and a distance increasing behavior), if fleeting - "heard" by the other cat and does not escalate should be allowed.

When interactions are not fun for the cat, this can be readily apparent. "Stand-offs," so called, because of both, how they appear, with cats fixed in gaze and place, and as this is one of the ways cats will resolve conflict. As such, they may not necessarily need intervention if they have the desired effect of one cat backing down. Responses are not paced on human time lines. Especially as sudden movements indicate all the wrong messages. Add to this, our need to remember, behavior happens on a continuum and a spectrum.

The cats do need to be able to communicate and respond to requests for stopping or distance. It is helpful if all observers can decipher what is being "said." More serious signs of stress include spitting, growls, yowls, flattened ears, whiskers back, tensed and tucked body posture and rippling muscles along the back. Increase in vocalizations indicate emphasis. Pressured vocalizations, such as growls, howling, moaning, or yowling, signal with purpose things are most likely to escalate. Where standoffs are accompanied by those vocalizations that are ratcheting from hissing to spitting and yowling, these can be signs that altercations are more likely to intensify.

Not fixed in place, emotions move with actions. If one cat's apotreptic communications dissuade the other from aggressively moving forward effectively and retreating, case may very well be closed. It is necessary for communication and relationship, that the cats are allowed to parse through the sequence. When the standoff is prolonged, as in more than 20-30 seconds, stares are fixed and intense, escalated vocalizations are heard, and the other cat is actively blocked by the other, intervention may be necessary. Determining when to step in can often be a case of less is more but not always.

If the other cat responds to a distance increasing behavior or warning signal, there is probably not the need for human intercession. Should you start to see signs of elevated stress in any cat that are not responded to or replaced by calm signals, you want to end the session as quickly as possible while remaining calm. A soft voice coaxing the cat to another room may help, so may shaking a treat bag or a favored fishing wand toy. Can you redirect tension with a toy or treat when you see it? This can be particularly effective at both changing emotional states to reset the stage and if necessary, helping to move a cat to another room. If you can toss a treat or two that is accepted at the end of a room change, even better.

Never, ever put yourself between two cats engaged aggressively. Similarly, do not handle or attempt to touch or pick up any cat that is in a heightened emotional state or otherwise aggressive. Displaced aggression can result.

The use of an appropriately sized object that is an extension of the body can help to block the visual stimuli and detract attention. A calmly introduced couch cushion, cardboard box panel or wall calendar between the two can give either cat the opportunity to move away. Do not pressure the moves with threats, pushing or chasing. Allowing for time for the cats to move on their own can be the most effective.

Fighting is never where the cats or guardians want to go. Management of environment and encounters can help to avoid their occurrence. When extreme vocalizing and rolling biting and scratching are underway, stopping the altercation with stomping or yelling may be necessary. Stop that intervention when the altercation stops. Such interventions which may be necessary for safety will add to negative associations, increase the probability of their repeating and increase thresholds to resolutions.

Do not lecture, scold, or reprimand either cat for their behavior after an aggressive event or its resolution, such actions can only add to the negative associations of the cats encountering each other and will make things worse here not better. (See the chapter on cat aggression for more on aggressive behavior.)

Remember the 4 P's: No Punishment, Interactive Play, Priority of Original Cat, and Raised Resting Places. Add the significance of time, understanding cat behavior, environment and resource distribution makes to a feline friendly integration. In accordance with conventional protocols, staggering introductions of each phase is recommended in addition to assessing each step of the process as the individual animals respond and tailoring accordingly.

Implementation of any protocol or intervention is not possible without human execution. Desensitization and counter conditioning strategies require skill to apply with care, in order not to push the subject animals past threshold. Such sensitivity required to integration protocols can be taxing and overwhelming to add to demands of everyday life:

> "Management and behavior modification strategies are tedious and time consuming for owners to implement with the necessary consistency" (De Porter, Bledsoe, et al. 2019).

Insuring compliance to achieve successful outcomes can be difficult for behaviorists to accomplish. Without evidence or data to determine which protocols or interventions are being put into place it is difficult to determine which are most effective for integration or for owners to implement.

Research from work with dogs and separation anxiety bear consideration: Takeuchi, Houpt, et al. (2000), compared treatment outcomes and owner compliance with 52 separation anxiety dogs and found owners that were given less than five instructions to be the most likely to follow them. Removing punishment, increasing exercise and giving a chew toy when leaving were most implemented. According to the owners, 62% of the dogs improved. Inconsistent applications of desensitization protocols has been shown to yield results. Butler,

Sargisson, et al. (2011) in a smaller study on separation anxiety dogs, found six of eight dogs improving even as compliance with treatment was uneven. The researchers concluded that it was "systematic desensitization" to be "a consistent factor in the improvement" and that the consistency of that "did not predict the speed of progress or final success."

So how would this work with cats? I followed up with 18 of my aggression in multi cat household cases and three cat/dog aggression cases (Nine responded). I asked which of the interventions recommended owners had found easiest to put in place and which were hardest. The responses would help determine not only which interventions owners were putting into place; they would also highlight which interventions applied, owners found to be the most effective.

Results to date showed: No punishment, keeping routines on schedules, more raised resting space, interactive play were rated easiest to follow. Switching rooms, separating pets, playing classical music and puzzle feeders were hardest/or not followed. Supervised interactions, priority of original pet and Valerian Root were split equally between easiest and hardest/not followed. Raised resting spaces and interactive play garnered the most in positive endorsements and owner compliance. 87.5% of respondents still had the pets, except for one. The one respondent (a cat/dog integration) who no longer had one of their pets (the cat) followed one of 11 recommendations made.

Watch for paws under the door or supreme indifference. Your new cat combination may be interested in interacting with each other or not. Chemistry is an unknown quantity in every individual relationship. Allow for this and do not force a friendship if that is not what is developing. Tolerance and time sharing of spaces can be just as acceptable in multicat households as affiliations. Juvenile cats or opposite genders are reported to bond more readily but this is not guaranteed.

At any step of this process, going backwards to go forwards can often help when things seem stalled or not proceeding. Each step of this process counts, for the cat, timing is calibrated on a different scale than ours. Trust and relationships based on positives may take many weeks or many months to develop. Progress in integration does not move at the linear slope we expect, we are more likely to see the plateaus; one day cats may be spaced at different sides of the room and the next they are allogrooming. The cats are the ones who decided it was time. We need to support that decision.

<u>References</u>

- Butler, R., Sargisson, R. & Elliffe, D. (2011). The efficacy of systematic desensitization for treating the separation-related problem behaviour of domestic dogs. *Applied Animal Behaviour Science*. 129. 10.1016/j.applanim.2010.11.001

- DePorter, T. L., Bledsoe, D. L., Beck, A., & Ollivier, E. (2019). Evaluation of the efficacy of an appeasing pheromone diffuser product vs placebo for management of feline aggression in multi-cat households: a pilot study. *Journal of Feline Medicine and Surgery*, *21*(4), 293–305.

- Takeuchi, Y., Houpt, K. A., & Scarlett, J. M. (2000). Evaluation of treatments for separation anxiety in dogs. *Journal of the American Veterinary Medical Association*, *217*(3), 342–345. https://doi.org/10.2460/javma.2000.217.342

SOCIALIZING FERAL CATS?

"Two dog-owners met one day to walk their dogs together. One owner had grown up in a small family that valued health, safety, and orderly, disciplined behaviour. The dog of this owner received regular veterinary care, two meals a day of low-fat dog food, and was walked on a leash. The other owner had grown up in a large community that valued conviviality, sharing of resources and close contact with the natural world. This dog (the owner's third - the first two had been killed by cars) had burrs in its coat, was fed generously but sporadically, and had never worn a collar in its life. Each owner, judging quality of life from very different viewpoints, felt sorry for the other's dog. " (Fraser, Weary, et al., 1997)

"Socialisation refers to the process by which an animal develops appropriate social behavior towards conspecifics. Typically, an infant animal first relates to its parents (usually the mother), then to littermates or siblings, next to peers, and finally to members of its species. To study the normal socialisation process, scientists have frequently interfered by taking an infant animal away from its mother and/or littermates (deprivation or isolation experiments), or by exposing the infant to an unnatural substitute or caretaker. This substitute has sometimes been an object, sometimes a member of another species, and sometimes a person. " (Turner, 2000)

For several years I was a colony care taker; trapping, arranging for vetting and neutering, providing recovery, releasing, and monitoring the feral and free-living cats that roamed the backyards of an urban neighborhood of West New York, New Jersey. That I lived in a two-family house with a deck for a feeding station, a backyard garden to set shelters in for inclement weather, a basement for separation and recovery after surgeries, made all the difference.

A lot goes into such work, including the art of trapping. To lure hungry street cats inside traps, they are baited with cat food, tuna, treats, or any number of tried attractants. Smart cats turn "trap smart" when they figure out how to get in, get the food and get out, all without releasing the catch. Or become "trap shy," after once trapped or observing those trapped, learning never ever to get near one again. An unsuspecting feral cat can panic to the point of self-injury in a steel trap abruptly closed off from freedom. Timing is crucial. Being able to cover the trap to reduce stimuli and get the cats inside them off for vetting is essential to reduce stress and limit injury. I watched set traps in the back garden to be close by to them once sprung.

Then there was trapping Kitty. Kitty was different than your average feral. A long-haired tuxedo cat, so called after the same character in the "Monsters" movie, who looked scary but was anything but. Kitty had the gentle nature of his namesake and never vocalized, no matter the situation. On his first trapping, he was beyond terrified. He trashed so badly, attempting escape, he bloodied his nose (a not uncommon injury for trapped and caged ferals). Kitty's reactivity and anxiety so scared the vets we took him to, they refused to do a thing with him. We released him, convinced we would never be able to trap him again but we did. The next veterinarian was less frightened or perhaps Kitty was less scary. They neutered, vaccinated, and did the routine tests for Feline Leukemia (FeLV) and Feline Immunodeficiency virus (FIV) on him.

Kitty turned out to be FIV positive, more scared than scary, and relatively easy to care for, recovering easily in the basement kennel. FIV affects more of the male population in cats as it is mainly spread through blood exchange. Intact male cats actively fight each other in conflicts over territory and access to females. Pregnant females who do have FIV can pass it to kittens in vitro. Unlike the human equivalent of this virus, FIV cats can live long and relatively healthy lives without treatment and males, once neutered, have little motivation or occasion to pass on the virus.

Returned to the rear garden, Kitty apparently made the decision that inside was better than out. He made those feelings known by sitting nightly on our deck and looking in. Every night. Kitty was never far from the back garden. When one of our house cats got out accidentally, he was the one who stayed at her side and walked her back up the stairs. A story I would not believe if I did not see it happening. As interested as Kitty was in what was going on at home, he was adamantly averse to being touched and would retreat from close contact. But he wanted something.

This was turning into a little match cat scenario and one the inside cats wanted no part of. Kitty was a constant presence on the deck for hours each day. The other colony cats came, had a meal, and left. Not Kitty. No matter the weather, he was there. So it went, until we sold that house and planned our move. The buyers were clear on not wanting any of those backyard cats around. What about Kitty?

Consensus on bringing Kitty along on that move was not forthcoming. My husband offered what he believed, as he later told me, an offer he was sure Kitty would refuse. Be able to interact with Kitty, be able to pet and pick him up and we would bring him with us when we left, he told me. I took the offer, knowing there had to be some way to bring Kitty on board with a plan he seemed to want as well. I would like to say that I intuited the inspiration to become less scary to the little monster but the idea came from a radio talk show I heard while driving. Read out loud to the shy or scared animal was the advice. Just sit there every day and just read out loud. Do not look at the animal, do not pressure them, do not reach for them when they start to react. Even when they come close. Let them come to you. It worked.

I was able to lure Kitty into the basement by setting up a trail of food and treats, closing off exits and keeping out of sight. Once contained, I set him up in a recovery room with a bed, litter box, food, water, and toys. I went down to his room every day, sat on the floor, and read out loud to him. A chapter or two a day from a book I was interested in. No horror stories, instead animal books, or literature, both better choices.

Wanting the plan to work, I followed it to the letter. As the days went on, I never looked up, even as he inched closer and closer. I kept reading out loud each day and I kept pretending he was invisible. When he was finally right at my side, so close I could almost feel his whiskers, I put a hand out without turning my head and felt him bunting. I stroked softly and left it there. I kept up with the reading adding only a little more stroking with time in between each day, treats added. The day came when I could pick him up. We took him with us when we moved.

Kitty became part of our home. He loved us unreservedly, barrier broken - he would come to us at the end of the day and solicit petting and follow the other cats in the house around. Like many a former feral, Kitty was trusting, warm and affectionate with his people but never comfortable around strangers. As time passed, I would learn more from the literature and its applications on how to best assimilate the feral, fearful, and shy into new environments. Time in the process, space, patience to let it unfold and consideration to the cat have always counted.

Ferals, street cats and shy cats, may and often do, have a history where interactions with humans are mainly or mostly negative experiences. In the world of trap, neuter, release, and monitor, cats are trapped and transported to clinics to be sterilized and vaccinated, then released. As much care as we might take with it, every step of that process is an overwhelming horror for the cat, tricked and caged against their will, taken from known environments and associates, subject to invasive procedures, forceful restraint, often erratic and inadequate pain medication, recovery time, and attention.

There is no way to prevent the stress only to minimize it. Depending on the experience and the length of exposure, the cat will regard traps along with certain or unknown humans as dangers to be avoided at all costs. Knowing this can motivate well intentioned cat rescuers to keep trapped cats caged longer when treating for illness or additional vaccines before release.

Keeping feral and free-living cats, deprived of liberty, trapped, and caged is overly and severely traumatic for them. They cannot know such measures are in their self-interest. Interventions we can provide such as pain medication, boxes to hide in, classical music, pet remedy plug-ins, valerian root sprinkled on resting surfaces, consistent schedules by thoughtful and informed caretakers can mitigate some of the inescapable stress while recovering and waiting for release but never remove it.

Kitty was the first feral cat I worked with on acclimating to humans but not the last. I would meet cats brought into homes who spent every hour they could under furniture, accosted by resident cats, coming out only under cover of darkness, injured and ill cats who would never have survived left on the streets and who resisted every minute of treatment no matter the intervention and who begged the question of which life was worth living. There were those huddled in the back of the cage they were kept in by well-meaning rescuers believing such a life was better than the one left behind.   The cat would not agree.

Others would follow, each with their own story on arriving in human environments. There are stories like Kitty's but not all feral cats want to be in our homes or will do better in them. For those, I have advocated for their return to their home territory and not our houses. The question mark in the title of this chapter emphasizes the point. One of the first readers of the draft of this book, provided with sample chapters (this was not one of them) and a table of contents objected strongly to a chapter on socializing feral cats. The title of the chapter was the same but without the question mark. That title alone prompted strong reactions to such material on the basis that such a process would be highly unethical or perhaps upset the natural order of things. How considered is such a point?

When I studied conservation biology in graduate school, much was made of what interventions were warranted where humans and wildlife were concerned. Basically, conservation biology being mission driven, posits that we can respond or intervene when we are the reason for the need to intervene, when humans are the agents of the situations we are faced with. This can be clearer in the sea, jungle, bush, forest, or any of the (fewer) places left without heavy human footprints.

Our interference in the natural order is now seen mostly everywhere; we are the ones who have introduced cats to environments around the world to serve as pest control on our journeys and then objected to certain impacts of their presence controlling pests, terminated them selectively, drowning those we see as surplus, imagined them complicit in supposed witchcraft, expelled them from our homes, fed colonies we have supported in expanding and then withdrew the feeding, trapped cats from stable colonies well adapted to their independent lives, neutered them without testing for existing conditions which can be greatly exacerbated by surgical procedures, not allowed for sufficient recovery or attended to overall surgical complications post neutering for any feral (no matter the condition), the list goes on.

Cats have lives worth living independently of people, when we let them. While humans have welcomed cats for centuries for their utility in controlling vermin, there can be equal contempt for their temperament, and particularly, their reproduction practices. Especially where female cats are concerned, who make their needs known much to misogynistic distaste.

Mistaken, though it may be, the cat's perceived solitary nature lessens their popular appeal. Domestication of cats and commensal living arrangements have increased the demographics of cat populations. In the natural environment, cats are prolific breeders even as life expectancy for kittens is extremely low. Born exceedingly fragile, one out of every four kittens is predicted to survive. A cat can live close to twenty years given the protections of human care. Outside the home, life expectancy will be less than five years, affected by calamity, predation, sickness, and resource scarcity. Multiple births in species often comes with low mortality but species can thrive where scarceness turns to plenty, even when fleeting, straining carrying capacity.

And where we have taken the world and agency from the feral cat, we may have given the best of both worlds to the barn, bodega, shop, or door step cats. Exposed to the comings and goings of different people along with their affections or disinterest, these cats live with us in an appropriately complex world. How then we have dramatically reduced the world of the inside only cat, limited to how we live in our too small rooms for them as humans and the company we keep. Is it any wonder that any new experience for them is overwhelming? This minute world is all and only what they have been socialized to, just as ferals and cats exposed to the larger outside world have had a vastly large field of experiences in their socialization. History and experience set the stage for the present and the future in our interactions with others.

Feral cat populations in urban areas are best controlled by well managed trap, neuter, return and monitor programs. Cats returned to their colony and or birth environments, with known resources and affiliates are well suited for welfare and survival. Not socialized to human contact and company, ferals are most likely markedly shy and reluctant to engage with people, depending on personality and experience. Our insistence on changing living situations is not the choice of the cat and is highly stressful. But to be fair, as the philosopher, Babette Babich, notes on what we can mistakenly think of the feral cat's experience: "There's always mice. They live in heaven." Hardly, the point she's making.

There are circumstances of sickness, safety and fate that can change that. We have all heard or experienced stories like Kitty's, of cats finding people, sitting on our doorsteps, coming in our homes on their own cat feet, or people finding cats and kittens on the road side, dumped, and abandoned, the list goes on. And for rescuers working with delicate populations; the sick or injured, juvenile or geriatric, those stories increase exponentially. And what stories would the cats tell?

Which cats are then the happiest cats? The ferals in streets and alleys, behind shops, in ship and train yards, living in barns or backyards? The inside and outside cats? The doorstep cats making the rounds in a neighborhood? Or the ones who never get to leave a home with its same four walls, furniture they are not allowed on, people and pets distrusting of them? The cats who get to hunt, roam, chase, forage, mate, den or choose friends on their own time and to their own liking? Or indoor cats who climb the living room curtains as kittens, have the best choices in wet cat food rotated per meal, food puzzles for dry food and treats, cat furniture with plush cushions to curl up on, valerian root rubbed sisal scratchers, classical music playing in the other room, a human who anticipates needs and adores them? Or ferals so cold in winter the tips of their ears freeze and die off from the cold, get hit by passing cars, bikes, or scooter, targeted by cat haters, sickened by poisoned rats, preyed on by coyotes and more? Or house cats whose boredom, frustration and sickness goes unnoticed, forbidden to rest on surfaces most comfortable to them, sprayed with water, neglected or ignored, litter boxes left to chance, whose food comes intermittently, is too soft for teeth and never is what satisfies or what they would have chosen? The list goes on.

Given the right circumstances, feral kittens, can be our biggest socialization success stories. If we can get to them early enough to socialize to humans and late enough for having spent adequate time with their mothers. Such is a difficult window to calibrate under even the best of circumstances and one rarely seen. When a shelter asked me to socialize two four-month-old feral kittens, we both knew that four months is far out to do this kind of work. The sensitive period to socialize kittens to humans (and other animals) is recognized as ending at seven weeks. When handling kittens before seven weeks;

how long handling lasts and how consistently it is done, can make all the difference in whether these kittens like humans or are fearful of them. Researcher Sharon Crowell-Davis notes that where cats are concerned the research on socialization is sparse:

> "...probably because of the misconception that they are not social. Karsh found that the best socialization to humans occurs if kittens are handled from the second to the seventh week of life. The presence of a calm mother is beneficial to the process. In order to facilitate the formation of socially stable groups in cats, research needs to be conducted on when socialization to their own species occurs, and what particular early experiences are necessary for the development of high rates of affiliative behaviour and social attachment, and low rates of aggression. Until such research is completed, it is reasonable to assume that the period of socialization of kittens to other cats also occurs between 2 to 7 weeks of age." (Crowell-Davis, 2007)

I and no doubt, others, have found that kittens (and cats) can be socialized to humans and other cats after the seven-week period, but, and this is a big but, it will much take longer and require much more finesse, creativity, and patience. Interactions with humans in early kittenhood has a more profound impact on attachment and trust of humans than in later kittenhood and adult life. Still, cats will and do interact with humans after seven weeks all the time. How positive or negative those experiences are for the cats and for that particular individual, will affect their welfare, how humans are referenced socially, and how they respond to them. It is the nature of the cat to take time to take the measure of the environment along with the players in it. The assessment by the individual cat is always done in their own "cat time." Absent easily shared language and ability to trust in the permanence of place and our intentions, the cat must fully assimilate and experience the introduced human environment as friendly, appealing and most of all safe.

Cat behavior expert, Dennis Turner, notes that shy cats not socialized to humans, are most often more wary of new experiences and require multiple positive experiences with new people to trust them. This means, the work you do with these cats must be about creating long lasting, intensive positive experiences with humans. Turner also found that these cats will react more strongly to a single negative experience and those negative experiences have staying power.

> "The former socialised cat generalizes positive experiences quickly; the latter must learn to trust the individual person (or family) and does not generalise its later positive experiences, but if anything, its negative ones." (Turner, 2000).

Scare these shy and fearful cats by going too fast or too close, too soon or handle in the wrong way and there is a whole lot of damage control to be done. Conversely, immerse ferals in safe surroundings of duration to nurture trust and strong relationships can take root. This process from feral to former feral entails fostering trust that the strange new world of living with humans is a good idea. You need to figure out the timing and the method to get the message across—how to make the relationship worthwhile from their own feline point of view.

As Turner points out, working with cats after the sensitive period is not truly "socializing" them as they have already been well socialized to their environment, including the animals that are a part of it. These cats already have had meaningful lives and a history that includes socialization to their environment, family members, affiliates, and preferred associates. Turner agrees with other experts who have proposed a more correct term for including human affinity as "social referencing," I would say we, as humans, need to figure out how to be more "socially relevant" to these cats.

Of further consideration, with forging relationships, is how attachment theory figures in. Recent attention has been paid to how strong the bonds are between affiliates in human dog relationships and human cat relationships. Attachment theory speaks to styles of interacting in relationships and these styles are typically characterized as secure, anxious, fearful, or disorganized. Thought to mirror the relationships juveniles had with their caretaker; the corresponding style defines the dynamic of interactions between parties. Fearful, shy, and feral cats may have and most typically do have, pressured histories where upbring is concerned. Food and resource scarcity, inadequate shelter, predators, heightened vigilance - all contribute to stressed maternal care and impact offspring.

I have seen over and over, former ferals socialized to and forming relationships with humans, demonstrate anxious or insecure attachment where they become "velcro cats" or spend most of their time hiding under the sofa, bed or anywhere they can find for safe harbor. Such behaviors clearly signal fearful attachment, when not enough time or attention has been paid to the expanded socialization process for each individual. Further formal research is warranted here, but any animal coming from a pressured environment and into a strange new world, needs careful and gradual assimilation and support.

As times passes and I work to desensitize, habituate and counter condition the feral kittens, who are now about seven months old, they actively solicit petting and contact, especially before eating and after play. They are the most comfortable with me since I have spent the most time working with them but when armed with soft voices and toys, my husband, the pet sitter, and new people can approach them.

Knowing history can help with shy and ferals but often their behavior will tell you what you need to know. Past food scarcity shows in swatting at others before feedings, gobbling a meal, and growling at anyone who comes by while eating. Give the best, highest and most satisfying food available to both assuage hunger and satisfy. Food is more than just nourishment. Feeding multiple cats at a distance from each other and keeping dry food available for free feeding in food puzzles can help with food insecurity and resource guarding. Flinching and shying away from sudden hand and foot movements tell a story of fear of humans in general or avoiding blows. Slow movements and keep hands and body where you can be seen. Extra time on hand feeding and petting can help. The list goes on.

Late socialization to humans can take months, many of them, but given the careful and considered approach, it can happen. Progress is often not linear, incremental increases in contact or distance can be seen in plateaus instead. One day, that cat that would never approach can bunt for pets or sit on any available lap. The cats decide when, as long as the necessary components and resources are in play. Time, distance, and attention to process are the magic ingredients for any cat, new, shy, or feral. Consider the following to add to the mix:

Go slow in every step to go fast with any cat. Especially feral cats. Do not rush them. Go slower than you think you need to be going. Much slower. Even slower than that. Keep the pressure off the cat to engage. The feral cat that is wary and hesitant of engaging is speaking volumes. We talk of distance increasing behaviors being communicated in body language by every species. If you are seeing such requests with any cat, listen. Cats wear a different watch telling a different time. All relationships take time, space, and trust to develop, those with cats take longer, and even longer with some shy and feral cats. Let them come to you.

Be a positive experience and person to interact with. Announce your presence and your intentions with a low friendly voice. No high-pitched baby talk. Offer short greetings or reassurances as you go forward on an approach or to perform a task. In general, cats prefer quieter and more melodic tones than dogs. After greeting, silence can be golden with these cats, resist the human urge to keep the conversation going and focus on body language and what the cats is "saying" in response.

Pay close attention to tucked body posture, head pulled back on the body, flattened ears, dilatated pupils, whiskers back and tightened muzzle, all telling you the cat is stressed. "Feigned sleep," so called because the cat appears to be sleeping with eyes closed, paws and tail tucked but is taut and tense, is an extreme stress behavior. In these scenarios, stopping and taking a break in routine is warranted to give the distance being asked for by the behavior and body language.

Keep caretakers consistent and routine feedings, cleaning and play sessions on schedule. In a world turned upside down with loss of liberty and familiars, schedule and routine offer control over events in this new environment where those events are out of their control.

- Sitting on their level and reading out loud will accustom cats to your presence. Pick neutral or soothing things to read. Your reading voice prevents unwelcome eye contact and guarantees a modulated tone. Read for a long enough period of time, one that is long enough to allow the cat to relax into the setting and narration. A chapter or two is a good marker. Read for the same amount each day. While you read, make sure to ignore the cat as they approach. Successes will be measured in the cat coming closer as you continue this practice.

- Cat greeting behavior is nose to nose. Offering an outstretched finger at or below muzzle level can approximate this behavior for the cat. Do this only when the cat has come close enough to you, as in within a foot, that you can see them of the corner of your eye. Should the cat pause at less than a foot and wait in contemplation, they are saying they still need more time. Give it to them.

- After a few days of close proximity, try patting the floor softly next to where you are sitting. Patting here calls attention to a good spot to settle on. Wait. Use sidelong gazes only. Pat again. If the cat does not approach, stop. Remove your hand, wait a day or two before trying this step again.

- Tossing treats to the cat or past them can help but the approach must be on the cat's time. Resist, no matter how delectable, what you have on offer, tossing the treat to where they are forced to advance past their comfort zone or flight distance. When and only when the cat approaches your finger and makes contact, let it happen. Are they still there? If so, stroke along the muzzle once or twice. Continue if they bunt or lean into the contact. Keep silence or a soft voice to avoid startling the cat. Keep reading every day.

- Classical music has been shown to lessen stressful behaviors in animals. A classical radio station will also offer the added benefit of associating the pleasing tones of human voices (the announcers) with classical melodies. Be mindful of the cat's sensitive hearing and keep the volume turned lower than you think they might prefer. If the cats have access to more than one room, only place the music in one room. Play the music for a set period either during the day or during the evening, not for 24 hours.

Give them something species specific, i.e., of cat relevance, rewarding to do and somewhere to do it. Resting places, scratchers, toys, feeders, and more are essential. Stable scratching posts or pads, covered with sisal or corrugated cardboard for somewhere to scratch, need to be in every room to allow for outlets for natural and necessary stretching and scratching behaviors. Puzzle feeders (for all dry food instead of bowls) allow cats free access and engagement with food foraging. Make sure puzzle feeders allow for rewarding natural behavior and not frustrate them. Avoid "slow feeders" and look for rolling feeders that simulate batting actions cats use with prey or large flat feeders with multiple access points to scoop out kibble pieces. The flat feeders have the added benefit of allowing for positioning so that multiple cats can access them without competing for space.

Multiple and varied raised and floor level resting spaces and hiding spots are needed for every cat's welfare and can mitigate some of the loss of complexity of the free-living cat's environment. Enriched environments contain boxes or beds with at least three raised sides (essential for feeling safe and warm) along with elevated resting spaces - for every cat, in every room. Add cat beds to towers or place beds or igloos on existing furniture like dressers or desks in the home. I add a table mat with skid proof backing under a cat bed or igloo to prevent slippage when the cats jump on or off them. Make sure they are easy to access and exit. For the shy and feral, this is vital to allow them to feel secure to hide or safely survey their new environment.

Toys for solitary object play can offset boredom and substitute some for predatory play activity, fur covered mice that rattle, crinkle balls, stuffed toys with feathers, crumpled paper balls are some choices to offer. Watch to see which are most engaged with and give cats time to decide on their choices. Then provide more that are similar and new ones.

Rubbing powdered valerian root, silvervine powder, or catnip onto the surfaces of cat furniture, such as scratching posts, cat beds or platforms can increase interest and attraction to those spaces and places we would like our cats to target and add additional sensory stimulation with these natural cat attractants.

Interactive play is one of the most beneficial things you can do for cats in this process and huge in creating positive relationships between human and cat, relieves stress and engages cats in intrinsically satisfying activities that mirror natural and necessary hunting behaviors. Toys and objects are great to play with but interactive play adds the dimension of unpredictable and dynamic to bonding moments. Cat dancers, fishing wand toys or even pieces of string are great play instruments. Experiment with different types of these toys to see which the cat most responds to. Put the fishing wand toy away and safely out of sight and cat access when finished with a play session.

Keep play sessions on schedule but make sure they happen even when schedules fall off. Consistency is huge in mitigating stress. Adding another task can be easier said than done with demanding days of our own, pairing interactive play sessions to feeding and our own daily routines can help in keeping them consistent. Toys for object play, as in crinkle balls, faux mice, stuffed toys, etc., should always remain out and freely available throughout the home, for solitary play.

Work on your play style by remembering to pass the object across or away from a cat's line of vision to engage them. Think of the movement of prey: reptiles and rodents which are more favored in cat hunts than birds. Slinking slowly, stopping, starting, zig zagging, etc. Mirror the behavior of these prey animals with your fishing wand toy as you interact in play. Start with slower movements as opposed to faster ones. Shy or uncertain cats and those who have not had the opportunity to play may not engage initially. Play with them anyway.

Start at a comfortable distance for the cat, determined by how relaxed they are dependent on your proximity. If you are seeing tension, take several steps back and assess. Keep your own reaction muted and avoid the jolly camper voice you might use with a puppy, instead use soft "elevator" voices or quiet, which work better for ferals and shy cats. Aim for at least two to four minutes of play at the same time in the morning or evening (or even better, both). For ferals and shy cats, a shorter session is best to start with – two minutes daily paced over a week, before adding more time can be less stressful.

Hand feeding is a good way to build trust. Once established, incorporate petting into the process to associate the positives of feeding with contact and closeness.

To do this, start with tossing bite sized treats (I like freeze dried salmon or chicken), the higher the value here, the better, towards the cat first. Toss to the cat or slightly behind them, you are not trying to press them closer to you, more to find good things in your presence and that you provide. Do this several times a day for several days. Measure the progress here by shortening the distance after several days.

Does the cat hesitate to take the treats? Or advance quickly? If you go back to tossing further away, do they take the treat even quicker? If treat taking is faster further away from you, go back to tossing at this distance and stay there for several days before advancing further. Listen to their behavior.

When the cat is up to treat taking closer to you, you can progress to either using a long-handled teaspoon for shyer cats and for more forward ones, directly to hand feeding with either squeeze treat tubes or dry treats or kibble. Start with hand feeding by placing a good number of high value treats on the flat of your palm which you are holding at or just below muzzle level. I like freeze dried salmon or chicken. Allow the cat the time to approach, remember that cats have poor vision up close, so several treats will help the cat to identify them with greater ease by sight and smell. Keep your hand flat and take time to wait to allow the cat to take the treats. Keep silent or your voice soft and acknowledge with a single syllable or their name.

Once the cat is consistently taking treats and you are hand feeding, gradually introduce petting using the hand that is not doing the feeding. Time the contact to only when the cats is actively taking the food. Because cats will pause between bites of hard food or chunks of some foods, this exercise is better suited to soft items like chicken or turkey baby food on a teaspoon or cat treats in squeeze tubes. I have also added baby food to plastic snack bags for homemade squeeze tubes. Make sure to offer the petting hand at the same level or lower than the cat's head, the side of the face is ideal.

Keep the duration of caresses short initially, three to four strokes along the side of the muzzle or behind and between the ears or back of the neck to start. Watch for reactions that indicate the contact is welcome - as in leaning in to your hand and reactions such as flattening ears or whiskers back which indicate that it is stressful. Remember to hold off on petting unless treats are offered at the same time and remember to also increase the number and frequency of treats offered.

Where petting is welcomed increase the strokes gradually, from two to four, for example, being vigilant and observant of how the cat is receiving them. You do not want to be parsimonious with treats here. Dry food continues to be fed throughout this process in puzzle feeders. Where you see signs of stress signaled, going back to the last step to the last step in the process helps. Go backwards to go forwards.

- Feral kittens may initially be more approachable than older cats and may still retain fear of humans or interact with humans as they would another cat with little fangs and claws. When your kitten rubs against your legs, make sure and exhale slowly. How approachable are they? Say the kitten's name first and then slowly lower yourself for petting to kitten level. Extend a finger out as an invitation, should the kitten retreat, turn away in place and softly say its name again and wait. Remember, this is a good time for reading out loud while at kitten level.

Handling or holding your kitten as much as you can is KEY to socializing to you and your family. If the kitten panics at being held, first offer treats in a squeeze tube to desensitize and counter condition the kitten to being held, progress after several sessions to adding in pets with one hand while offering treats with the other. For the kitten that can be handled or minimally objects, hold the kitten against your body and not away from you where it will feel insecure and may seek to struggle. Make sure to use two hands. Support the kitten from under by holding your hand underneath the body with your hand spread between the front legs. The chest of the kitten should be resting on your palm. Depending on the size of the kitten you may or may not need to support the lower body with your forearm. Cup your other hand around the body. Stroking should be confined only to the head. If the kitten is squirmy, try stroking and talking softly to the kitten for a minute and once the squirming stops release the kitten immediately (squirmy kittens will also benefit from classical music in this scenario). Wait at least three minutes and begin the process again. Practice holding and handling your kitten several times throughout the day.

As much fun as playing attack games with your kitten may be, it is never a good idea to play this way. Baby kitten claws and teeth are all fun and games when they are little but it leads to playing this same way with adult cat claws and fangs which gets them in all sorts of trouble when we started it in the first place. Never a good idea. Be assured this is play but leave this kind of playing for the cats only. When the kitten attempts to play attack, swap in a stuffed animal instead or use a short word or quick intake of breath, immediately. Meaning at the exact moment of contact say "Oww!" or gasp (not both) and hold still. Use one short sharp syllable and no movement – providing feedback and taking the fun out of the chase.

The key here is timing. The very second you feel the bite, use the response above and the very millisecond the cat stops, use a softer voice in praise to reinforce the stopping and keep the encounter positive. That praise after the cat pauses/stops is important and rewards the stopping not the biting. Cats are extremely sound sensitive due to their exquisite hearing and do not like loud or discordant noises which is why the feedback is so effective at getting them to stop the behavior. Your reaction will startle the cat. Remember in that very second -when the cat reacts, to immediately say "Good kitty" in a soft voice and stroke along the side of the muzzle to reinforce that stopping the bite is the wanted behavior. That's it, do not lecture the cat the afterwards as this only confuses a cat for doing what you asked for. Do not forget to provide the kitten with the kind of interactive play best suited for kitty and human to retrain.

Litter boxes may be a new thing for ferals and plants can be targeted for elimination instead. Begin by relocating plants out of reach initially. Make sure to have sufficient litter boxes. Boxes which are kept clean, uncovered and with sufficient and fine-grained litter to encourage use. Research which plants are toxic to cats. Reconsider indoor plants and flower arrangements with cats in the house. If this is not acceptable, make plants themselves unattractive by using double sided tape affixed in a lattice pattern over the top of the pots or on to inch wide by foot long cardboard strips to lay over the opening. One step and the sticky tape touching the paw is a deterrent and makes the plant an unwelcome place. You can also crumble aluminum foil around the base of a plant or use moth balls. Do know, moth balls smell and the foil once removed is no longer effective.

There are additional considerations when bringing ferals into homes with resident cats, neither will take kindly to the introductions with exceptions. Please see the chapters on assimilating cats in multi cat households and introducing kittens to older cats in addition to this chapter.

In our sometimes dogcentric society, would that the hard luck story of these street kittens had more of the "rescue" appeal that dogs enjoy. If it did, finding homes for cats like these would be easier. Yet once again, the truth is, not everybody likes cats. They hiss when scared and scratch when cornered or defending themselves. Their breeding habits are smelly and noisy. They mostly do not come up to you like dogs do and beg for attention and affection and we hardly encourage or invite them to.

Then again, street cats and ferals are too scared to do most of what dogs do. Too scared of what we might do to them and we blame them for that. But some cats are in desperate need of rescuing, living on the streets and in back yards and vacant lots without shelter, too cold in winter and too hot in summer and then there is rain. They are often hungry and scrounging for food. And the streets are dangerous. A second too close to a passing car can be crippling and fatal. Humans can be dangerous for a street cat, risky to be around, they are not welcome in every backyard, things get thrown at them, and the human bullies can find them and hurt them or even worse.

Winter is almost here and there are too many cats on the streets and snow, ice, and oil slick streets with puddles of toxic chemicals to keep car radiators working, engines that provide heat and turn deadly when cars move, cars and bikes and dogs and lack of shelter, so many dangers. And now there are two more kittens who have started to believe that trusting humans is a good and safe thing to do, ready for a family of their own that can go slowly with them and be OK if they startle at first a lot and then a little as time goes by. The promises we make.

<u>References</u>

- Crowell-Davis, S.L. (2007). Cat Behaviour: Social Organization, Communication And Development. In: Rochlitz, I. (eds) *The Welfare Of Cats*. Animal Welfare, vol 3. Springer, Dordrecht. https://doi.org/10.1007/978-1-4020-3227-1_1

- Fraser, D., Weary, D. M., Pajor, E. A., & Milligan, B. N. (1997). A scientific conception of animal welfare that reflects ethical concerns. *Animal Welfare*, 6, 187-205.

- Turner, D.C. (2000). The human cat relationship. In D.C. Turner & P. Bateson (Eds), *The Domestic Cat the biology of its behaviour*, (2nd ed., pp. 194-197). Cambridge, UK: Cambridge.

HOW TO PET A CAT

When I began teaching the pet care technician course, the program was taught without cats or dogs in the classroom. This was a six-month certificate course, so that meant six months of no live animals, even as there was an applied portion putting learners into internships with actual cats and dogs.

Learning the basics of feline and canine behavior along with low stress, species specific approaches and handling for cats and dogs was a necessary part of the pet care technician curriculum. Handling in pet care services can be intrusive and invasive, especially absent relationships and trust. Working with animals can often mean insufficient time to develop both. When force and restraint are used, the pet now has a history to draw from of fear and apprehension and can avidly avoid future contact. Our companion animals may tolerate unacceptable interactions to a point, a latitude an unknown animal may not give us, offering in its place, what they may consider a necessary defense. Becoming versed in low stress handling and restraint can help to make the difference for welfare and in future interactions for both animals. But first you have to learn it.

Most of the educational material geared to interacting with companion animals is aimed at children. Mainly because children (young boys in particular) are the recipient of most dog bites, cautions on approaching dogs make up most of what can be easily accessed for safe pet interactions. Resources for cat interactions are now gaining in popularity but are more limited. Aside from simplistic phrasing, the content is just as valid for adults. Content for both cats and dogs can be rich in what not to do and pass over the what to do instead portion. Not knowing what to do instead does not prevent self-interpretation, often mistakenly based on our own human likings. We are drawn to interrelating with other species as we would our own –

hugs, cuddles and holding animals upside down, as we would a human baby, as if all creatures can easily relate to being treasured like our own human intimates. It cannot be so. But surely, we think, the animals must know we just cannot resist their cuteness, mean well, and are captivated by them. They do not know.

Any instructor takes what the curriculum offers and works to enhance it. Visual materials are important references to help illustrate foundational concepts, as is modeling and practice with stuffed animals before applying to living subjects. But first you need the visuals. To get the right resources in front of learners, I spent a lot of time searching for supplementary educational videos and photos to help learning in canine and feline body language along with how human body language and handling impacts both. At the time, my search had limited results. Especially if the results were in the force-free, science-based camps. Results for handling cats, were practically non-existent.

Needing more to work with, out of necessity, I made my own videos with my own animals. And needing a platform to show it from, I posted my videos online. My students were not the only interested viewers. At this writing, my YouTube video on how to pet a cat has 33,000 views and my video on how to pet a dog has 1,500 views (suggesting we may think how to pet dogs is a sure thing -the dogs would not agree). My video on how to pick up a cat was an even bigger hit with 173,000 views and how to pick up a dog with 9,000 (more suggesting we have dogs figured out over cats).

I also revised the no pets in the classroom policy. When learners had progressed in the course to a point, I brought Daisy on certain days to class to work on dog behavior, handling, walking and dog park trips and Cinnamon, on other days, to work on cat behavior and handling (she stayed inside all day).

When the cat or the dog came to class, it was soon apparent that most students felt confident interacting with Daisy and at the dog park. Whether or not the mechanics were right we worked on them. When it came to Cinnamon, there was always more trepidation when she arrived for the day. Learning the basics with cats must go at "cat speed" and be measured in how comfortable the cat is around humans to begin with. Most companion cats, unlike dogs, are not well socialized to a wide variety of people. The housecat can spend an entire lifetime, years, in the same rooms with the same people, with little opportunity to encounter new people or changes in environment. It's a small world. Being a prey species as well as predator, cats will instinctively avoid new people. Individual cats that are more forward and confident may welcome strangers, but such cats are less common. Cats that have larger roaming areas and spend more time around different people are more easily approachable. Shop cats like bodega cats are often well socialized. In my classroom, once petting fundamentals were solid, petting the bodega cat was the next assignment. But those petting fundamentals needed to be solid.

After an hour or two to decompress in the carrier on my desk, Cinnamon spent the rest of the in her cat bed sleeping, taking treats, and watching with carrier opened. A long-haired calico with a tipped ear, she was a topic in herself to quiz learners on. What could just a look at her reveal? What did that tipped ear tell you? That calico coat? (Cutting off part of a cat's ear during sterilization is routinely done with feral cats to indicate they have been spayed and to avoid trapping them again. We don't see this with owned cats, such an alteration indicates some of a cat's neuter status and history. Calico cats are 99% likely to be female.)

With Cinnamon I could demonstrate how a person should first observe the individual cat standing sideways, what was happening with head position, ears, whiskers, eyes? Next approach also sideways to gauge what the body language was saying, and next assess possible consent: Did she turn towards me or away from me? Did she drop and roll on to her side ("social roll")? Reviewing followed with which touch cats, and she liked best; stroking behind the ears, back of the head at the base, and the sides of the muzzle were always preferred. How to pick up a cat, which cat holds and why technique mattered came last.

Every species and every individual animal has a certain and preferred way to make contact. Ritualized greeting behaviors communicate neutrality, affiliation, and reassurance. The natural behavior of cats interacting with cats shows respectful distance, blinking, social rolling, tail up, nose touch and flank rubs recognized as affiliative or friendly communiques. Touch can help to create or affirm social bonds but only once a foundation has been laid of safety leading to trust. Along with ears, whiskers, and eyes in neutral positions, how the tail is held can signal intent. A study done on a Roman cat colony showed that advancing toward another cat with the "tail up" was more often used by lower ranking cats towards more dominant individuals and might serve as a signal of amiable intentions, a recognition of social order and to hinder hostile behavior. Female cats were found to be more likely to initiate the greeting sequence with their tails up and rubbing towards males while the males were more likely to initiate nose to nose greeting towards females.

We can observe our own or local friendly cats raising their tails as they come closer to each other and briefly touching nose to nose. Close affiliates may rub along the flank, twine tails or rub ("bunt") heads as salutations. Social rolling can further connections or invite them. Affiliations are also signaled with a slow blink, a tail tipped toward a familiar cat and any number of welcoming vocalizations. These behaviors may be combined, they may be fleeting or repeated. Such observations inform preludes to engagement when we use a finger outstretched at muzzle level towards a cat, brief muzzle caresses on consent and avoidance of body stroking initially. Cat greeting rituals are both seemingly and simultaneously, languorous and brief, you must pay close attention to see them happening but the more you observe cats greeting each other, the easier it will be to recognize, and to identify what degrees of friendship or standing are being signaled.

If cats precede contact with each other with certain necessary behaviors, we need to follow suit. Greeting, determining consent, and careful petting if welcomed, needs to go before any attempts at handling or picking up any cat. Cinnamon, like all cats, was not necessarily a fan of strangers handling her. Being held and held incorrectly is resisted by most cats. Holding cats upside down may appeal to humans who favor holding cats as if they were human infants but this position triggers a righting reflex in cats. That wriggling to turn over is innate. Cats lose control over their own stability and control over their freedom when held and restrained. Our own emotion plays into the mix as well. If we are anxious and stiffen, this is communicated immediately to the cat, giving them even more to be concerned about.

No matter the speed I broke it down, it appeared, even the students professing great affection and comfort with cats, appeared hesitant and unsure of the right approach when it came time to touch Cinnamon. Difficulty with reading cat body language and fearfulness of a cat's reactivity surge when presented with handling an unknown cat. Concern can be a valuable part of learning, if attention and retention to what is being seen goes along with it. Understanding and putting into practice a proper feline greeting ritual is the necessary first part of that right approach.

Like most cats, Cinnamon could sense a human's discomfort. Her reaction to human tension, was to slightly tense her own body, curling upon herself opossum like.

Another study, this one done on cat responses to human petting, found that cats prefer to be pet/stroked most in the temporal or ear region and least at the base of the tail. The research also noted stroking in the perioral or lip/chin region was preferred to be confined to the chin and avoided the lip corners. Additional research continues to show cats prefer touch in these areas and mirrored in how they interact with each other. Following these findings means feline friendly petting is kept to our touching to these areas as well, after we ask if the cat is interested in being pet in the first place. More recent studies confer the reminder:

> "Based around a "CAT" acronym, guidelines focused on providing the cat with choice and control ("C"), paying attention ("A") to the cats' behaviour and body language and limiting touch ("T"), primarily to their temporal regions" (Haywood, Ripari, et al., 2021)

Close consideration when making initial contact with any new cat will determine consent by the cat's response. Are you at a distance that increases the cat's vigilance or is their posture relaxed? Are you standing off to the side or looming in front and over? Can they see your hand approach or does it swope in quickly? Do they lean in to touch? Bunt? Social Roll? Move away? Do their ears flatten or remain neutral? Does their body remain relaxed or stiffen? Do they want to leave or attempt escape? Do they look away or at you? Does their tail flick or stay still? Do the whiskers flatten against the muzzle or remain out to the side?

Social rolling, leaning in to touch, bunting on to a hand or rubbing against an outstretched finger, relaxed body, ears upright, focus on the person and not scanning ways away from them, are some signs the interaction is welcomed. If touch is not welcomed, give the cat distance. And time. If, and when a green light is given, confine the touching to the head and face initially and keeps caresses or stroking limited. Remember, cats will rub along the side of a preferred feline associate and twine tails, both actions being momentary ones. This may repeat but again, the duration is a short one. For humans this means we need to limit strokes along the sides or length of the body to one or two, and give them, only if the cat you know appreciates flank touching. The same goes for petting at the base of the tail.

For red lights or stop signals, pay close attention to: cats that move away from touch, sharply turn a head towards the touch, freeze, stop purring or moving against touch, lip licking, tail flicking, ears flattening, skin that ripples or a break to groom – all are signs that cats display when an interaction is no longer welcome or to increase distance.

To sum up and add more steps to know for more successful cat petting:

- Prior to engagement, exhale slowly to relax your own body and release tension. Cats are expert at reading our body language. We want them to read us as soft and non-threatening. Tense and anxious is threatening.

- Start by softly announcing your presence. When you enter the room or approach the area where the cat is, say hello and use the cat's name in your greeting. Use a modulated tone and not a higher, louder voice better suited to babies or dogs. Approach the cat from the side, as in your body turned sideways to the cat not full frontal. Avoid direct eye contact as you get closer. Pause on arrival, take a few moments to let them become accustomed to you being next to them.

- Do not bend over or loom over the cat. Stand or sit alongside or at an angle next to the cat or bring yourself down or close to cat level. Patting the ground next to you gently can get the cat's attention and invite them to station on that spot. Cats often prefer quiet during these encounters, especially if the interaction is a new one and not speaking allows you to focus more completely on how the cat is receiving the attention. If the cat is a familiar one, try saying their name again softly in greeting.

- Now look at the cat and at the same time slowly blink in the cat's direction. Do this twice, keeping a gentle smile and silence generates the right energy here. Watch for a blink back to know your technique is working. Some cats take longer to warm up to this but all appreciate the effort.

- For initial forays with a new cat or to heighten interactions with a known cat, treats or a toy can set the stage further that your presence is a positive one. If you are bringing either, let the cat investigate the toy or take the treat placing or tossing it closer to the cat or just behind them. Resist the urge to have them take it from your hand. Cats are short sighted and one or two treats are hard to locate in your palm or in your fingers.

- Next, offer your hand with an outstretched finger from below or at muzzle level. Remember how cats greet each other and initiate contact with nose-to-nose touches and channel this by keeping your finger in the cat's visual range which allows the cat to track where it is coming from. Make sure to pause so the cat may sniff, push, or bunt against it. A sniff, contact or dropping down to social roll is a good invite to go further.

- Keep contact to where the cats contact each other. Confine your petting to three or four strokes along the side of the muzzle, behind the ear and or between the ears. Under the chin, front of the chest, the back, sides, and base of the neck can be welcome areas, but maybe not on a first petting session with a new cat.

- When it comes to ears, with a known cat, and if the touch is welcome- you can try ear slides – starting from the base of the ear, gently slide your thumb and the base of your pointer or index finger over the ear, only one finger is moving. Do one or two ear slides initially to gauge response.

- No matter what, keep all touching short at first in duration, do not over pet and avoid full length body stroking. Remember to keep it to three or four caresses or strokes per area initially unless you know this cat well or the cat bunts your hand or nuzzles (rubs against you with the nose first followed by the side of the muzzle) in return.

- Always keep in mind that each cat is an individual and may prefer one area of contact over another.

- Pay attention to and consider the cat's response to differing areas, touches, and approaches. How does the cat react to all of this? Do they respond with cat enthusiasm as in coming towards you, is their tail up? Is there more bunting or head butts? Are they purring or are your hearing short little trilling meows? Are they staying in their original position but social rolling? Bunting, nuzzling and/or purring? These responses clearly indicate the cat's pleasure at the interaction and their desire for you to continue.

- Does the cat remain in place but show some interest in the interaction as in a somewhat lifted head and chin but the rest of their body remains still and close? Do they not move away or move at all but allow the petting? Hesitation and tolerance can mean here that you want to keep that petting session on the short side, stay silent and done after three strokes to two areas as in side of the muzzle and behind the ears. Remember, always, pay attention to cat body language. It can change moment to moment.

- The cat that is saying "NO" loudly and clearly is either recoiling from your touch, has airplane ears, looks quickly towards touch as in guarding, looks away without tilting the head in acceptance, dilated pupils, airplane or flattened ears, whiskers pulled back, tail flicking, tightened muzzles or eye area, lip licking and rippling muscles, swats at you, tries to leave and has body tension. You may see any or all these signs. Each means STOP. Please listen, all signs that petting is not welcome at that moment. If you see any of these it is time to stop and try your approach on a different occasion. Help future interactions by looking to things like interactive play, reading softly out loud and enriching the cat's environment (see the chapter on either in this book).

- Again – remember that consent acronym, CAT - providing the cat with choice and control by checking in for their reaction, paying attention to the cat's body language in response to touch, and being mindful of where your hands are always on the cat's body.

- Do not pet alongside the flanks or at the base of the tail during initial cat encounters no matter how well you think it is going.

- Belly rubs are not welcomed by most cats. Individual cats with a developed history with their person may be the exceptions that prove this rule.

- Tolerance is different from consent; cats may allow a touch but not want it. Pay close attention to what the cat looks like when any approach is contemplated, carried out and how the cat feels about it. Wait to find out. Wait even longer than that.

- Petting is not always what a cat wants. Cats may seek out affiliates for social proximity or straight forward physical contact, resting or sleeping together. These are not highly mobile interactions. Human affiliates can be approached when working at a desk, watching television, lying down, or sleeping. Enjoy the closeness, wanting to be next to you is major. If you lay a hand lightly on or alongside the cat, does the response say the contact is wanted? If not move your hand further away. Wanting to be close to you is a significant step in cat relationships. Let it develop.

- Grooming each other ("allogrooming") is another way that cats will maintain relationships and bond socially. This too is mostly concentrated on and behind the head and not the body or the tail. Continued strokes and pets can likewise be a stand-in here for this behavior The same areas can be focused on when brushing cats.

- Brushing needs to be introduced gradually and with positive associations. Not all cats like brushing and most cats do not need it. Cats being superb at self-grooming can do without our assistance in that department. Certain flat faced breeds or geriatric and ill cats may benefit from brushing if the brusher is mindful of painful areas and pays close attention to technique and cat response.

- When grooming or continued petting, only engage when the cat is relaxed. A tense cat wants to be left alone.

- Introduce the brush by showing it to the cat and letting them sniff it. Follow with a high value treat or soft voiced praise.

- Stroking the area to brush first, gauge the cat's reaction for comfort.

- The cat's natural behaviors are always our touchstones to follow when interacting with them. Cats groom in short strokes, a little at a time and often.

- If the cat resists or shows any signs of stress. Stop.

Treats are always a good way to end any interaction. A soft "Thank You" and "I love you" never hurt.

References

- Cafazzo S, Natoli E. (2009) The social function of tail up in the domestic cat (Felis silvestris catus). *Behavioural Processes*. (1)60-6

- Haywood, C., Ripari, L., Puzzo, J., Foreman-Worsley, R., & Finka, L. (2021). Providing Humans With Practical, Best Practise Handling Guidelines During Human-Cat Interactions Increases Cats' Affiliative Behaviour and Reduces Aggression and Signs of Conflict. *Frontiers in Veterinary Science*. 8. 10.3389/fvets.2021.714143.

- Soennichsen, S., & Chamove, A.S. (2002). Responses of cats to petting by humans. *Anthrozoös*, 15, 258 - 265.

- Shelley-Grielen, F.   (2012). How to pet a cat. *YouTube* https://youtu.be/-MU7VYmwtSk Retrieved 11/6/2022.

- Shelley-Grielen, F. (2012). How to pick up a cat. *YouTube*. https://youtu.be/mVbsjijbQhI Retrieved 11/6/2022.

- Shelley-Grielen, F. (2016) How to pick up a cat from the floor or a raised surface. *YouTube*. https://youtu.be/4O13sqnk-0U Retrieved 6/30/23.

HELPING YOUR CAT TO USE THE LITTER BOX AND WHY THEY AREN'T

"Chauncey absolutely, positively, adamantly refuses to use the litter box."

"No matter what we do, she just won't go in her box. We want to keep her but we are running out of patience."

"I have tried it all. Everything. You name it. Odin goes everywhere but in the litter box."

When I moved house with my cats, Ed, and Sarah, some years ago, we went from living by ourselves in an apartment, to moving together with my new husband, into a house. Ed, plain to see by his behavior, hated everything about the moving experience, he cried piteously and non-stop in the carrier on the way to our new home. He had only lived in that apartment and besides, far as he was concerned, anything about a carrier was sure to lead to something he would not welcome. Who could blame him?

Ed appeared especially horrified by the new set up of strange and unfamiliar spaces and smells. To make matter worse, we did not move straightway into the house. We were full scale renovating, so we moved our bedroom temporarily into the basement. Ed was so traumatized by this arrangement, the first night we got into bed, he jumped up to join us and with fixed gaze and a long meow, squatted and urinated right next to me. I had yet to read up on all this but to my mind there was no mistaking that the purpose of that pee mail was to communicate how very, very stressed this cat was about the situation he had been forced into. That and maybe - could I please do something about it?

Litter box avoidance gets cats in trouble. Trouble they pay for with owner surrender and desertion. Litter box issues or "inappropriate elimination" (now better known as "house soiling" or "unacceptable elimination," as not using the box is appropriate here, as far as the cat is concerned) have been cited as the number one reason for owner surrender at animal shelters, and, one of the top reasons cat owners consult with a behaviorist. No matter how well established it may be that maintaining faithful litter box use begins with frequent scooping, no matter how closely adhered to, at times, there is more going on, enough to keep a cat out of the litter box often because not using the litter box in certain scenarios is about stress and not clean litter and multiple cat boxes.

Cats care about litter boxes and cleanliness. Fastidious in matters of hygiene, cats spend 15 to 30% of their day grooming. Even when it comes to toileting, waste is covered and at a distance from most territory and nesting sites. So why is litter box avoidance such a problem in the home? Whatever allowances we might make for behaviors unique to other species, whatever engages or captivates us, that is of other animals' worlds, anything that relates to elimination is not among them. Waste, is mostly to be avoided, unsanitary and offensive. We also give all sorts of motivations and meaning to the leaving of waste products, mostly imagining it to signify all sorts of intentional insults and complaints. That it can be anything more than that is given little regard, but there is uncharted territory to survey in waste for species not our own.

Such a predicament is frustrating for both humans and cats. First steps in figuring out what is going on should always start with the medical to rule out possible urinary tract infections, bladder stones, arthritis, etc., all of which can contribute to litter box avoidance (de-clawing and other surgeries may also be problematic, more on this below). After obtaining a clean bill of health, going back to litter box usage 101, is in order along with tackling the reasons behind the behavior.

Key to understanding the cat's elimination requirements or any animal behavior is learning about the natural history and behavior of the animal as a species and as an individual. In their natural environment, cats select separate areas for urination and defecation and prefer unsoiled areas for "squat urination." Spraying, more often seen in male cats than females, is a communication for cats to convey information to others. Alarm, availability for reproduction, and territory, are the most understood messages conveyed. When spraying does happen, cats, unlike dogs, will not "overmark" or spray over another cat's urine, seeking to locate unmarked locations instead, spraying urine higher or lower than where another cat has sprayed. While individual preference dictate, male cats have been found to have larger roaming areas in the natural environment than females do (male cats can roam up to four acres and females up to one acre).

Getting a cat to use a litter box often works so well because it approximates what they would normally use in a free-living situation; a clean, safe location with a soft, fine substrate (ground covering) to cover waste in, away from where they eat and drink. It makes perfect cat sense that a clean, uncovered litter box, of the right size, with the right sort of litter (fine textured and unscented), in the right place, is essential.

One absolute universal cat litter box law is clean litter. The cat's natural behavior and numerous studies support cat preferences for clean litter. Rule number one: The cleaner the litter the better. At minimum, scoop your cat's litter box daily, twice a day is optimal, especially if you have more than one cat at home using the litter box(es). I like keeping litter boxes in the bathroom where their condition is both easily observed and easily accessed to clean.

Do a complete change of litter at least once a week. If you are using clumping litter, be vigilant about thorough scooping -those clumps smell, take up valuable litter box space and are hard to maneuver around. A good box rinse every month or when changing litter along with replacing plastic litter boxes, which are porous and can hold odors, every three months, will keep litter boxes to a feline clean standard.

Types of litter boxes and litter material can also come into play. Most of the products on the market are designed to appeal to human animals. Cats do not ask for scented litters, we do. Scented litters may not necessarily be the best choice for felines, especially older ones. Choose unscented litter and opt for finer grained litters over larger pelleted ones. Cats look for and prefer finer substrates or surfaces to eliminate and bury waste in, softer, finer litter as opposed to clumping or gravel type litters are more in keeping with cat preference and a kinder choice for cats, especially declawed cats, as manipulating is often painful for these cats.

While scented litters developed to appeal to humans can be off-putting, other litters have been formulated to entice cats to use them. Not all cat attracting litters are created equal.   Scientists in a 2019 study, blind tested a plant (corn) based litter with an added attractant compared to one without an attractant and found the cats preferred using the one with attractant added. The attractant is not specifically named in the study but a popular cat attracting litter on the market uses timothy grass, a grass used for guinea pig housing. Adding a generous sprinkle of an attractant marketed to encourage litter box use or such as timothy grass, to the top of your current fine-grained litter may also add to litter box attraction.

Depth of clean litter can be a factor, enough to bury waste easily and not too much to get lost in. Give enough litter to achieve good coverage especially with avid cat excavation efforts. Sufficient litter has been shown to make a difference to using the litter box, especially where defecation is concerned.

Young kittens may find litter to be the greatest of sandboxes to play in. Less litter in the box and timothy grass can be helpful here and with former ferals that may need more time and training on litter box usage especially those that have used the great outdoors as their litter box. Cover soil surface areas of indoor plants to prevent opportunistic use.

Avoid liners as the plastic may interfere with natural burying actions used.

Have you recently switched brands or types of litter? Thinking of what your cat has preferred in the past can give clues to preferences and help in replicating it.

Is your litter box large enough and uncovered? Is your cat trying to use the box and not being sufficiently contained, either by box design or physical limitations? Make sure that box is long enough - research shows that cats prefer a box at least one and a half times the length of the cat. Your cat needs to maneuver and turn in a litter box and size here matters, think long and wide enough to make necessary posturing easier once in the box. Can your cat both turn around and get easily into and out of the box or is that swinging door getting in the way?

Does the litter box have a cover or a door? This is another human preference - cats do not need "privacy" here. As prey and predator, they actually do want to see who is around when eliminating. Removing the door or the cover adds to better sight lines for your cat when it matters to them and adds usable space to move in. Trying this simple fix may help quickly or within days. Higher walls on a litter box can help contain litter and over spraying but older and arthritic cats can have difficulty accessing boxes with higher walls at the entrance. Modifying this by cutting down the entrance side of the box can make entering the litter box easier, encouraging use. In addition to the huge selection of cat boxes available in stores and online, look at plastic storage containers, the kind that slide under a bed or for a closet, which come in a variety of sizes and can be retrofitted to become the perfect longer, bigger, higher walled on the sides, cat box.

Robotic boxes with motor components to remove waste are often covered and not large enough to adequately contain cats, litter, waste trays, and the motor. As much of an appeal as these might be in relieving **us** of scooping duties are they really what your cats want or keeping the box clean where the cat and their superior sense of smell is concerned? That waste has not gone anywhere as far as the cat is concerned. They can still smell it, even if we can't. Or the other tradeoffs of the off-putting effect that the noise, cover and size can have on the cat? That cats do use these boxes means they tolerate them and not that they like them. Added stress from having to put up with a less than desirable litter situation can also result in use falling off, displacement behaviors or stress showing in other areas where cats are concerned.

Litter box location needs to be considered. It may be tempting to hide that box in a closet or to centralize food and water bowls along with the litter box but these choices are worth rethinking. Cats using a litter box are in a vulnerable position, they want to see what is happening around them and not hide away from sight. Avoid placing the box in a place where the cat might feel trapped by either other cats or passing humans and cannot easily access. Walkways and heavily trafficked areas should also be avoided along with noisy areas such as next to washers, dryers, or televisions. When given choices to use boxes in the middle of a room or against a wall, cats in research studies, overwhelming chose the box against the wall.

In a natural environment, cats do not eat and drink where they eliminate, relocate litter boxes away from food and water bowls.

Is there something else going on that your cat is objecting to? House soiling can also be a means to communicate distress. Stress hormones and pheromones transmit information in both urine and feces that are apparent to other cats and often not understood by humans. Think of recent changes in environment: moves, losses or additions of people, pets, caretakers, schedules, etc., that may be troublesome for your cat and how you can address them with mitigating enrichment.

Remove any SOS signals by replacing, removing, or adequately cleaning those spots where the litter box has not been used. Removal and replacement can have the most successful outcomes in this regard, as any chemicals left behind are discarded with the object. This may not always be possible. Products used to eliminate those chemicals matter. Cleaners need to be a good enzymatic solution. Repeated applications of these formulations may be necessary to be effective. Adding enzymatic cleaners to the wash with soiled linens and clothing can remove protein deposits left on fabrics. Make sure to read labels for instructions on use, one popular product advises "shake before use," how often does that happen?

Cleaning with standard cleaning products to remove odors we may object to are fine for us but do not remove the amino acids left behind that our cats can still identify (and an enzymatic cleaner can remove) and may continue to target. Removing those deposits properly can help.

Multiple cat households can be another issue for some cats. A second box plus one can help and is frequently recommended by experts for this. The rule of thumb is one for each cat in the home plus one. Apartment dwellers who are already dealing with a space premium may not find this realistic, but adding that additional box (and keeping both boxes routinely scooped) can often make all the difference, especially where territoriality issues may surface around the box or with new cat integration where the newcomer's use of the box may be prevented by the resident cat. Help to address more concerns by ensuring cleanliness, type and size of box, litter, ease of access, location, etc. (we cannot say this enough), just as carefully for the additional litter boxes as you have with the first box. All the reasons to make that first box attractive for use, count for each litter box. Make sure to add supplemental boxes in places far enough from the first box to be and feel separate.

When addressing this, no matter how often you have gone back and tried a solution with litter box basics, it is always helpful to start again from the beginning.

But what about when inappropriate (for us, the cats have reasons) elimination or litter box aversion is more than just about the box? A closer look behind the scenes can help the both of you.

Human beings mainly communicate visually and verbally. Smell is not high up on our list, but for cats and dogs, it's a whole other story. Olfaction is a number one sense for dogs and falling only slightly behind hearing, a number two sense for cats. In the feline and canine world, pheromones, odor, and scent work to convey and process vital details about individuals and their environment.

Urine marking and "middening" defecation (intentional placement of feces) are definite expressions of information to be shared. Cats who are urine marking are usually intact males, this is thought to relate to reproductive and territory concerns, as is middening. Cats may begin spraying, intact male or not, when "intruders" are involved whether the intruder be a visiting outdoor cat, raccoon, or new addition to the family. This sort of signaling is mostly to delineate territory, especially when placed next to an exit door.

Even knowing that marking and middening are deliberate communications, we do not come close to knowing the full particulars that cats and dogs are transmitting through "smell-o-vison". Urinating and defecating in other areas, like an owner's bed has a definite stress /frustration/insistence aspect and may relate to a traumatic event (as perceived by the cat) such as a move, loss, bringing in a new cat or person, mistreatment or other significant change in environment or routine. It bears repeating for emphasis that urine and feces do not mean the same things to cats as they mean to us. While we may think such an act is an insult, it is anything but in this scenario. Limiting it to territory is surely too simple an explanation in every scenario. The cats and dogs that encounter the purposeful placement and scents left behind are certainly fully aware of their significance, for us, it's a bit more to untangle.

Even having determined a definite component of communication in some instances of house soiling or unacceptable elimination, we still need to puzzle out what is being communicated or "said." Some experts have supposed that cats view human caretakers as maternal stand-in's. Cats continue to meow to us past kittenhood, such behavior changes between cats as they mature as cats are not known to meow to each other as frequently as they do to us. Meowing to humans is often seen when requesting a meal, activity, or access, all of which they recognize we are in charge of providing. Verbal beings such as we are, it is easy to read meows as requests, chemical signals left behind and out of the litter box are a lot more work, especially without some needed equipment. Humans are missing the Jacobson's organ, cats and certain other animals have near their nasal cavity. This organ picks up moisture heavy odor particles and pheromones, and is best seen in action in cats with "smell tasting" or the "flehmen" response, when they take in a chemical stream through the organ, jaw is slightly parted and eyes may momentarily lose focus.

While, we may not know the chemical components of the urine or feces when elimination is purposefully left out of the box, we do know that chemical signals are an additional way animals do share pertinent information, including through pheromones that influence behavior on physical and emotional levels. In addition to those pheromones used to influence reproduction, alarm pheromones are also known to be used by cats. What other messages in depth or nuance or we missing?

It is very possible that elimination in certain instances might be an attempt to "communicate" alarm sort of messages through scent if we were able to process it. Scientists often use a non-invasive method to determine stress levels in animals by measuring the cortisol (stress hormones) levels in urine and feces. Levels of cortisol or pheromones in the leavings at what we perceive as unacceptable locations might have a lot to "say" if we were able to puzzle it out. No doubt, cats themselves, easily decipher this and other information when encountering the urine and feces of other cats and actively use waste products as part of how they communicate all the time in outdoor living situations.

While we may find such leavings objectionable, this communication is not directed towards us in a vengeful or adversarial way. Rather, it might be meant to share an urgent concern that the cat has about what is happening around them. So urgent, perhaps, that they may feel the situation to be so uncomfortable and untenable that they are pressed to communicate this to their human in a location that has the most of our own scent deposited on it, our beds, clothes, or shoes.

Research looking at scenarios where litter box aversion has been remedied, offer insights into what factors may help. The cat's welfare is directly and forcefully impacted by routine and environmental events. A ground breaking study done in 2011 found that disruption to routine resulted in sickness behaviors (which are defined as vomiting, diarrhea, decreased food or water intake, elimination outside the litter box, lethargy, fever, decreases in grooming and decreases in social interaction) in healthy cats and that providing an enriched environment to sick cats resulted in a significant decrease in the number of sickness behaviors and/or symptoms exhibited. The study found that routine and schedules, including keeping the time the same every single day for each feeding was paramount to stress reduction. Other instrumental factors were providing for the same caregiver, playing classical music (no rap or heavy metal please), offering playtime including the interactive kind, keeping clean litter boxes in the same locations, and avoiding manual restraint. Applying these interventions in the home, may have even greater benefits.

A separate study published in May of 2017 by the *Journal of Feline Medical Surgery* compared the behaviors, including inappropriate elimination, excessive grooming, and aggression, of cats that had been de-clawed compared to cats that not been de-clawed. The de-clawed cats significantly demonstrated more of these behaviors. Sixty three percent of the de-clawed cats were found to have bone fragments left behind after surgery in their digits. These cats were more likely to have back pain, inappropriate elimination, biting and aggression. De-clawed cats without retained bone fragments were found to have increased biting and inappropriate elimination. Pain is not easily addressed but the use of gravel type litters or clumping litters is not a good idea for de-clawed cats. These cats have compromised abilities to manipulate litter so the cleanest, softest possible litter is the kindest in a large enough, uncovered box.

Side effects of neutering and their stressors can also affect litter box use. Pediatric spay and neuter limits the development of the cat's sex organs. Theories have existed that the opening of the penis, or the "lumen," remaining underdeveloped might lead to blockages. Two studies done in the 1990's discounted this. More are needed to further examine continued claims. A more recent study done in 2017, found castration to decrease the density of collagen fibers in the corpus spongiosum of the cat's penis:

> "This suggests that the compliance of the periurethral region is reduced, and these changes could be a predisposing factor for urethral obstructive disease" (Borges, Pereira-Sampaio, et al. 2017)

This is a significant finding. It is important to note that age at the time of castration was not studied as a factor and needs to be further investigated.

Think again, a traumatic event, move, change in routine or schedule, loss, introducing a new cat, dog or person into an existing cat household can also generate house soiling and litter box issues. The work is in figuring out what is stressing the cat so very much that this is what they feel they must do in their cry for help to reach us to do something about it.

Start with trying to determine first what has changed and what change would be most upsetting from the cat point of view. Again, and again and again, no matter how much has been implemented in the past, a two-pronged approach, where litter box basics as noted above needs to be reviewed and implemented at the same time as working on behavior, for the most effective solution.

Once the stressor is identified, remediation, mitigation, and enrichment, need to happen. Allow for a period of latency, for a time when the cat will continue or attempt to continue the prior behavior while you are making changes or appear not to be influenced by the changes you are making. Cats wear the proverbial different watch then we do. It is important to remember that change is a process not an event. In those cases where a new pet, whether cat or dog is being introduced, review introduction strategies, including revisiting them from step one (see the chapter on successful cat integration or the one on cat dog integration). New persons in the household can also be perceived as threatening and upsetting the existing order the cat has come to trust. Newcomers can build relationships by taking over feeding routines and interactive play sessions, both highly valued activities and events to associate with the change and person. Evaluating and redoing environment and enrichment protocols can also benefit.

Vital to keep in mind, is being aware that part of changing your cat's behavior, is changing your behavior with your cat. Should your behavior include punishment, no matter what the form, even verbal with scolding, angry asides, or body language, you want to remove it. And spraying water or shaking a can of pennies or rocks? Stop doing that, definitely out. Punishment creates fear and the need for defensive behavior, increases stress and makes behavior problems worse not better, especially with cats. For what will make things better — adding in the following changes may help:

Do apply the management of litter box basics reviewed above initially along with addressing the causes of stress. With a cat that is targeting the bed, placing the litter box on the bed might shift placement, but is definitely not palatable from a human standpoint. In this scenario, new linens, mattress pads or even a new mattress to remove any residual odor that we might not be able to detect but a cat can, may help. Placing unwelcome objects from a cat point of view on the bed will also help to deter the cat from reaching the bed. Temporarily prohibiting access to the bed or bedroom might be a more workable strategy.

Another example of a management strategy for outdoor intruders is blocking the view to cut down on visual stimulation, think of taping paper over window panes. Additional strategies would have to be employed to deter the outdoor visitor as well because even if the indoor cat cannot see the intruder, they can smell them. Mothballs, placed outside, are an excellent deterrent. Motion activated lights are another option. Although not always effective, a plug-in pheromone diffuser that has been tested specifically for marking behavior, can be tried. Add in overall soothing scents such as catnip and valerian root sprinkled around which have been shown to be preferred scents for cats and are beneficial and stress reducing.

Make sure the appropriate cat furnishings are available in every room, such as beds with at least three sides to offer containment, both floor level and raised resting spaces, such as shelves, towers or beds placed on existing furniture for safety and retreat. Multi cat households benefit immensely from this, and so do multi species households. Cat things and places to do them are important, as in crinkle tubes and boxes that come in the mail. Do not forget sturdy scratch boards or posts rubbed with catnip or valerian root (because they need to scratch) and plenty of toys (fur mice that rattle, crinkle balls and more are a must) with the most important aspect of play, again, being with you -unpredictable, bonding and way more engaging.

An enriched environment is more than essential for these cats to alleviate stress and allow for necessary and natural behaviors that are intrinsically rewarding. Schedule and routine are paramount to lessening stress and offering control over events cat cannot provide for themselves. Consider and institute set times for meals and interactive play. Schedules instituted in the 2011 study were key factors found to lessen litter box avoidance and sickness behaviors. This is an important take home. Pairing routines of cat care to our own routines are an effective way of getting schedules on the consistent calendar. Another key factor was constant caretakers. Keep trusted caretakers steady, including pet sitters and make sure pet sitters (and caretakers) make the morning feeding an early one to satisfy the cat's crepuscular (most active before dusk and dawn) nature. For multiple cats, you want to make sure resources are adequately distributed throughout the environment, including your attention and careful integration strategies observed for all manners of newcomers.

Provide opportunities for satisfying cat activities like foraging and hunting with puzzle feeders for dry food meals instead of food bowls. Puzzle feeders that allow for natural behaviors such as batting with rolling feeders or scooping with larger tray puzzle feeders simulate movements cats use in hunting. Avoid feeders that are designed to slow eating as these are more likely to frustrate the cat than satisfy behavioral needs. Keeping puzzle feeders full allows cats' intrinsically satisfying interaction and problem solving in feeding and control over accessing food.

Offer routine daily interactive play with their humans with fishing wand toys. And keep fishing wand toys out of sight when they are not being used. Pairing play sessions with feeding or litter box cleaning can help our own routines and keep these activities on schedule. Most importantly, do play with your cat even if the time is off but be careful not to create an unpredictable schedule instead. These intermittent schedules of reinforcement, so called, are the most effective at getting a response but also the most frustrating.

Petting and grooming can be welcomed by cats when they consent to it and when it is offered in that cat friendly fashion or as another cat would. Approach your cat from the side, gaze lowered, offer a slow blink, extend a finger a muzzle level and observe the cat's response. Consent can be indicated by blinking in return, tilting the head towards the finger, or rubbing against it. Short gentle strokes along the side of the muzzle, under the chin and behind the ears are the most welcome. Avoid full length body stroking.

Review the enrichment chapter in this book for expanded ideas on the cat friendly home and make sure to institute them. The key here is looking at the whole entire picture from a human and cat point of view to begin to solve and address things that are troublesome. That a cat would think we have the power to solve the problem gives us some big shoes to fill, it is also amazing and motivation enough for us to do just that.

References

- Borges N.C., Pereira-Sampaio M.A., Pereira V.A., Abidu-Figueiredo M., Chagas M.A. (2017). Effects of castration on penile extracellular matrix morphology in domestic cats. *Journal of Feline Medicine and Surgery*. 19(12):1261-1266. doi:10.1177/1098612X16689405

- Bradshaw, J.W.S., Casey, R.A., Brown, S.L. (2012). Undesired behaviour in the domestic cat. In *The Behaviour of the Domestic Cat* (pp. 190-205). Oxfordshire, UK: CABI.

- Ellis, J.J. McGowan, R.T.S. Martin, F. (2017) Does previous use affect litter box appeal in multi-cat households? *Behavioural Processes*, 141, 284–290.

- Frayne, J., Murray, S. M., Croney, C., Flickinger, E., Edwards, M., & Shoveller, A. K. (2019). The Behavioural Effects of Innovative Litter Developed to Attract Cats. *Animals*, 9(9), 683.

- Martell-Moran, NK, Solan M., Townshend H.G.G. (2017). Pain and adverse behavior in declawed cats. *Journal of Feline Medicine and Surgery*, (published online May 2017)

- Stella, J.L., Lord, L.K., Buffington, C.A.T. (2011). Sickness behaviors in response to unusual external events in healthy cats and cats with feline interstitial cystitis. *Journal of the American Veterinary Medical Association*, 238, 1, 67-73.

SCRATCH THIS – NOT THAT

My first Manhattan apartment was one of those five-flight walk-ups with a bath tub in the kitchen. The "bathroom" was little bigger than a closet and papered with that very popular '80s throwback of grass cloth. Nubby in texture and pattern it worked for the space and the cat loved it for scratching. He added his own unique pattern of ripped paper, up stretched front leg claw height, to every wall. Not his fault. It was all he had for good scratching and the satisfaction of sinking his claws into that slubbed surface! Scratching posts were not on my radar as a must have when I took the apartment or when the cat moved in. My high school years consisted of living with indoor-outdoor cats that, as most cats, who can access the outdoors will do, confined their scratching to outdoors and not in. I did get scratching posts, after the bathroom redesign, more than one. Still, grass cloth wallpaper will be a forever reminder for me that cats being cats need to scratch.

Scratching is a natural and necessary behavior for cats. It is interactive and dynamic; beneficial for stretching the upper half of the cat's body, releases energy, removes the older, outer sheath of claws and releases pheromones. Whether or not scratching is a visual signal for other cats continues to be debated as cats are not observed to focus on visual placement of scratch marks left behind by others. These are the functional aspects of scratching we tend to focus on when discussing scratching, the aspects that supply a necessary biological component to welfare but what about welfare's necessary emotional aspects?

Scratching feels good. Bodies are designed to repeat behavior that is rewarded intrinsically or through external reinforcers. The components of scratching reward, that pleasurable and intense muscle stretch, that stripping of outgrown claw sheath, that sinking feeling of the claws tactile push and pull against a resistant but not too resistant surface the cat alone has decided is perfect for the job. Cats need to scratch for more than one reason and we mostly would not like them to.

That cats scratch surfaces owners consider inappropriate is a common behavior problem. Furniture, fabric, curtains, carpet, doorways are the most frequently targeted. Texture, optimal resistance, and interplay of the object to be scratched, all factors which are most probably more significant to the cat than its location. Looking at the properties and characteristics of what cats like to scratch can be helpful in replacing those objects with similar properties and characteristic in things we would like them to scratch instead. Along with how to attract cats to those replacement objects.

Just as with my grass cloth wall paper, we often are not aware of the scratching problem to us, or the need to them, until after the fact. Substitution may not be our first response. More likely we try to stop the behavior going as far as surrendering cats or even more drastically de-clawing them. Two studies looked at how cat owners were dealing with unwanted scratching behavior as in what was working and what was not.

A 2018 study found owners reporting that many of their cats were daily scratchers targeting mostly at least two different pieces of furniture or carpet. According to the survey respondents, close to 70% of the cats had a scratch post or pole and over 50% had a scratch pad available. To stop cats from unwanted scratching, 69% of owners indicating yelling at the cat, 37% sprayed water, 15% spanked, and 13% rattled a can. Deterrence accomplished, with 26 % clapping, noise making and moving the cat. To encourage alternate scratching locations and objects, 57% of owners relocated the cat near the preferred item, 51% praised appropriate scratching, 44% applied catnip and 36% moved the cats' paws over the items they would like the cat to scratch. How effective were these methods? From the study:

"Interestingly, cats appeared to use the designated scratching item less if owners placed them near the item. Without further data it is difficult to explain this observation. One possible explanation could be that scratching behavior often occurs within the context of waking or stretching, or after social tension, and that being lifted and carried interrupted the behavior chain. The majority of owners yelled or sprayed water to interrupt scratching, although these strategies did not affect the expression of the undesired behavior in this study or the study by Wilson et al. Previous studies in dogs have confirmed the inferiority of punishment-based methods to modify behaviors." (Moesta, Keys, et al. 2018).

A similar study done in 2022, found less objectionable scratching reported in cats that had access to the outdoors (where there are probably far better items and places to scratch than indoors) and less scratching when sisal and flat scratchers were furnished to cats. Unlike the 2018 study, this study did find that rewarding the use of scratching appropriate items reinforced their use. Every animal responds to positive reinforcement, timing and reinforcers factor in. Applying attractants (pheromone sprays or catnip) and furnishing additional items to scratch also increased use. (See more on why I like Valerian Root for this below.) Punishment and interruption of inappropriate scratching increased its occurrence. What we can increase with punishment is fear, loss of trust and new places to perform forbidden behaviors.

The limiting factor of the indoor environment is linked to scratching in both studies, there is less to do inside, accompanied by the effects of mitigating its constraints with enrichment. Cats kept within the home can be stressed, bored, and frustrated from lack of opportunities to perform the natural and necessary set of behaviors, of which scratching is just one. Optimizing the house cat's home can have obvious benefits:

> "When cats are kept indoors, the current findings suggest that providing enrichment may serve to reduce problematic scratching. This is similar to previous research that showed the positive influence of exposing cats to a variety of enrichment items in managing other common behavior problems, such as aggression and inappropriate elimination." (Cisneros, Litwin, et al., 2022)

Owners in both studies did recognize the needs of their cats to scratch and offered them places to do it. Whether what was on offer was what the cat wanted to scratch or in the right place are further areas to explore. Add to this that every cat is an individual and while sisal rope may seem the front runner in what a cat likes to scratch in commercial cat scratchers, some cats may prefer corrugated cardboard and others fabric. There is clearly a cat reason our upholstered furniture and carpets are such a draw. Fabric, particularly fabrics with textures that provide a degree of resistance to drag claws through, can be highly desirable for cats but not a surface owners want to encourage scratching. Recognizing the appeal of fabric and or other surfaces, coupled with location preferences, and looking for alternates with attractants added, can heighten the attraction and use of where we do want cats to scratch.

In practice, I have found when introducing a cat to any new cat furniture or surface you would like them to be interested in, rubbing the items with valerian root, catnip or silvervine can be helpful to entice them to use the object. Some cats immediately gravitate to the attractant and others take a minute, but they all, with their highly evolved scenting abilities, notice the attractant and will investigate given time. Valerian root, a member of the catnip family and a proven cat attractant, is easy to find whether online or in local stores. Drugstores and health food stores, often stock it for people using the plant for its reputed help with occasional sleeplessness. It comes in a fine powder (dried catnip is flakey and gritty) in capsule form. That fine powder lends well to being rubbed into surfaces. Topping off weekly an additional capsule's worth for those surfaces you want cats to scratch can reinforce interest and add the stimulation of the scent along with scratching.

Stability is an important component of a scratching surface, as is being able to stretch to use it. Scratching posts or flat scratchers that move around are too flimsy to get a good scratch in on. There is a reason cat towers with sturdy pillars covered with sisal and carpet are a draw, these stand firm against the scratch, allowing for full body stretching during the prolonged scratch bout. For flat scratchers – corrugated cardboard has quite the appeal in satisfying tearability quality, engages a different posture in the stretch and scratch, working a different set of muscles and a different rhythm to the movement. Every room needs two cat scratching stations (or one that has both) to allow for upright and flat scratching.

Location of scratch targets may matter a lot less than what is being targeted to scratch. The upholstery and solidity of the living room couch may be the draw and not where it is situated. Preventing scratching by covering the surface of the couch can eliminate that object being clawed but the cat will still need to scratch and will still orient towards it. Here is where placing the alternate object comes in. Such strategy can only be successful or moderately successful when the alternate object is just as good if not better than the original object. To replace the side object of affectionate attention or our couch example, now covered to be inaccessible, I have experimented with commercial incline corrugated cardboard scratchers, sisal posts and carpet pads placed against the former scratch site. None worked, until a custom two by four wooden post, tightly wrapped in valerian rubbed sisal and mounted on a sturdy platform was put in place. They loved it.

Try these steps to get them to scratch this and not that:

- Always remember, scratching is a necessary behavior for cats.

- Punishment is not effective in modifying this behavior. And if punishment creates distrust and fear, it does, it also lessens the effect of reinforcement and praise. Punishing a cat is not just cruel, it is counterproductive in changing behavior. Do not yell, scold, spray water, yell at or move, etc., a scratching cat.

- Prevent access to the undesirable surface by covering it or moving it. Pheromone sprays to reduce aggression can be helpful sprayed on undesirable surfaces. But be careful not to spray those same deflecting sprays on the surfaces of the objects you want the cat to scratch or if you are placing the alternate object next to a sprayed surface. For this reason, I do not use these particular sprays and only use attractants on the surface I would like the cat to scratch.

- Experiment with multiple types of scratchers and allow for sufficient time for the cat to acclimate to the scratcher to use it. This includes trying out multiple scratchers with different shapes and heights; incline, wider, longer, flatter, taller, etc. Make sure those shapes and heights accommodate cat length. Scratching stations need to be sturdy and both wide enough, stable, and able to withstand the force of the cats' movements to encourage use. Coverings matter and must, must, must have tactile appeal and be fun for cats to scratch. Sisal rope, carpet and corrugated cardboard can be good choices.

- Providing multiple desirable surfaces for use and allowing for time to acclimate to them is necessary to determine preferences. Every room should have an upright place to scratch and a flat place to scratch.

- Use cat attractants to rub on the surface of scratching stations. Valerian root, silvervine or catnip are proven attractants. Tatarian honeysuckle is another known cat attractant; however, this plant is an invasive species and cannot be sold or propagated in much of the Northeastern United States. Check your state. For most all states, valerian root, silvervine and catnip are freely available with the fine powder of valerian and silvervine being ideal for being retained by the surface of the scratchers. Replace the powder weekly.

- Praise scratching appropriate objects when you see it. Make sure your praise is brief and delivered in a calm, soft and loving tone using the cat's name. Loud and excited is great for dogs and disruptive for cats.

- Enrichment is not just about adding raised resting spaces, scratching posts and other appropriate objects to interact with. Routine interactive play and owner attention are needed for welfare and may also help reduce unwanted behaviors.

References

- Cisneros, A., Litwin, D., Niel, L., & Stellato, A. C. (2022). Unwanted Scratching Behavior in Cats: Influence of Management Strategies and Cat and Owner Characteristics. *Animals*, *12*(19), 2551.

- Moesta, A., Keys, D., & Crowell-Davis, S. (2018). Survey of cat owners on features of, and preventative measures for, feline scratching of inappropriate objects: A pilot study. *Journal of feline medicine and surgery*, *20*(10), 891-899.

PICA AND CATS – HOW TO HELP WHAT'S EATING THEM AND WHY

Grigio and I go way back, she is a former feral from the backyards of Brooklyn, one of two sisters living with human caretakers. Bonded to Pinot (Cute right? They came that way.), her littermate, she is fortunate that she has never been separated from her. Her rehoming history is a different story. Pinot and Grigio were brought to me for late socialization from an animal shelter, rehomed into a situation that did not meet their needs and next dumped. Left behind a restaurant at a highway rest stop, 100 miles from their adopter. Found and rehomed stably for years with periods of adjustment, that home has just lost footing and they are on their way to yet another place. In all this bouncing around, Grigio has developed an appetite for plastic. She eats it.

Theories on why cats might pick plastic include components used in production such as tallow or chemicals that might mimic pheromones. Possible, but at certain times, her desires had expanded to include portions of certain worn articles of her late guardian's clothing, especially when he was late to come home. Pinot, may deal with her comfort eating in a different way, becoming fonder of extra food and puts on weight instead. For both these cats, something is helped somehow by what they are eating and why.

"Pica" is defined as an atypical desire for and eating of substances not normally eaten. This behavior is seen in humans as well as cats. It has been found to satisfy nutritional deficiencies when chalk, ashes, or bones are consumed for phosphorus or when clay or dirt is consumed for iron. Grass chewing in cats is not strictly pica as cats often exhibit a natural preference for grass when it is available and ingesting it may benefit digestion. Pica also refers to eating non-nutritive substances and is commonly seen in cats that suckle or consume fabric, plastic, or foreign objects. When dealing with pica, as with any behavioral concern, ruling out the medical needs to come first. Even when the basis is thought to be behavioral, be aware there are cases where pica can contribute to dangerous intestinal blockage and more.

Various theories as to the motivation for the behavior range from medical conditions, to a need for more fiber in the diet, early weaning, emotional upset, stress or lack of choice and control. As an indicator of stress, pica is a sign of poor welfare for the cat. It may also be argued, that at times pica might serve to enhance welfare as opposed to detracting from it. Also, that not every close contact with an inedible object is pica.

Cats are meat eaters and hunters, "obligate carnivores" who depend on those amino acids found only in meat for complete nutrition. Plastic, fabric, and other inedible objects offer zero nutritive value, so why do some cats consume them? Start with a closer look at what might be going on with fabric suckling. Usually seen along with kneading with the forepaws, "making biscuits" can be an endearing action evidencing contentment by some observers. Conversely, the same performance may concern other cat owners and even has been termed as "infantile behavior" by certain experts, along with grouping this as a behavior consistent with pica. However, where the fabric is not being ingested, the behavior is not consistent with pica and really should not be a concern for the cat's well-being.

Kneading and fabric suckling soft or woolly surfaces may be a comfort or soothing behavior or invoke pleasing memories. These actions may be in themselves, an intrinsically rewarding behavior for the cat. If the motivation is self-soothing or pleasurable, and not harmful, it should not be denied to the cat. To follow then, what constitutes a "problem" behavior when looking at pica and cats and what interventions are useful?

Cat experts John Bradshaw, Peter Neville and Diana Sawyer looked at pica behaviors in a study of 152 cats, most of whom were Siamese and Burmese. They found the subject cats demonstrated a clear preference for wool, followed by cotton and then synthetic fabrics. Rubber or plastic materials led the choice in foreign objects. Cotton was ingested more than any other material, perhaps due to availability over clear preference. Of note to look at, are certain circumstances with the study cats, the majority of which had been rehomed. Pica occurred within four months of rehoming in 81% of the cases. The authors' state:

> "this data strongly suggests that rehoming is one factor which may trigger pica." (Bradshaw, Neville, et al.1997)

Additional research, led by Isabelle Deomntigny-Bedard, looking at pica and chewing behaviors, surveyed cat owners of 91 cats performing pica compared to owners of 35 cats who did not perform the behavior. In this study, cats who chewed on an inedible object without ingesting it were not counted, as this behavior was not considered pica. The pica cats exhibited significantly more vomiting along with other digestive signs than the non-pica cats. These are inedible objects after all. Further, shape or malleability play into partiality for what is consumed apparently, shoelaces, thread, and plastic were favored over fabrics. Significant findings included noting that fewer cats in the pica group had continuous free access ("ad libitum") to food compared with the non-pica cats where more cats did have continuous access to food. The authors raise perhaps the obvious questions:

> "does hunger play a role in pica behavior? Is ad libitum feeding protective of pica?" (Demontigny-Bedard, Beauchamp, et al. 2016)

Such research is valuable as we begin to investigate associations with environmental forces, stress, welfare concerns and pica. There are more cats out there with pica concerns and not enough data about the behavior or its motivations. More studies are necessary to look at what other factors are significant for pica along with what interventions might be the most successful.

In reviewing each of the studies, stress and lack of choice and control can also be implicated as primary factors, whether due to change in home environment or in being denied access to food. Stress occasioned behaviors may serve in some way to allow the animal to cope with the pressure. They also have an impact on overall welfare. It is inherently difficult, if not impossible, to tease out from observation what the benefit of pica is for the cat that performs it. Hunger, anxiety, and frustration, may each or all, play a part. It is possible and even probable, the act of suckling, chewing, or ingesting an inedible object in the performance of the act, the very time spent in the suckling, chewing, or ingesting - is rewarding or soothing. Perhaps the fabric consumed is wonderfully imbued with and redolent of the scent of the owner and this itself is rewarding. Conceivably the prolonged suckling, chewing, or ingesting is reminiscent of the intimacy and sustenance found in nursing from the mother. The cat alone knows the why, we can only suppose, work from educated guesses and begin to try for mitigating stress and providing alternate outlets which may relieve the need to perform the behavior.

Not every cat that experiences stressful events will perform pica but this does not mean that these cats are not stressed. Whether internalizing stress or acting on it is more harmful needs further investigation. We do know that pica is a behavior that may be harmful to a cat even as it may be stress relieving in its execution. Managing the cat's environment to limit access to those inedible objects targeted is the first step in working with this behavior, relieving the stress, and offering opportunities for other behaviors are the next steps.

The following plan of action may be most effective when implemented and given time to work. As always, remember that cats wear a different watch than we do and need sufficient time to trust the improved environment:

Manage the environment: Limit the opportunity to consume inedible objects. If the cat consumes plastic, do not leave any lying around. No plastic bags, packaging material, containers, etc. Similarly, if the cat is chewing socks, shirts or shoelaces make sure these are inaccessible. Be a detective in figuring out what the cat is targeting and how to remove those objects from the environment. Lessening the occurrence of the behavior is the first step in managing it. The next step needs to be addressing whatever function or relief the action is providing.

Remove any punishment: **whether in** your tone of voice, body language, or the simplest admonition to more aversive strategies such as shaking rocks in cans or spray bottles. Punishment is not effective with cats as it increases stress, exacerbates an existing situation, and creates a negative association with the human involved and not just the action. Additionally, no punishment is effective as a deterrent, unless it is performed at the exact moment of the behavior, and if it is not continually increased which impacts welfare.

Modify behavior through learning, interventions, interactions, and environment: Know that to modify or change a behavior you need to offer a behavior that is equally satisfying or rewarding and to allow for the place, learning and opportunity to perform that behavior.

Offer cats an enriched home environment with the appropriate cat furniture in each room. Each cat should have at least one raised resting space, floor level cat bed with at least three raised sides, place to hide (from cardboard boxes to cat igloos), scratching post in every room, cat attractant scents like valerian root and catnip sprinkled on cat surfaces rubbed into cat toys for solitary object play, and classical music at soft levels played on occasion.

Utilize fountains for water sources and place away from food Cat fountains, properly maintained, can increase a cat's water intake, an important component of cat health. In a natural environment, cats will seek out fresh water sources away from prey they have consumed.

Cats are not social eaters. Use separate plates to feed each cat. Feed multiple cats the wet food portion of their diet in separate rooms or at a distance from each other. Place food plates away from walls and not in corners. The prey and predator aspect of the cat means more comfort in being able to survey the environment while eating.

No more dry food in bowls. Look to a cat's natural hunting behavior and provide rolling or tray puzzle food feeders for the entire portion of the cat's dry food diet, to engage cats in those behaviors which stimulate cognition, problem solving and are intrinsically rewarding. Avoid slow feeders that can irritate rather than satisfy. Refill feeders daily and check the supply at bedtime, to insure food is available for overnight engagement and predawn hunger pangs and zoomies. Puzzle feeders with freely available food, may also serve to address hunger or choice issues with food, both of which were implicated, as a probable cause of pica in one of the studies mentioned.

Engage cats in interactive play sessions with fishing wand toys to stimulate and encourage play activity with all the helpful endorphins and neural firings of joyful and stress reliving activity. Experiment with different toys to find the most popular. Remember to draw objects away from and across the cat's line of vision to best approximate prey movements of rodents or reptiles. Keep play sessions on or close to schedule and keep fishing wand toys safely out of sight when not in use. Routine and schedule offer a sense of control to captive and domestic animals who cannot determine when significant and necessary things and events will be provisioned. Feeding, husbandry, and interactive play on schedule has been shown to reduce sickness related behaviors in cats. Providing this to every cat can be significant.

We know looking at and instituting at ways to improve welfare and lessen stress contributes to the quality of life of all cats. Grigio has benefited from these interventions. I would like to think she is happier for it. We will continue to keep plastic out of sight now and in the future. Whatever taste there is in that filmy stuff that drew her or whatever comfort association may have developed back then no doubt remains. Supplanting that association with safer alternatives that still satisfy the need behind the motivation is the goal. We think we got there.

References

- Bradshaw, J.W.S., P.F. Neville and D. Sawyer. (1997). Factors affecting pica in the domestic cat. *Applied Animal Behaviour Science*, 52, 373-379.

- Demontigny-Bedard, I., G. Beauchamp, M.C. Belanger and D. Frank, (2016). Characterization of pica and chewing behaviors in privately owned cats: a case controlled study. *Journal of Feline Medicine and Surgery*, 18(8), 652-7.

OWWW, MY CAT BIT ME! WHAT DO I DO? CAT AGGRESSION TOWARDS HUMANS

Victoria, is an older kitten and former feral street cat with a small biting issue, more along the lines of light nibbles as opposed to bites, nibbles given as play signals and seeking attention. What she may envision as a friendly fun feline invite does not translate to human scale. Victoria doesn't know that, so I need to break it down for her, into small parts, with the appropriate feedback, redirection, and behavior modification, for who she is as an individual. Victoria has a minor case of wobbly cat syndrome or cerebellar hypoplasia (CH).

Kittens born to cats infected with feline distemper can have part of the brain (cerebellum) affected or underdeveloped. The cerebellum affects movement, balance, and learning and development. For most cats with this condition, life is otherwise unaffected (Never declaw any cat especially a CH cat, for whom, carpeted, non-slip mats and carpeting can aid in traction, raised food bowls can also help.) and we mainly see the effects in lack of coordination or wobbliness. With Victoria, you see her condition mostly in teetering in tight circles at mealtimes, splayed toes in place, and sometimes failed jumps to higher surfaces. Victoria is also markedly slower to respond to redirection compared to other cats. She stays focused longer than most on one thing. Victoria needs repetition, emphasis, and several loops in redirection. She is also a Velcro cat, with a level of insecure attachment who is an avid audience.

When working with biting behaviors, it is helpful to remember that all cats bite as part of a normal set of behaviors, whether grooming, in play or defensively. As with any behavior, these do not happen in a vacuum, they are colored by individual history, physiology, responses to another's action or interactions. Where biting is between cats, the messages being sent are often loud and clear to the cats (Victoria's biting issues are nonexistent with the rest of the cats at home), but when it comes to cats biting humans, we need to first look more closely to what is happening for both perspective and remediation.

Aggressive behavior can get cats kicked out of homes. The top three behavioral reasons for owner surrender of cats to animal shelters are house soiling, problems with other pets and aggression towards people according to a 2000 study in *The Journal of Applied Animal Welfare Science*. A more recent 2023 study noted behavior as the most common reason for return with cats adopted from a shelter. The latest statistics supplied by the ASPCA reports that of the 3.2 million cats that entered the shelter system each year, 2.1 million of these cats will be adopted and 530,000 of them will be euthanized. Retention and, or returns, are not included in that set of statistics.

Formal discussions surrounding aggression and pets stem from reported incidents. While medical attention may be warranted in certain cases, minor occurrences, with little to no injury, often go uncounted, the more severe of bites can be the ones to receive media attention and account for more hospital visits. Children under the age of 12, and boys more than girls, are the recipient of most dog bites to the face or torso. Middle aged women, with injuries commonly to the hand or wrist, are the reported recipient of the majority of cat bites. Understanding how a bite can happen and how to avoid being bitten, is the best strategy for prevention. Ruling out any medical concerns the cat may have, is also needed, as pain and or illness can contribute to aggressive behavior.

Human tolerance and safety limit acceptance of inappropriate elimination and aggression, even as such behaviors suggest and are evidence that the cat's basic behavioral and perceived needs may not be understood and/or sufficiently met. Unmet needs affects the cat's welfare. Despite this, when it comes to litter box avoidance, scratching, and biting behaviors, caretakers can respond with various punishments, including scolding, noisemaking, and water bottle spraying. All a context and environment ripe to foster defensive behavior, including aggression.

The most widely accepted definition of "aggression" is action with intent to cause harm. "Violence" is further defined as being a form of aggression where the intended harm is severe or fatal. When it comes to human beings, we can further define aggressive behavior into "physical aggression" or "verbal aggression." For all animals, threats and warnings are not aggression, as they are projected purposefully and can serve to prevent action intended to cause harm from happening if they are communicated effectively - that is "heard" and responded to by another party. While most aggression is based in fear, there are also different underlying emotions and motivations that can accompany it. Behaviorists will often categorize the behavior accordingly into several types of aggression that exist with cats; defensive aggression caused by lack of socialization or in response to actual threat, pain aggression, play aggression and fear aggression. (Anxiety is thought to be the cause for abnormal or problematic aggression.)

It is vitally important to realize that we tend to overuse "aggression," especially when talking about animals, to the point where the word has become a catchall for every behavior we may think is negative or are not comfortable with. This sort of thinking colors how we feel about the animal's actions, and in such quick thinking, dismisses the animal's underlying emotions, history, context of the event, and motivations. Empathy is easily lost in such a response, and owners can tend to over react as a result.

A look at two studies shows insights into what is happening around a bite event. Animal Behavior experts, Ramos and Mills, studied cat aggression directed at humans in the Sao Paulo region of Brazil and found that the top two reasons for owners reporting aggression happened after study cats were "petted or put on to the lap" followed by "when playing." While it is important to know when to pay greater attention to what humans are doing with cats to avoid biting, the difficulty with this study, and similar studies, is that "aggression" is not well defined. It is not clear what the exact behaviors in question are. We may be reading about preliminary distance increasing behaviors, such as hissing, up to and including last ditch efforts, such as biting.

A hiss is a preliminary distance increasing behavior and not overt aggression. That hiss has probably been preceded by even more preliminary communications, such as flattened ears, retracted whiskers, dilated pupils, stiffening, tucked and tense body. There is often a lot being said that we are missing. Surrounding cats have not missed any of it. We need to be able to tell the difference between warnings, threats, and aggression along with accepting that aggression is a necessary and normal response when a cat is threatened or in a dangerous situation. And we need to be able to recognize de-escalations. Actions and emotions change congruently. In addition to a wide range of warning vocalizations that start with hissing and progress in meaning and intensity, cats will employ highly ritualized threat displays and depend heavily on standoffs, back downs, and retreats to avoid actual physical contact. All behavior happens on a spectrum and is fluidly responsive, dependent on the response received. Aggressive behaviors fall into the camp of distance increasing behaviors, if the hiss accomplishes greater distance or retreat from the other party, the hissing can stop, the body tension can lessen and fixed gaze can be punctuated by "I mean no harm" blinks.

Body position also matters. Serious fights are launched when the cat rolls on to the back or side to be able to use all sets of claws and fangs. Fighting is biologically costly and cats only fight when a standoff is unsuccessful, they are attacked or cornered and when all other options to avert fighting have been exhausted. No surprise, cats that have not been socialized around people or have been improperly handled by people will be more traumatized by human encounters. And depending on how they are handled, cats will more readily utilize aggression for self-defense. Being lunged at, pinned, and immobilized by scruffing does not open peaceful dialogue. Low stress restraint and handling are more ethical, humane and offer the cat way less to fight back against.

A separate study published in May of 2017 by the *Journal of Feline Medical Surgery* compared behaviors, including unacceptable elimination, excessive grooming, and aggression, of cats that had been de-clawed compared to cats that not been de-clawed. The de-clawed cats significantly demonstrated more of these behaviors. Sixty three percent of the de-clawed cats were found to have bone fragments left in their digits. These cats were more likely to have back pain, inappropriate elimination, biting and aggression. De-clawed cats without retained bone fragments were found to have increased biting and inappropriate elimination. It is worth the emphasis to note that, in addition to the mutilation of removing digits and residual lasting pain, de-clawing leaves the cat with only teeth as a defense. While painful cats have much to object to with environmental and human stressors in this study, we also are not made aware of exactly how aggression is defined or what proceeds its appearance.

Cats need to use their mouths for more than eating, not having hands to manipulate, hold or touch things with. Feline mouths may also be used to explore, groom (oneself or others), and communicate. The mouth is also used to eat, vocalize, sniff-taste or perform "flehmen" behaviors. When we get up to biting or mouthing as a form of communication, the next question to ask is what is the point or function of the behavior? What is the cat saying and why?

Painful cats can be aggressive in attempts to defend themselves from dangerous situations and environments. Understanding what a painful cat looks like, securing treatment for medical concerns, knowing how to approach and handle these cats will lessen the stress. Language and how it predisposes us to our own understandings of others matters. Veterinary terminology labels a cat "fractious" that reacts aggressively to handling as if the cat came into the practice in a bad mood as opposed to an appropriate response to being manipulated so she is immobile against her will, injected with sharp objects, subject to the insertion of foreign objects in her rectum, etc., etc., etc.

The fractious cat is not a personality type, rather a reaction to a history or intense fear of capture, loss of control, over restraint and aggressive handling. Awareness of these impacts has increased. The late Sophia Yin pioneered low stress restraint and handling practices for veterinary professionals through her work which included books (*"Low Stress Handling Restraint and Behavior Modification of Cats and Dogs"* is foundational), seminars, workshops, and certifications. The American Association of Feline Practitioners launched "Cat Friendly Practices" to address the need for humane handling of stressed and painful cats. Efforts continue in other venues with "Fear Free" programs. Make sure to patronize those veterinarians who institute these practices and most importantly, be able to observe and provide needed social support to your cat at clinic visit by remaining with the cat throughout exams and procedures. Research continues to find owner presence as defusing stress for cats in veterinary settings. Knowing more on cat body language and handling as a cat guardian, improves these interactions at home and helps to make sure those practices are being put to work in other environments.

A closer look at aggression towards people in the home includes play aggression and fear aggression:

Play aggression. We see play aggression when typical cat play behavior such as chasing, pouncing, the use of feet, etc., happens out of context or inappropriately. In the Ramos and Mills study, aggression in the context of play was cited as one of two top areas of concern. In genuine play, when one cat alerts another cat that they do not like what is happening by vocalizations or freezing, the other cat stops what they are doing. With play aggression the stop signals are ignored and can sometimes trigger even rougher play. Stop signals are often repeated for emphasis when unheeded and play is terminated without the appropriate response.

There is a strong element of frustration in play aggression, a ratcheting insistence either for the level of interaction to be reciprocated or for play to commence or continue. Communication has gone off track. In true play aggression, the play behaviors are not typical, welcome, or appropriate. When looking at this, it is key to note that single kittens raised in human households have not benefited from the socialization of acceptable behaviors with other cats. Even with cats raised with other kittens and who had benefited from such socialization, it is very important to rule out any rough and tumble play that is appropriate cat to cat being engaged in. Or, key point, that the cat may have learned is acceptable with humans. If the cat "asks" often enough and is answered, has been taught to play with a human hand or foot under the covers or to chase or be chased, than this is then how the cat has been taught to play and it is not fair to call it play aggression just because the cat may initiate it and not the human. More on retraining this below. Unpredictable and inconsistent interactive play sessions with caretakers can also prompt play aggression where the cat's insistence on play is not being met.

It is also important to rule out attention seeking behavior as being play aggression. Attention seeking behavior like grabbing, swatting, or biting usually works for cats because the human, whose attention the cat wants, pays attention. Such behavior can also be motivated by hunger, frustration, or a caretaker's lack of attention to schedules or routines. The difference between attention seeking behavior and play aggression is that it is possible to redirect the cat from the behaviors. The cat wants attention, petting, food, and the contact they have initiated, is to gain a human response to those requests and not to further engage in play fighting.

Fear aggression in cats can be identified by behavioral signs such as, withdrawal and passive and active avoidance. This cat wants distance in a big way. Fearfully aggressive cats hiss, yowl, arch their backs, and raise their hair on end. fearful cats with head pulled close to the body, eyes open with pupils fully dilated, ears back and flattened, and whiskers pulled back. Such cats may be trembling or shaking, still or crawling, lying on the abdomen or crouched over all paws. The tail will be close to the body. Vocalizing for fearful cats may be little or sounds that range from mournful meowing to growling or yowling. If the cat can hide or take flight they will. If the cat is pursued with no escape possible, the cat will stop moving, pull his head over the body, crouch, growl and roll over with the feet over the belly. This posture is a posture of true defense, where the cat has exhausted all efforts at escape from the threat. In this position the attacked cat can use all set of claws and teeth to defend herself.

The defensive cat appearance is like the fearful cat but with the head positioned lower on the body. The coat is fuller, with hair is raised on end ("piloerection") and the cat is crouched on top of all four paws. Defensive postures can make the cat look bigger with piloerection and posture that can give way easily to attack. Vocalizations for the defensive cat include hissing in addition to plaintive meows, growling or yowling as arousal progresses. The offensively aggressive cat will have constricted pupils and may spit, swat, scratch, bite, etc.

Fear aggression is typically in response to trauma, as in a direct threat or a response to a person, environment or situation that is threatening. Fear aggression can also be in response to people, situations, or environments that have been menacing in the past. It is important to remember that those aggressive behaviors that seek to defend any animal against threats and dangers are hard wired and necessary for survival. Removing the threats and danger are the necessary first steps in changing the behavioral response.

With any form of aggression, it is essential to first and foremost, steer clear of situations or circumstances that trigger aggressive behavior and fear responses. It is important when identifying what the trigger is, to focus the view from a perspective that is significant to the cat. The scope needs to include, the natural history of cats, the history of the individual cat, their personality, and the context of the environment.

First manage the environment. Immediate or antecedents to the behavior of concern are significant in what they inform and prompt in response. For instance, if a cat reacts to having his body stroked by scratching, stop stroking the cat's body. If the cat hisses at the dog when the dog chases the cat, control the dog's behavior so she does not chase the cat. If the cat sees feral cat colonies out of a window and is stressed by them, lower the window shade so the cat cannot see the cats and work on rebuffing such close contact, if possible, outside the home. If the cat does not like when children pull on her tail, stop the children from pulling on the tail. And if the cat does not like children because in the past they have pulled on or her tail –whether you are aware of it or not– keep children away from the cat. And so on.

Never ever fight with an aggressive cat. Humans can severely damage, if not kill cats, and hurting a cat will only increase fear and aggression in a cat. If the cat is displaying threatening behaviors such as flattened or "airplane" ears, hunched and rigid posture, growling, sustained, hissing, or yowling and it is possible to distract the cat, use a high-level distraction like a fishing wand toy or in last ditch efforts, a laser pointer to redirect the cat (this is the **only** advisable use for a laser pointer while being careful to avoid any contact with the cat's eyes). The cat as predator is hardwired to hone in on prey movements and lasers and wand toys are great substitutes. This can work remarkably quickly in changing emotional states if employed correctly and in time. If a situation has already escalated to repeated displays or contact, the best response is to get up slowly and calmly and walk away. If possible, leave the cat in the room where the situation occurred to settle. Do not yell. Do not escalate your tone of voice. Maintain a calm and neutral cadence to your speech. If necessary, place an object as an extension of the body, such as a pillow between person and cat. Do not use your hands, arms, or legs. Never pull on a cat that is physically engaged as it will exacerbate the situation. In extreme cases, an application of seltzer water will cause the cat to disengage. Highly aroused cats can also be contained in a towel. Exercise extreme caution and remain calm (your excitement will add to the reactivity) when applying these techniques.

The second tenet to remedy aggression is to remove all punishment (including anything the cat finds punishing such as spraying with water bottles, shaking cans full of rocks, scolding, abusive tones or language or physical force of any kind).

This is not saying to remove all negative reinforcement -negative reinforcement introduces something that rewards the cessation of a behavior -using double-sided tape on a couch to deter cat scratching is an example. The cat does not like the sticky tape and is rewarded by not scratching it.   Positive reinforcement can be complimented in this example, by placing a scratching post rubbed with a cat attractant such as catnip or valerian root next to the couch. This in turn, reinforces and rewards the behavior of where to scratch instead along with the intrinsic satisfaction of scratching a great scented scratching surface. These associations created are with the environment, the tape on the couch, the scented scratching post and not a human. With punishment doled out by humans there is no reward system. The end of punishment is not a reward, it is an associative learning process that signals the punishment has terminated.

The third principle is, modifying the behavior by working directly with the cat and the environment to enable the cat to learn different responses and to benefit from an enriched environment that offsets the need for those responses.

Taking a wholistic approach to welfare is needed for the whole cat. Cats are social animals, requiring attention, affection, and interaction to thrive. Greater social interaction may be afforded to man's best friend with multiple daily dog walks, training, fetch, tug, romps, agility sports and other forms of interactive play. Cats can basically be ignored unless they actively solicit our attention, and their attention getting ways can backfire. As cute as knocking things off the mantle may be, a reprimand probably rewards that behavior, while a bunt or gentle paw on is more likely to earn a caress in return. No matter how they ask, they are still seeking connection.

Indoor cats are fiercely dependent on a complex and appropriately enriched environment for maximum welfare, and that enriched environment includes the interactions with the humans living in it. Changing the physical environment can often be the first and easiest step we can take in lowering anxiety, arousal, and aggression.

- Is this guarding or defensive behavior related to pain? Are there areas of the cat's body or approaches which are more objectionable? Is the cat de-clawed? Are medical concerns present? Have they been addressed?

- Breathe and relax. We may not be aware that our first human inclination in a stressful situation is to tense and hold our breath. Be physically aware of your body language and breathing and take a moment to inhale, exhale and roll your shoulders to release your own body tension. Remember that our cats are supreme masters at reading our body language so the more comfortable we are the more comfortable they are and vice versa.

- Become familiar with what your cat is "saying." A cat that is tail thumping or swishing, looking away, holding ears back, is rigid, muscles rippling, hissing, growling or yowling is adamantly asking for whatever is going on in an interaction to stop. These sorts of behaviors are called "distance increasing behaviors" because they are exactly that. When you see a cat asking for space, give it to them. Cats, like most animals, go through a whole set of communicative behaviors as requests and warnings, they never "just" do anything. Routinely ignoring a cat's requests can cause a highly stressed cat to skip steps in asking for something to stop.

- Be very careful with cats to completely avoid punishment (including spraying with water bottles, shaking cans full of rocks, scolding, abusive tones or language or physical force of any kind) as it is never or rarely applied with any degree of timing (any reinforcer requires application at the microsecond of an event to be effective), greatly stresses cats and teaches only that they should be fearful of us. Using the appropriate positive and reinforcing feedback carefully employed with the correct timing and added praise for the correct response can be truly helpful.

Teaching cats what we do want, can change their behavior, understanding what their behavior means, teaches us.

- Positive reinforcement in behavior work is not confined to dogs nor is it confined to acknowledging/marking and rewarding alternate behavior we have trained for. Undervalued and often overlooked is noticing/marking "good" behaviors we would both like to see more of and which benefit cats. The cat lying next to you and contentedly purring can be benefited by a well-placed caress behind the ears or on the side of the muzzle or a soft hand laid alongside them and the behavior named and praised "Good Sleep" (for instance) as soon as you notice. Sitting and looking at you, blinking, watching out the window, etc., all count to be acknowledged and reinforced. Marking naturally occurring behaviors and responding to them puts them on repeat.

- When it comes to petting cats or picking them up, being mindful of how cats interact with each other and the most appropriate way to handle them is kinder to cats and us. Never scruff cats. This is extraordinarily painful for them. Mother cats may carry very young kittens this way but this is only appropriate for that mother and those kittens. We are neither and they know it.

- When picking up cats, remember that cats have a "righting" reflex, so turning them upside down as you would a baby is highly stressful and will cause them to struggle to regain a safe position. It is also important to hold the cat securely (not tightly) against the body and not dangle the cat, equally precarious for the cat.

For petting, use the approach cats use when greeting each other; confining your stroking to along the sides of the muzzle, behind the ears and between the ears. This is the safest and most feline friendly approach. Several studies have been done confirming that while some cats may like petting at the base of the tail, they are often well known to the individual doing the petting or in the minority.

Social media is a new manner of hell for cats with too many of all the wrong things and not nearly enough of the right things for what to do with cats. Dangling a cat does not indicate their temperament, it is highly stressful, uncomfortable and can hurt them. Neither does dressing them up, challenging them, putting them in "funny" situations for viral potential. Do not do it.

Please. There is nothing cute or funny about it for the cat. No matter how many likes or hearts you see.

- When it comes to playing with your cat, remember to do it. Interactive play is one of the most beneficial activities to engage in with cats to develop affiliative relationships, build trust in new relationships, provide for natural predatory behaviors and overall enrichment. Kittens and young cats will play with far greater frequency than adult cats. Confined spaces and lack of engagement with any cat, especially a young cat, can be frustrating and "invitations" to play such as swiping a paw as you pass, biting ankles or hands are expressions of desire and need.

Fishing wand toys are the most effective ones, experiment with different types. Shorter sessions as in one to three minutes are good to start with. Make sure to drag the toy away from or across the cat's line of vision so the cat can follow it. Adjust your speed so it is not too slow or too fast, with a little faster being better than slower- studies suggest cats are less able to focus on objects that move very slowly. Insure the play object retreats from the cat as prey would rather than advances towards them. Rodents and birds try for escape from predators rather than attacking them. The best feedback for knowing the right movement and speed is whether it solicits a response from the cat. With cats that do not respond at all and who have not "learned" to play, work up to engaging the cat by presenting the opportunities for play on a consistent basis. When even a preparatory reaction is displayed –the cat monitors the movement of the toy, follows the human when they take the toy out to play, etc., it is important for the human not to overact by speaking loudly or getting excited and scare the cat.

On longer play sessions with cats that are playing or even with those that are warming up to playing, try switching types of fishing wand toys. (All fishing wand toys need to be hidden away when not in use). Demonstrating play for the reluctant participant needs to be paired with no to little verbalization, especially not those jolly, high raised voices dogs respond to.

Cats like our elevator voices or silence. Even when thinking about play and watching humans play, there is benefit from stimulating mirror neurons in the cat's brain. This exercise intensifies bonding, repairs relationships, is fun, and satisfying for both human and cat.

- Resist playing with the cat on demand in this intervention or on an ongoing basis. While your cat's requests are more than beguiling and we want to make them happy, complying is setting up an expectation that interactive play can be asked for and supplied at the cat's initiation. While you may be able to respond to some requests, it is highly unlikely you can fulfill every appeal, which can be highly frustrating for the cat. Biting in these contexts, can be play aggression as in insistence to play which is not forthcoming. Don't let it get there.

  Give instead, control in structure and routine of regular scheduled interactive play sessions. Schedules you set. Routine playtimes also relieves the cat of the pressure and stress associated with needing to ask for play and interaction that is sometimes provided and sometimes withheld. Regular play at set times is rewarding for the fun of it and can help dissuade that asking seen in jumping out at you, chasing you or using your hands as toys. Adding standard times for interactive play of three minutes (yes, it can be longer if you are so inclined to keep those times regular) tacked onto our schedules as in before a meal and before bed can also help to make these practices easier to keep for human routines as well.

- Leave a variety of object cat toys out in every room for cats to interact with. Tunnels, "fur" mice that rattle, tennis balls and other toys the cat displays a preference for.

- If a cat is biting defensively due to being petted or held incorrectly it is the person's fault and not the cats. The cat should be released immediately and the person needs to learn how to hold and pet correctly so the cat does not have to defend themselves again.

If there is a history of petting or holding incorrectly, the cat will remember. Take some time to not initiate any contact at all. More time than you think necessary. Next, gradually, by taking baby steps in approaching your cat sideways, avoiding direct eye contact, blinking softly and silently, and offering tentative proper pets that are consented to starting with the cat coming forward to an outstretched finger (see more in the chapter on how to pet a cat), begin to reestablish trust.

- If a cat bites in play (which humans are going to retrain with interactive play) or in affection, immediately, meaning at the exact moment of contact, say "Owww!" and hold still. Modify your tone so that it is not overly aggressive or sharp. The cat is offering affiliation, recognize that you are providing brief instantaneous feedback in return, this is not to be punishing.

Where "owww" has become meaningless - you say it and the cat keeps biting or rabbit kicking, a gasp or sharp intake of breath can be tried, and remember the next steps count: Use one short sharp syllable and **no movement** –providing feedback and taking the fun out of the chase. The key here is timing and stillness, the second you feel the contact use the response above and the very millisecond the cat pauses and/or stops, use a softer voice in praise to reinforce the cessation and keep the encounter positive. Cats are extremely sound sensitive due to their exquisite hearing and do not like loud or discordant noises which is why the feedback is so effective at getting them to stop the behavior.

I ask not to use the word "No" because body language is powerful and "No" is a loaded word for us. "Owww" is better depending on how you say it and for some cats. In certain situations, when we use "owww" or any word without the marking and reinforcing it loses meaning. A gasp or short intake of breath and pause can be truly effective as feedback, especially when we follow with marking the exact moment the cat stops/pauses with a "Thanks" or "Yes" (for stopping/pausing - remember?). Add in a stroke behind the ear.

Your reaction will startle the cat, remember in that very second when the cat pauses to immediately acknowledge and reward, for instance, say "Good kitty" in a soft voice just as soon as there is a pause/stop and stroke along the side of the muzzle to reinforce that stopping the bite is the wanted behavior. That's it. Do not lecture the cat the afterwards as this only confuses a cat for doing what you asked for.

- If the biting/scratching has been happening for a while, this is a behavior that has similarly been reinforced, intentionally or otherwise for a while too. While new behaviors are being introduced the old ones will take time to diminish. Frustrated petitioning for play to continue on cat schedule, can be subject to "extinction burst" or trying very hard this thing that has worked in the past before letting it go. Learning alternate behaviors that are lasting takes consistency and repetition. If biting persists, work on not reacting with stillness and silence (more on this below).

- Enrich the environment by providing complexity and things for cats to do without you. Domestic cats living indoors are subject to when and what their owner decides they should eat, where they sleep, when they get to interact and with what, where on what they eliminate and what they are able to do, or not do, to fulfill their natural behaviors. Such complexity depends on enrichment that is species specific. This goes to adding resources that can replicate characteristics and components of the cat and their natural living environment such as, for example, a prey and predator's need for raised resting and hiding spaces to safely survey their surroundings. Similarly, also keeping feeding times for wet food on schedule, with all food freely available in puzzle feeders, provides a sense of control over events and allow for positive anticipation without frustration.

Cats need cat furniture, from cat beds to igloos to cardboard shipping boxes placed on the floor. Cat trees and cat shelves are ideal ways to add vertical resting spaces. Add a window ledge or access to raised surface next to an interesting window so the cat can benefit from the visual stimulation of outdoors. Make sure windows are screened and the view is not on neighboring cats which may cause anxiety.

Always carefully consider placement when adding spaces from a cat point of view. Towers and shelves are ideal against a solid backing, placed next to a window, accessible and not in locations that leave the cat overly exposed such as the middle of a room -unless we are talking a shipping box. Those are great anywhere, ask any cat. Being able to easily get to and away from vertical resting spaces raises use and appeal. Towers and shelves may not fit every living space, adding a cat house to the top of a dresser or a credenza is an under used great way to integrate a hiding space onto an existing vertical space in the home.

- Cats need to scratch both to stretch muscles and to strip nails as they grow. Scratching posts and boards are something fun to do and allow for necessary stretching, encourage the right places to scratch, and are satisfying. Place one in more than one location. Scratching stations need to be stable during use without wobbling. Sisal covered scratchers have been shown to be strongly preferred by some cats and those corrugated cardboard scratchers that offer an incline are particularly attractive to cats. Both sisal and cardboard can be rubbed with catnip or valerian root initially and on a weekly basis to provide enrichment with a scent attractant cats are drawn to.

- Classical music has been shown to have positive benefits for both cats and dogs. Leaving a classical radio station on with the attendant soothing voices of the human announcers has an added benefit of counter conditioning.

- Consider catnip or valerian root to add value to toys and beds. Both plants are members of the mint family and enrich interaction and heighten attraction with toys, beds, etc. You can purchase catnip and valerian root in dried form. Remember to crush catnip between your fingers when using it to better release scent. Sprinkle and/or rub in to apply to scratching posts, raised resting spots, beds and toys.

Key to remember:

When we take away punishment, we are not just modifying the cat's behavior we are modifying our own. That can take some mental stretching for us; especially with biting or scratching, when we are asked to provide feedback on the action, and mark/note and reinforce the moment it stops or pauses. We need to remind ourselves we are reinforcing the stopping and not rewarding the biting or scratching. And that scolding, lecturing, or more, are not energy, words, and gestures, cats can benefit from. Mostly what they do is put an animal further and further into a corner where they are forced into being defensive and fearful to protect themselves. They will never, ever understand that you are telling them what they did was wrong and why, only that you are threatening and hurting them.

References

- ASPCA, Facts about US Animal Shelters
https://www.aspca.org/helping-people-pets/shelter-intake-and-surrender/pet-statistics
Retrieved March 2, 2024

- Ellis S.L.H., Thompson H., Guijaro C., Zulch, H. E. (2015) The influence of body region, handler familiarity and order of region handled on the domestic cat's response to being stroked. *Applied Animal Behaviour Science*, 173: 60-67.

- Martell-Moran, N.K., Solan M., Townshend H.G.G. (2017). Pain and adverse behavior in de-clawed cats. *Journal of Feline Medicine and Surgery*, (published online May 2017)

- Mundschau, V., & Suchak, M. (2023). When and Why Cats Are Returned to Shelters. *Animals*, *13*(2), 243.
- Ramos, D., Mills, D.S. (2009). Human directed aggression in Brazilian domestic cats: owner reported prevalence, context and risk factors. *Journal of Feline Medicine and Surgery*, 11: 835-841

- Salman, M.D., Hutchison, J., Ruch-Gallie, R., Kogan, L. New, Jr. College, J.C., Kass, P.H., Scarlett , J.M. (2000) Behavioral Reasons for

Relinquishment of Dogs and Cats to 12 Shelters. *Journal of Applied Animal Welfare Science,* 3(2), 93–10.

- Soennichsen, S., Chamove, A.S. (2002). Responses of cats to petting by humans. *Anthrozoos* 15:258–265.

PLAY WITH YOUR CAT

"*Some of the cat's psychological needs may also be met through social interactions with the owner, such as play and displays of affection.*" - Bradshaw, Casey, et al., 2012

"*One element that does appear to be common to all types of play in the cat is that cats seem to enjoy playing. Although the essentially private nature of this element may prevent it from being studied objectively, its potential importance must not be ignored. The fact that play is a pleasurable activity may explain not only the high frequency of play but also changes in preference for types of play activity. Social play in kittens may be accompanied by a form of pleasurable or positive feedback which would result in its frequent repetition.*" - West, 1974

What we love about cats, what occasioned their utility for humans, why their domestication was of great benefit to us, is also what some of us just do not like about them. The sleek, indolent, and gorgeous cat, adorable kitten, fabulous feline, sharing the couch with you, sleeping in your bed, is one of the most fearsome predators on earth. An obligate carnivore, so called because their nutritional needs can only be met through meat, hunts a wide range of prey in a natural environment. Cats prey on what flies, swims, crawls, or walks. And they are hard wired to love the thrill of the chase. The cat is successful in less than 20% of hunting attempts but will, in a natural environment, forage successfully for several small meals a day. When cats have secured their meals, they are known to "play" with their captives, often to death. Cats learn hunting behaviors through play and social learning (observation). How to dispense prey may elude the cat not taught this finishing skill but the "fun" of playing with your food is intrinsic. Techniques in administration of the "killing bite" is demonstrated by maternal caregivers bringing home live prey to kittens to observe this in action. Something to think about when our indoor-outdoor cats bring home their catches – in hopes of teaching us too?

The cat's biological need for hunting for sustenance can take up to ten hours a day in the life of a cat living without humans to supplement meals. Cats are solitary hunters and due to their size, meals obtained are small and therefore need to be frequent ones. Cats who are dependent on humans, will spend less time on stalking, foraging behaviors but still pursue quarry up to three hours daily.

Even as habitat loss and collisions with man-made structures are the main reasons for bird fatalities, feral and outdoor cats contribute to these losses through "compensatory predation;" injured, weak, or older victims are targeted and easily picked off by cats. Concerns for wildlife predation by cats, can keep owned cats indoors with added concerns to include dangers to cats themselves, such as being preyed upon, vehicular collisions, exposure to parasites, and pathogens in the environment. Life expectancy for indoor cats versus outdoor cats is vast. Cats can live 20 years or longer in the home but rarely exceed five years in free living settings.

In a study on indoor-outdoor cats, which were routinely bringing home captures, the numbers brought back to the home decreased by 36% when a higher quality, as in no grain, high meat protein content was introduced to their diet. Such a finding suggests hunting may fulfill unmet nutritional needs in some commercial diets and prompts a review on what kibble we are feeding our own cats. In the same study, when five to ten minutes of interactive play with a fishing wand toy and a feather mouse toy by the owner was provided, the number of prey victims declined by 25% (Cecchetti, Crowley, et al. 2021). Further evidence, that the need to fulfill behavioral needs can be satisfied at home, and that added benefits come in keeping cats happier and intensifying social bonds.

Yet another study, comparing play behavior activity with cats with outdoor access compared to cats without, found indoor only cats that had not been exposed to hunting, compared to cats that did hunt outdoors, were more likely to move toward and play with a ball and interact with a fishing wand toy more often than indoor-outdoor cats. Does this suggest greater unfulfilled needs to hunt when stuck inside?

And that willingness to play can satisfy them? The researchers also looked at how effective the cats were in catching the fishing wand toy and found that those cats which had spent more time with their mothers were more successful in catching the fishing wand toy (Pyari, Uccheddu, et al. 2021). No surprise, remember - how to dispense with prey is taught by mother cats bringing back live game to kittens. Cats who are not exposed to this can hunt successfully but are often unskilled in delivering "the killing bite" and will "play" with their catch, often fatally, instead. Something we may not want to bear witness to, but happens, when the inside only cat finds the occasional and unfortunate mouse or insect.

When cats are deprived of opportunities to indulge natural behaviors, frustration, stress, and displacement behaviors can occur. Interactive play provided by cat guardians can give cats opportunities to perform necessary behaviors increasing welfare, reducing stress, boredom, and frustration. Interactive play also keeps cats safe indoors, forms positive associations in cat introductions, and increases trust and bonding in cat human relationships. In research done on owner practices with the frequency of interactive play sessions, owners reported that when play occurred for five or more minutes, fewer behavior issues were observed than for those reporting play sessions of one minute (Strickler, Shull, 2014). Further investigation is warranted on the age of the cats and how types of play factor into cat satisfaction, as in, if a one minute of fully engaging interactive play compares to five minutes of lackluster, going-through-the-motions interactive what counts as play.

There is no one reason for play in animals. Intrinsically reinforcing, play behavior, for many species is often thought to serve multiple purposes. Play develops skills needed for hunting, exercises the body, facilitates social bonds and boundaries and is fundamentally rewarding. Play is more than anything else, a fun thing to do. While believed to serve to sharpen and maintain predatory skills, hunting is no doubt just as fun for cats. One study looked at 101 hunting events by cats with only 32 of those events as successful. Another study looking at cats that had success in hunting showed those cats:

> "cats captured an average of 2.4 prey items during 7 days of roaming". (Lloyd, Hernandez, et al. 2013)

Hungry cats are known to catch more prey than well fed ones but hunting, roaming and being interested in prey for preys' sake appear universal. And interest is hard wired not just by need but by desire and intrinsic reward of the process.

Aided by extreme focus, stealth, and patience, cats remain opportunistic hunters, they will catch what is easiest. The study cats made off with mostly reptiles, namely lizards, as quarry. The cat's stalk, wait and pounce method of hunting is better suited to catching reptiles and small mammals (voles numbered highly in the study after lizards) as opposed to birds that if able to, can fly away. Cats were domesticated, after all, by humans for their devoted inclination and success in patrol and preying on rodent and vermin in grain stores and manmade structures.

Play behaviors work to sharpen skills used to overcome and outwit intended kills, but skills are also gained from play in social interactions that cement bonds and affiliation. Skill gained when the rules of play are followed. Lack of relevant socialization impacts initial learning but there is an almost magic in play at any stage. Play's low impact cause and effect has benefits past sensitive periods and can be rehabilitating post traumatic events.

Watching animals at play can yield the close observer tenets from the universal basic play guidebook:

- A game is being played only as long as everyone knows and agrees it is. Humans may announce an invitation or intent to play with words, dogs with play bows, and cats with side steps and pounces, but no matter what the species, the request must be met with acceptance. Everybody has to want to play.

- Everyone must collaborate on making sure the game stays a game and not a fight. Movements in play are softer and more fluid, performed with less severity and force. Some moves are off limits, inhibited, and care is taken not to hurt the other player. This so called "self-handicapping," means that each is aware of how they are affecting the other and is careful to keep things light to stay in the game.

- Attention to each player's reactions is constantly monitored and responded to. Teasing has no place in a game. Neither does playing too rough. That bite was too hard? A little yelp or meow told you so. Not listening? That yelp got higher and the meow turned into a growl. If you stopped and things got softer, we can keep playing, if not, we are done and if you keep it up, we do not play again.

For cats, play behavior is highest in frequency as kittens and begins to drop off at four months of age. Cats may play vigorously for up to four years but play can continue throughout cat life. It follows that kittens benefit from more frequent and longer interactive play sessions. While older kittens and adult cats may indulge in less play, they may still be interested in opportunities for play, if it is presented. Individuality, environment, and how play is on offer, no doubt affects each cat's appetite for play.

All this is to say that playing with your cat is not just what a cat wants but what a cat needs. We have established that the cat's biology and welfare requirements, support that all cats need to play. Objects to play with also are essential to enriching cat homelife. Cat toys, as is toy mice that rattle, crinkle balls, and the like to bat and chase, boxes to jump in and out of, raised spaces to zoom to, tubes to run through, and more, supply objects for solitary play. Individual cats have preferences, fur mice can be a huge hit for some and the bottle cap from your water bottle may send another cat over the moon. Providing a variety of things to choose from lets the cat tell you what they like best.

And then there's collaboration. Human and feline housemates offer needed social play opportunities. When a cat plays on their own there is predictability to the play - when you are at the other end of the wand toy, the interaction is dynamic. That toy you are manipulating at the other end of the fishing wand can be the unpredictable "prey" object they are designed to enjoy pursuing. These sessions can have the most impact on your cat's behavior and your relationship with them. Interactive play, with its fluid and changeable exchange is one of the most significant opportunities to increase positive associations around new cat (and people) introductions, changes in environments and developing the human animal bond. The top daily interactive play starts with a fishing wand toy.

Remember:

- Draw or drag the toy away from or across the cat's line of vision to best engage them. Think of the feathers or fur at the end of the wand as standing in for a mouse, bird, etc., and how that animal might move when a cat might have the best opportunity to catch them. Still or bursts of movement, low and slow, and then fast and sudden changes of direction. Be careful not to "attack" the cat with the toy. Mice and birds retreat from and do not advance on their predators.

- Play can stand in for the thrill of the chase for cats, but they need to be the hunter not the hunted. Vary your movements. Gauge the more enticing moves by observing the cat's response and interest. Keep it interesting.

- Cats have been shown to truly benefit from schedule and routine and through them the sense of control they afford. Keep play regular and as close to on schedule as possible. Linking a play session to before a meal can help to put those sessions on our own observed routines and can add to a cat's satisfaction of "catching" breakfast or dinner.

- Incorporate two interactive play sessions, one in the morning, and one in the late afternoon or evening into everyone's routine. A third, before bedtime, is optimal and especially beneficial for younger or stressed cats.

- Keeping play sessions on the regular schedule you set is key. Effects of play can have positive impacts on welfare and requests to continue sessions or for additional sessions can follow. Expanding session time or adding extra sessions is all good, only if they can also be put into your established interactive play routine.

- Intermittent reinforcement is one of the most trying experiences and imparting stress is exactly the opposite of where you want to go with this. Making play an "on demand" event is only a good thing if you can meet every single request. Not very realistic, leading to an unreasonable and unsatisfying expectation for the cat when those demands and requests go unmet. Play aggression, often based in frustration in thwarted attempts to initiate play or displacement behaviors can result. Kinder, more effective, and better to keep routines they can trust and depend on instead.

- Remember, for cats, time matters. Keep as best efforts as possible on set times and duration. Use a signal or word to signal the play session is over when you are done. Put the toy away and bridge the transition to ending play with petting, brushing or a treat in a different room or throw some treats or a valued object toy to a different area of the room you are in.

- While a long play session can be a good thing, keeping the cat's interest and keeping it doable for the human is key. Know that less can be more- two to five minutes in the morning and evening are good amounts of time to spend in play. The most important thing is that these sessions happen, even if the time they do, slips every now and then.

- Experiment with different types of wand toys to see which your cat loves best (I see the feathered ones getting the most response but every cat is different). You can vary between two separate fishing wands during a session to keep up flagging interests for either one of you.

- Some cats may not engage initially. Keep going. Notice which sorts of play movements you perform that garner the most reactions and which type of toy your cat shows more interest in, no matter how slight.

- Keep your reaction muted - no happy camper puppy voices please. Keep enthusiasm in your motivations to continue but cheer on cats by using your "elevator voice" and thinking encouragement instead of speaking it.

- Interactive play sessions can engage more than one cat. Position yourself between the cats and alternate attention between them both. This can have truly positive effects with assimilations. Cats being hunters, focus on the play/prey aspect of the toy at the end of the wand and not the opposing cat. With newcomer cats, begin playing with the original cat in a separate room at first. With two people, play with both cats independently at the same time.

- Keep the toy out of sight and in a place where the cat cannot access it when you are not using it for safety and novelty concerns.

- Avoid laser pointers- they can be dangerous if misdirected, are inherently frustrating and offer nothing to catch ever. If you must use them, let it be sparingly (they are good for redirecting cat squabbles for instance) and have them disappear under a door and throw a treat where the beam ends.

References

- Bradshaw, J. W., Casey, R.A., Brown, S.L. (2012). *The behaviour of the domestic cat*. Cabi, Oxfordshire.

- Cecchetti, S. L., Crowley, C. E. D., Goodwin, R. A. McDonald. (2021) Provision of High Meat Content Food and Object Play Reduce Predation of Wild Animals by Domestic Cats *Felis catus, Current Biology,* Vol. 31, 5, p 1107-1111. https://doi.org/10.1016/j.cub.2020.12.044.

- Lloyd, K.A.T., Hernandez, S.M., Carroll, J.P., Abernathy, K.J., Marshall, G.J. (2013) Quantifying free-roaming domestic cat predation using animal-borne video cameras, *Biological Conservation,* Vol 160, p 183-189, ISSN 0006-3207.

-McGregor H., Legge S., Jones M.E., Johnson C.N. (2015) Feral Cats Are Better Killers in Open Habitats, Revealed by Animal-Borne Video. *PLoS ONE* 10(8): e0133915. https://doi.org/10.1371/journal.pone.0133915

- Pyari, M.S., Uccheddu, S., Lenkei, R., Pongrácz, P. (2021). Inexperienced but still interested – Indoor-only cats are more inclined for predatory play than cats with outdoor access, *Applied*

Animal Behaviour Science, Vol. 241, 105373,
https://doi.org/10.1016/j.applanim.2021.105373.

- Strickler, B. L., Shull, E.A. (2014). An owner survey of toys, activities, and behavior problems in indoor cats. *Journal of Veterinary Behavior*. Vol 9-5, p 207-214

-West, M. (1974), Social Play in the Domestic Cat, *American Zoologist*, Vol 14, 1, p 427–436, https://doi.org/10.1093/icb/14.1.427

LOW STRESS WAYS OF GETTING YOUR CAT IN THE CARRIER

You already know the drill when it comes to cats and their carriers. For the most part, they just do not like them. Period. For most cats, the carrier is far from a transport to a positive experience: they are trapped, held, restrained, or grabbed against their will. Pushed into one, moved about encased through all sorts of noise, motion, and scents to a strange new place with no familiar sights, sounds or smells. Most likely to be delivered to the veterinarian's office, for being once again restrained, held too tightly by strangers wielding needles, poking, and prodding. To the cat's way of thinking, this is surely not the best association for time spent in a positive situation.

For the cat that has spent time in a carrier, the learning experience often results in carrier avoidance at all costs, both for what it means and where it leads. While humans may know that cat carriers are safe havens for moving cats around, the only time your cat may agree is when it is time to leave the vet's office.

As a rule, your cat, being the homebody and highly territorial creature they are, would prefer staying put to traveling, whether to a new home, for the holidays, or to the vet. Even so, a visit to the vet is an occasional necessity. And what about when your travels may be extended and you choose to bring along your feline? Or your ticket is one way rather than round trip? Feline friendly, low stress, and pain free methods to help get your cat in the carrier for time like these can make all the difference to how you and your cat start the journey.

Researchers tested whether cat carrier training could make a difference to cats being transported to the vet's office. Cats were divided into two groups and both groups went for individual carrier transport via car to the vet (low stress handling methods were used for all cats at the vet office). Next, one group received cat carrier training several times a week for the next six weeks while the other group did not. Both groups were again transported to the vet. The cats were evaluated throughout using the Cat Stress Score. The study found all the cats to range from "weakly tense" to "fearful, stiff." When the scores were averaged between both groups, the cats when confined to their carriers during training, in the car or during the exam, the consensus was "very tense."

Obviously, no cats was a fan of the carrier. But training and handling mattered. The difference was in after training, during the car ride to the vet, the cats who had cat carrier training were less tense. Additionally, cats who had cat carrier training had shorter exam times, indicating less resistance – also worth remarking, exams were carried out in the carriers for both groups.

Cats, as with many animals, will never like restraint or being confined. Where we can lower stress for the inevitable when the use of the carrier is a must, guidelines for remedying negative associations can help. Along with thoughts on what to do in those situations when all else fails:

- A new carrier, where nothing bad has ever happened, to start your new cat carrier friendly experience can help. Choose wisely. Carriers need to be well-made to stay together with movement and animal contained. Make sure the carrier can be easily opened and accessed from the top. And separated into two distinct parts if it is rigid (more on this with vet visits below). Soft sided or rigid, become familiar yourself with assembling and closing the carrier before introducing it to your cat.

- Avoid "cat bags" that immobilize a cat, backpacks or carriers intended for human babies. Being restrained in this sort of transport is not only highly stressful but dangerous if a stressed cat escapes.

- When carrying the cat carrier, support it against your body or with your hands so it is not swaying of swinging freely which can further terrify the cat inside. And "no" to rolling carriers for the same reason.

- Patronize a cat friendly veterinarian who will perform most, if not all, of your cat's exam by leaving the cat in the carrier, making your cat's visit less stressful and you more likely to return for more. The vet exam in the carrier is another good reason that the carrier you are using, rigid or soft, opens easily from the top or sides. Add a cat bed with shallow raised sides to the carrier so the cat can remain comforted in the carrier for their exam.

- Be adamant about staying with your cat for necessary social support in clinic and exam room setting. Avoid multiple room changes at these spaces. Studies show these practices will reduce stress. The less stress you can associate with a veterinary visit, the more likely you will return, the more you benefit the cat's welfare, and the less stress is associated with the carrier that brought the cat there.

- In the time you have before leaving for a move, trip, or vet visit, or to prepare the carrier in general, make the carrier a more attractive place by beginning to leave it out and open in either a secure and elevated resting space such as the top of a dresser or a corner of which ever room your cat spends the most time.

- Make sure your cat has independent access to enter and leave the carrier. Remove the top half or leave the soft sided carrier open. Over time you will progress to closing the top of the carrier but leaving the door open. It is best to remove the door to the carrier or to keep it securely propped open at all times during that step.

- Placing a small, oval cat bed with raised sides in the carrier will add to the comfort of the open ended, enclosed space. Supplement the number of cat beds in your cat's general living space for a more cat friendly environment and to serve as transport to the carrier (details follow).

- Sprinkle catnip or valerian root around and inside the cat bed in the carrier and around the entry outside of carrier. Position several favored cat toys strategically around the carrier to raise its appeal.

Take a few moments each day, in the morning and evening, to begin playing with your cat's favorite interactive toy next to the carrier. Start at a distance away from the carrier – think ten to six feet and gradually decrease that distance as in gaining six inches every other day. Keep these interactive play sessions on schedule. Interactive play is a huge draw for cats, putting it on routine gives a sense of control that it will happen on a regular basis.

Offer treats close to the carrier. To be enticing, use treats that are high value like dried salmon as opposed to those manufactured kitty treats, and progress to leaving them as close to inside the carrier as possible. If your cat does not eat the treats this is an indication that the cat is still not comfortable with the steps in the process. Try placing the treats a further distance away from the carrier. Remember, go slow to go fast with cats.

Once you have your cat comfortable around the carrier, onto the steps to getting kitty inside it. Always remember, the fewer and smaller your movements here, the better:

- Consider offering some valerian root sprinkled over the meal your cat will have before your trip

- Have the carrier open, in place and ready to receive the cat before you pick up your cat.

- One of the least stressful and easy ways to put a cat in a carrier is by picking up the cat (burrito style) in their cat bed and sliding cat into carrier (if you are using this approach, make sure there is no bed already in the carrier). If you can pick up your cat and the cat is agreeable to going in the carrier, lift up the cat, keeping them close to your body, and deposit in the carrier in one fluid motion – important, if you have never done this, first practice with a stuffed animal or doll.

- Place a worn article of your clothing in the carrier. A recently worn tee shirt or sock will go miles towards making your cat more at ease when it is in the carrier. The familiar scent of you on the garment will offset some of the stress. Dig in the bottom of your hamper for the most "comforting" shirt/sock scent wise.

*For the approaches, where you have not sufficiently worked with your cat to acclimate to the carrier, here are some strategies. Know that utilizing these approaches will **not** make the carrier a friendly place for your cat and should be last ditch only:*

Last ditch option one: Use the smallest space, possible such as bathroom. With a resistant cat, turn the carrier on its end with the door facing upwards. This way you are lowering the cat back legs first into the carrier. Know that the next time you use this approach with the resistant cat, means a cat that is also wise to what is coming.

Last ditch option two: The next option is to position the carrier by advancing it toward the cat in such a way as the cat has no other option but to enter it. Only, if you know how (do not do this if you are not well versed in doing so because you can really hurt the animal) pick up the cat using a towel. *NO "Scruffing" or grabbing a cat by the nape of the neck* to immobilize the animal which is painful and inhumane (yes, we know mother cats carry infant kittens this way. You are not a mother cat and the cat knows and feels the difference). Towel wraps are more effective and when employed by experienced handlers with minimal restraint, work well with even the most extreme fractious, worst-case cat. Speed, handling and fluidity are key here.

- Used a last-ditch option? Be prepared for how not to go there the next time and take the time to find a great review online on cats and carriers from Catalyst.

- Once your cat is in the carrier it helps to drape a towel or button-down shirt (not too long, you do not want to trip on it when you are carrying it) over the carrier to cut down on the visual stimulation. (Consider also using one of the towels from the bottom of your hamper for this as well.)

- If you are working with a cat that is absolutely panicked with the prospect of the carrier no matter what, consider utilizing a product that will help soothe your pet such as Valerian Root, Rescue Remedy or Pet Remedy that you can sprinkle or spray onto the cat bed in the carrier. Do **not** spray the cat.

- Layering the bottom of the carrier with newspaper will also aid in absorbing urine should the cat become frightened enough to lose control of his bladder in the carrier.

Individual cats react differently to being contained in carriers. Some shut down in fear, some panic and some will huddle towards the back and some will howl piteously, not for effect but as an expression of emotion. Speak softly and reassuringly to soothe your cat with one of two sentences. Do not act as if the sky is falling, this will really scare the cat even more. Poking your fingers through the grid in the carrier door (do not do this if you need to release any openings such as latches, clasps, zippers, etc.) and making contact can also help.

The first step of converting a new carrier to a den can be an easy one, the rest follow with practice. Do not force speed with any of these steps. Do whatever time permits you to do. Good to remember, go slow to go fast with cats.

With any of these next steps, if your cat is becoming stressed by the approaches, as in tense, rigid, looks away, ears flatten, whiskers back, eyes dilate, tails flicks, hisses, leaves, etc., stop and give both of you a break for the rest of the day and the day following. Begin again by going back to what was last working.

References

- Pratsch, L., Mohr, N., Palme, R., Rost, J., Troxler, J., Arhant, C. (2018) Carrier training cats reduces stress on transport to a veterinary practice, *Applied Animal Behaviour Science*, (206), 64-74. ISSN 0168-1591, https://doi.org/10.1016/j.applanim.2018.05.025.

CLASSICAL MUSIC AND CATS – ALWAYS A GOOD IDEA

I used to spend a part of each work week, stopping into the sites where my students interned for evaluations or prospective sites. Those pet services we worked with were animal shelters, grooming salons, doggy daycares, veterinary clinics and offices. On one visit to a doggy daycare, I found classical music playing and relaxing dogs. Nice. I asked the manager if he was listening to the music to soothe the dogs. "No," he told me "I listen because like it myself, but I do notice that the dogs are much calmer when the radio is on."

Classical music with its soothing melodic tones and harmonies is now more readily accepted in pet care as being a positive for the animals that get to listen to it but do those special pet versions being marketed really make a difference? The answer may be "yes," "no" and "it depends on what kind." In addition to being able to observe a positive impact, it makes sense that the right sort of classical music needs to be played from the animal perspective to be effective. Research and practice shows soothing music may have a universal component for all species, including our own. Factors such as lower pitch and regular tempo are typically associated with more pleasant music while high pitched and irregular tempo are usually perceived as disturbing.

A study on the sort of music cats prefer by researchers, Drs. Snowdon, Teie and Savage, addresses cat specific music by testing how well cats respond to certain music. Here, the music was created to cater to what the researchers believed the cats would like to be listening to. The scientists composed pieces of music that were in cat frequency ranges and compared the cats' responses to music in human frequency range.

Of course, for any of this to matter, the animal in question needs to be able to hear it. Humans hear in an approximate range of 64-23,000 hertz (which is a measurement of the cycles per second of a particular sound wave) compared to dogs who hear in an approximate range of 67-45,000 hertz and cats, who beat out both species, hearing in an approximate range of 45-64,0000. Just look at those feline ears, 32 muscles in the cat ear compare to 18 muscles in the dog ear and 6 in the human ear. Shaped for maximum functionality, cat ears can also swivel and rotate independently 180 degrees to orient more efficiently to sounds.

And it is with the hearing of things that most animals aside from humans are listening at frequencies we are not aware of. When looking at the sensory capacities of other animals it is vitally important to know that the measurements we are working with are all educated guesswork on our parts, with significantly huge gaps missing in our knowledge. We need to remember that when we have tried to determine the range of an animal's hearing, we cannot ask the animal what it hears and receive a response. The cat being tested is not sitting there raising a paw to tell us which ear they heard the sound in. We never definitively know. We first train hungry animals to respond to a sound by rewarding them with food and measure which sounds they respond to. In fact, we cannot know what they are hearing only what levels they respond to in our tests. It is an important distinction and a valuable one to remember in the work we do with animals.

In the study, the researchers also added into their compositions what they thought would evoke a friendly response in the cats using purring or the sucking sounds cats make when nursing to regulate tempo. The human music used for comparison, pieces generally perceived as pleasing to people (Faure's Elegie and Bach's Air on a G String), had tempos similar to the human resting pulse rate, which is intimated in the study as possibly contributing to its relaxing qualities. It is confusing why the researchers choose purring and sucking as tempos to set their "cat music" to rather than cat resting pulse rate, especially as the cat music is compared to human music set to a tempo of the human resting pulse rate. The cat's resting pulse rate is significantly higher than ours, 120-140 beats per minute as compared to a human's which is 60-100 beats per minute.

The researchers found the study cats responded significantly more actively and positively to the cat music with younger cats in the study showing an even greater response. Positive responses were listed as turning the head toward the speaker where the music was coming from, rubbing against it, sniffing and purring.

It is also vitally important in the work we do looking at animals that we look at behavior in contexts that are meaningful and salient for the animals in question. The cat music study here begins with positing a relevant range of what a cat can hear as a starting point for what music might be more or less meaningful and perhaps loses focus by mixing in tempos of sounds made during comfort behaviors, like nursing and purring. Context matters. When the cats in question respond positively to the music (again, worth noting that more of the younger cats in question respond) it is difficult if not impossible to parse out what the cats are in fact responding to, the tempo of the purring or nursing sounds and those attendant comfort associations or the frequency or otherwise.

In a separate study on dogs, Patricia Simonet, recorded the sounds dogs make when "laughing" at play, a particular in and out pant, almost akin to huffing or chuffing. Dr. Simonet then played the dog laugh recording as background noise in various shelter environments and observed responses. She found similar impacts, with results showing the shelter dogs demonstrating positive behavior responses to the dog laughing sounds.

Whether it is studying animals, working with them, or living with them it is of great importance for us to always remember that other species perceive the world we share together, in sometimes markedly different ways even as we may be sharing the same physical spaces. We are not all watching the same movie. Soundtracks and color scopes depend on who is doing the processing when taking it in. Paying close attention to how our animals react to their environments in the ways we have come to learn are positive or negative for them continue to be the most telling for us as to how they "feel" about them.

Drawing from the prior studies on the effects of classical music on animals and personal observations, I recommend the use of classical music with pets I work with, specifically classical music found on the radio. Radio stations mostly play music for background effect, so arias and discordant pieces are often ruled out. Radio stations also have dulcet toned announcers softly speaking about the music, a bonus in desensitizing and counterconditioning for certain pets.

Classical music continues to have value for cats and other animals and this is most demonstrated by the behaviors they present to the appropriate music. The behavior tells, we just need to be clear what it is telling.

References

- Snowdon, C.T., Teie, D., Savage, M. (2015). Cats prefer species-appropriate music. *Applied Animal Behaviour Science*, 166, 106-111.

- Simonet, P., Versteeg, D., Storie, D. (2005) Dog-laughter: Recorded playback reduces stress related behavior in shelter dogs. *Proceedings of the 7th International Conference on Environmental Enrichment* July 31 – August 5, 2005.

WHAT HAPPENS WHEN ADOPTIONS DON'T WORK?

Not that long ago I received a phone call from a woman I had never met who was traveling southbound on the New Jersey turnpike. She was calling from a rest stop somewhere in southern New Jersey. Not being familiar with the area, she could not tell me which rest stop she was at. But she had something important to say. She was calling to say she had found "my" cats. Right there at the rest stop. She did not sound happy about either the discovery or speaking to the person she thought might be responsible. I had lost no cats and I live in New York City, many, many miles from that turnpike. Hearing this, I was confused, concerned, taken aback -all three. My pet greeting behavior on arriving home includes personal interaction with each animal and all pets were present and accounted for.

"My cats are all home," I said. Why, did she think the cats were mine? The traveler explained that she had been taking her dog for a relief walk and found the cats in question behind a bush, huddled closely together in the back of a carrier with the door ajar. She went on to say that it did not look as if the cats had left the carrier and they appeared clearly terrified, all huddled into themselves and each other. She had called me because the carrier had a sticker pasted on the end, one with my name on it. The sort of sticker a veterinary clinic places on a carrier with the identifying details of the pet along with the owner information. (It later became apparent my information was on that carrier for a prior spay surgery I had paid for and facilitated.)

I started thinking about the cats I had worked with, rehomings, and carriers I had given away. I asked what the cats looked like: were they long haired? Was one grey and one orange and white? Yes, to both. I realized which cats these must be and who had "adopted" them. I knew who these cats were.

Two years prior, I had worked with two fourish-month-old feral street kittens at the request of a Soho animal shelter. Four months is old to socialize kittens to humans but time and consideration are the magic ingredients in working with any cat. For ferals it is really just much, much more of each. Another four months of time working with these siblings, spay surgeries, and recovery time and these girls had graduated or so I thought, to being successfully adopted or "rehomed." But rehoming is a tough road, one not so easily traveled.

When it comes to finding a new home for any pet, rehoming is a bumpy ride and soft landings are hoped for but not guaranteed, with the brunt borne mostly by the pets shuffled around in the process along with the rescuers and the fosterers involved with them. The biggest fear for those working with animals, those who do the caring, feeding, vetting, socializing, rehabbing, taming, training, and treating, is that the standard of care the pets will receive in the future will not equal the care they have been receiving from the rescuer, fosterer, clinic, or shelter. In fact, it will not be. It cannot be.

The care received can never be the same when the individual care giver is changed (and this is no small factor in stressing an animal—a humane and consistent caregiver is vital to an animal's welfare). The relationship nurtured and developed in fostering is abruptly terminated. The humans understand why the break must happen, the animal has no advance warning, no gradual transition, no say in the decision. If even this change is a stressful one, how then can this ever work? Finding new places for companion animals to live without the companions they have come to trust, developed relationships with?

What makes rehoming work, the theory or hope behind it, is based on the animal being first, placed into supportive care in shelter or foster and next a permanent environment, all where a universal standard of care is adhered to which will benefit the animal's welfare and, that after an inevitable adjustment period (which the new guardian must successfully facilitate). the animal will thrive. And when a supportive environment and guardian come together for the animal so does a "forever" home.

But what happens when this is not the case? What about when the reasons industry people fear most come to fruition: The guardian who says they want the pet and then realizes the cat or dog is not who they expected, the bond they hoped for has not developed fast enough with either them or a resident pet or at all, decides after the fact, they would rather not consistently buy and open smelly cans of pet food, deal with pet messes, change a cat box or walk a dog several times a day or come home early from a night out to do any of these? And what about any of the horror stories we would rather not hear about but that happen every day whether we hear about them or not? What happens when the universe drops the very same animals you thought were in that forever home right back in your world? What then?

In the work that I do, people without pets will often apologize to me for not sharing their home with pets, as if they are at fault for knowing their limitations. I praise those who know that pet care is beyond them and who avoid taking on an animal they cannot care for. Sometimes the best choices for all involved are the ones we do not make. And what of the limitations realized too late for those who have taken on the responsibility of the life of another who find themselves not able to provide care? What is the thinking for the person doing the abandoning?

Rescues and shelter routinely ask that animals be returned if a placement is not successful. And while human beings are "diagnostically oriented" a/k/a judgmental, animal care workers need to reframe situations on an ongoing basis as a survival strategy in the field. Reframe and go on. Not knowing this, might the now reluctant adopter's thought of an animal worker's possible judgment of a return be enough to dissuade contacting them? Can what one person think, if they do think it at all, be reason enough to deprive a cat or dog from a dedicated process for humane care? Can a person truly reasonably believe that somehow by leaving these cats in an open area they will be able to fend for themselves? Maybe in a Disney movie. In real life, dropping off domesticated animals in new and foreign territory is not a chance to strike out and make a life for themselves. Even should one persist in the belief of a benevolent universe and laws of nature, one needs to know these laws do not include former Brooklyn street kittens turned house cats being able to figure out life at a New Jersey rest stop on 1-95.

The truth is cats need to be taught survival skills and be in familiar territory to employ them. Without learning to hunt and more importantly, how to kill prey from their mother, these cats would have little to no chance of catching a meal. If they were not so very, very terrified of this strange place they found themselves, they might be able to scavenge food from the trash but being obligate carnivores (requiring meat for their nutritional needs) they would be hard pressed to find adequate fare. More pressing would be the need for these cats to be comfortable enough to venture out to try and eat at all. And without eating they would enter dangerous waters. Cats need to eat frequently to survive. In starvation modes they process too much fat for the liver to handle effectively and this "fatty liver" disease once started is hard to reverse and treat. And of course, they desperately needed water.

"Were they OK?", I asked the caller. She was not sure. The day was close to ninety degrees and the days before it had been insufferably hot as well. Two grown cats in one carrier crouched together, not moving. I asked her if she could just close the door, hoping to reassure any fear in approaching the carrier or the cats. I said that since they had not left the carrier, they themselves were no doubt terrified and would surely stay in the back of the carrier and were sure to avoid coming close to anyone closing the door. She was not afraid of the cats she assured me, rather she was upset that they had been abandoned here.

If there is some sort of fortune in this story, with the universe bringing these cats back to me for whatever saving we had fallen short of doing in the first place, whatever flaws in placement, communication, ego, their fortune also included being found by a former veterinary technician who cared deeply for animals and for our responsibilities to them. She assured me that she would not leave the cats at the rest stop and without any other options she would find a shelter to bring them to.

Shelters are terribly overloaded with cats and many of the cats entering shelters will never leave them. The ASPCA estimates that of the 3.4 million cats that enter shelters every year, 1.4 million will be put to death and that the percentage of cats entering the shelters who will be adopted are about 37% with 41% being euthanized. The number of stray cats being returned to owners is less than 5%. Rescue organizations and individuals work to make a difference in these daunting statistics that translate into too many cats and not enough people to care for them.

And in this one story of so many others, the shelter did not admit two more cats to add to the numbers. This one story of rehomed cats dumped on the side of the road at a rest stop far from the place they had known as home has a happy ending for the meantime. Haunted by what "we" had done with these cats, I tried everyone I could think of to help, any favors to call in, any begging I could do, anyone left that cared for animals that had not already helped and rehomed already.

I made countless phone calls. Who had room? Who would answer the phone? Somehow, with more fortune and people working on the side of the angels, these cats came to yet another foster home. With hundreds of miles of driving and coordinating of all the major and minor components of figuring out how to get the cats from a rest stop in New Jersey to someone who cared enough to be persuaded to help and would wait for someone else who cared enough and would drive those many, many miles and pick them up and set them up with all the requisite cat furniture. Another foster, another relationship to make, more trust to ask for from two animals that have been so far twice betrayed.

When I wrote about working to socialize these, then feral kittens two years prior, I wrote about the relationship of trust I developed with these cats that had no trust for humans. I also wrote of how this work entailed a promise of sorts in which to make it work. So far, the cats are the ones doing their part in keeping the bargain.

GOODBYES

"I think we need to set aside the notion that there is a Right Time- some moral target that we need to hit precisely. Our goal is not to pinpoint but to find a golden mean between too soon and too late, between premature and overdue. Working with a veterinarian, we can seek to understand our animal's illness or injury, to know what kind of deterioration or changes they might experience…Still am I certain about my timing? Not at all.

"Why do I balk at the assurance that "They will let you know"? Because this places responsibility on the animal and removes responsibility from us. Our animal may indeed give us signs that they are suffering (refusing to eat, withdrawing into themselves), but it is us who must read the signs – and as should be abundantly clear from all that has come before here, the signs can be obscure. It is not easy to interpret pain in animals, we don't know their behavioral language, unless we take the sustained work necessary to understand it. And our interpretations of their "signs" is, as often as not, clouded by our own interests, presuppositions, and ignorance. And, yes, by our love for them." (Pierce, 2012)

The routines, the fabric of the day, we structure around the animals we live with, sits with enormous gaping holes when one leaves. What to do with those now extra minutes turned to hours? To those days as they add up? What about the schedules once so needed, tacked on to our own, the walk the dog and or feed the cat as soon as you get up? The get up on time because they are waiting for you and know it is time. And only after that's all been done, can you take the shower and make the coffee. The feeding, scooping litter, throwing their ball, time to play, all the each, and every, so many little things. The interactions, looks, pets, sleeping next to you, with you, having each in every part of home life surrounds you. Until it doesn't. And after they're gone? Who wants this extra time, who even knows what to do with it. We want them back instead. A bargain that will never happen.

These losses, are never just that are they? The cats or dogs we euthanize do not just go to sleep in our homes and never wake up. We decide when, after how much or how little interventions, and we do it around schedules of our own and the people who can arrive or be available to administer the injections.

In a bereavement group, I went to one time, there was a handout on the benefits on writing tributes helping with grief. I only attended that one time because the leader of that group checked her watch once too often. Still, it was good to know the remembering and the telling can help.

Every pet has a story. Live long enough and with enough pets and the stories can run into each other. We share them now on our feeds and posts and threads and share in other people's losses. We cry all over again reading them, this pain touches us just the same as if they were ours. And in a way they are. I am not the first pet owner to despair over this. And I will not be the last. This group shares the unsupported grief, the guilt over engineering the end of their lives, the very idea and carrying through of it, that is either too soon or not soon enough, the hearing of "it's just a cat/dog," when it is anything but just that. The well-meaning advice of "get another one" when all it is, is exactly the one we no longer have that is not another one that we want back.

For My Julie

I put away my Julie's bowl for the last time just the other day.

She will not be waiting for me to serve her breakfast or dinner any longer. There is now no little calico waiting for me when I come home. I will be able to sleep through the whole of the night without her paw gently tapping once, and then again, my eyes to see if they will open. I will now have all the space next to my pillow and can shower in complete privacy. I will be able to write on my computer with two hands, without the paw to remind my hand that petting needed to be uninterrupted. I will be able to move about more freely. My lap and side unencumbered by her presence. When I am on the patio there will be no one crying piteously from the other side of the screen door or practicing that guaranteed cue - hooking a claw in the screen and looking at me to make the door open.

How can I tell you about my Julie now that she is gone? That her mother found our backyard a good place to have kittens and that Julie was the only one we managed to catch? That she was small enough to fit in a shirt pocket, with a runny nose and eyes, a scrappy little street cat made for the harder life who found a softer place to land.

When we brought her to our vet to check out, he gave us a piece of advice we never took: don't pay attention to her around the other cats, he said, ignore her and let them bond with her. Don't make her special. It will be hard, he said, but let them accept her first. We must have tried for an hour or two to take that advice but we never succeeded – we were helpless with that ball of fur that was all mayhem and love.

Ed and Sarah, the then resident cats, did accept her. In their time and only after that. Sarah sat on her first and Ed ignored her. Daisy, being a puppy, wanted only to play with Julie in the most puppy way which made for plenty of muzzle swipes in return. Whoever came and went in the fur kid family, Julie was always the baby girl with the big personality, and she was always the star. It makes sense if you think people have karma with other people in their lives that they would have it with the animals in their lives too. We had karma with her.

She was 17. We knew this day would come but it came too fast. And how we let her go was the hardest part. When we saw the struggling, the weakness, the loss of appetite we made that decision to end needless suffering –this better death for someone we loved dearly. We would have a vet come to our home and put her softly to sleep.

That was what we wanted for her but that was not what happened. The anesthetic the vet administered didn't have the desired effect, she fought for escape – the vet remarked that he had probably not given enough for her muscles in the state they were in to absorb properly. It was a weekend and this stopover visit for the vet also turned out in him not having brought any additional supplies, no heavy carry bag. He had no more to give, so she did not get that additional needed dose. She railed and struggled and panicked as the effects took ineffective hold. The struggle took over for the veterinarian, he had never seen this he kept saying, he was not sure how to go forward but decided to inject the fatal dose into her heart but had difficulty finding where to place his needle. We were frozen in horror and pain in not being able to do anything to prevent the very bad death we had prayed, and payed, and wished this not to happen for this one we loved.

That next shot that somehow penetrated to stop her heart also did not work. The minutes dragged and her heart kept beating and beating and then there was another shot and still she stayed. Tormented. Dying. Terrified with everything left that could feel, Julie did not go gently into that good night, she raged, raged against the dying of the light.

As the days have passed since losing Julie, we are left with the memories of her in our lives and how her life ended. That the hand that delivered those drugs should have known the right amount to give or given more as soon as needed, was for her the difference between a peaceful passing and one where there was pain and fear. Julie did not have the best karma with vets, she had botched attempts at care in clinical settings over the years and yet, if it is true that as we have heard tell of a vet school saying – the one that goes "everyone makes mistakes, make sure to make a new one each time" than perhaps her karma with the people in her life can add that she is the last mistake of this kind for the next cat or dog visited by that practitioner. And for this, she helps many.

One of my sisters says that after life you can still tell them how you feel. Still make up to them. I want it to be so. So, here it goes then: I'm sorry Julie, please forgive us, we love you. Be free little rock star.

For My Cinnamon

I keep going in the other room and looking for Cinnamon. I lost her yesterday. She is not on her perch in the cat cubby on my dresser. I see her in the corner of my eye around every corner and in every space, I go into. This home is less home with her missing from it. Everyone is off, kittens included. I never, ever, again get to stroke her softer than softest fluff long haired calico/tortie coat. Never will feel her body stretch out on my lap when I watch a movie, sit next to me on the couch as I read or scroll the phone, bunt her head into my arm as she lay alongside me in bed or turn and see her on the chair behind me as I worked at my computer. No more of that gorgeous girl to come home to, to wake up with, to sprinkle valerian root for.

We went through the moves, the losses of other loves, humans and non. Spent the pandemic together, so much time suspended with time but we had each other in all of it. Fourteen years of shared history and now I never get to see her again. All the memories, what they marked, what they meant to me I get to keep and what they meant to her I can never fully know but they are lost to me too. The veterinarian who came yesterday, who so carefully and lovingly, gave her those lethal injections, she really was the best at what she did. I have been here before too many times. You get to know the differences, the skilled clinicians, the unskilled and the ones who should not be doing this at all. Cinnamon's vet worked with consummate skill, softness, and empathy. And she knew cats. That is saying an awful lot.

Cinnamon had feline injection site sarcoma. The irony of a disease we have given her in efforts to spare her one of the most fatal diseases we share among mammalian species. The cancer waited inside her for years. Only emerging months ago. How such a cancer could manifest so many years after initial vaccination, increase in size so rapidly, is medical mystery, but the literature says it happens. What the journal articles leave off is contributing factors, extenuating circumstances. How could we know with certainty any of that?

Stress we do know about. How strong its effects are on health and welfare. We know already, stress is huge for cats, idiopathic conditions present and bringing down stress levels can resolve many. We have studies on that. For Cinnamon, I think, and I will never know for certain, that it was two moves in the last year for this older girl, that lowered resistances allowing this malignancy to develop. Cats despise moving and no matter our best attempts to lessen the impacts, prepare the new spaces, they would rather stay where they were instead.

I like and cite often, Francis Galton's observation that the cat's extreme adhesion to the home is the foundation of their domesticity over their affection for humans. I would add that they want us to stay with them in that place, but yes, it is the home that is where their heart is. Cinnamon never really took to this last move. She only left the bedroom one time and that was enough for her. To be fair, that mass has affected her mobility greatly. She managed with it, jumped to cat trees, got to perches in good speed. But maybe those moves around one room was enough with such a mass to carry.

I watched her closely through this sickness, from the onset as the growth became more and more apparent. Cancer progress in its own speed dependent. And in some place in her cat awareness, her consciousness, she must have known what was happening in her body too. Not the name, but the force of it, the very place it took. That she ate, purred, groomed, stretched, sought contact, and bunted repeatedly through all of this with cancer, what did that mean about the worth of her life to her with it? She lived stricken for as long as we could bear. I can never know if the decision I made to call the hour is one she would have agreed with. The tumor enlarged and enlarged on her back leg and one can only imagine the insidious progression inside her body. She had lost weight, an older cat, loss of muscle mass comes with age, but these invader growths takes calories from the animal to spread and take over space of their own. Overcoming, this type of cancer, it is known, has limited success, even had we put her through leg amputation and the heightened risks of surgery at her age. That sarcoma is hard to contain no matter what, legacy remains, going into the abdomen, its ropy tendrils keep moving.

I am aware of quality-of-life scores and except for lessened mobility, with a leg she still uses, much of what would be in keeping for a senior geriatric cat, Cinnamon scored highly. I know about Kessler and Turner's Body Stress Scores, developed through observations of owned and unowned cats at British catteries and the Feline Grimace Scale, developed through observations of cats in hospital care. No cats were observed at home in these schemes. Perhaps those observations deserve finer tuning. I have read Yin, Heath, Leyhausen, Rochlitz, Rodan, Bradshaw and Ellis, among others. Mostly I know this cat, the relationship we have, the years of reading each other. I do know that towards the end when I tried to move her and there is pressure on the ventral abdomen or I hold the "bad" leg in the slightest off position, she now gives me a low growl. That hurts her. Perhaps the time I have deprived her of is not as much as I think it is.

Cinnamon came to us as an older kitten or younger cat. A former ear-tipped feral, she was one or two, when we came upon her at one of those big box store adoption events in suburban New Jersey. I had just lost Sarah, a 12-year-old cat to lymphoma. I so loved Sarah. She was another long-haired calico I had adopted from the city shelter on 110th Street. I arrived at the shelter with every intention of asking for a kitten only to be told that they were none to be had. All the cats sitting in the kennels with such looks of bewilderment struck me. Sarah was one. I took her home. She got sick not soon after coming home with me, an upper repository virus, no doubt from shelter stress. I nursed her through it.

Back then, adopters signed a paper promising to have a pet spayed rather than the shelters performing every such surgery. When I took Sarah to have her spayed, the veterinarian called me during the procedure, she had already been spayed. They sent her home bandaged and we soldiered on. When Sarah got sick, that cancer moved too with alarming speed. I tried everything the vet gave me, every medication, every procedure, all of it. I forced her to eat. She got sicker with all of it and never better. When I finally did have Sarah euthanized, I cried desperately over how I had waited too long and what I had forced upon her.

Cinnamon looked so much like Sarah, that was why I wanted her that day at Petco. Just to be honest, not really her, I wanted my Sarah back. In the beginning, as I got to know Cinnamon and spend time with her, I would tell her "You're not Sarah," not in recrimination, more of a wonder. She was though, like Sarah, soft, kind and incredibly gentle. She never hissed or objected to the other cats or the dog. One of those rarer cats that got along with everyone as soon as she arrived. She was such a wonder of a cat that I began to bring her with me to the pet care technician course so learners could observe cat behavior, anatomy and learn how to handle a cat.

Cinnamon (and Daisy, my dog) both had their own time cards at school. No paychecks 'though. When I would bring Cinnamon for the day, I would leave her in the cat carrier on top of my desk for the first hour or two, I wanted her to be able to settle. I placed an oval cat bed in the carrier so she might be more comfortable. I would play classical music. When I did open the carrier, the first thing I would ask students was what could they tell me about this cat, just by looking at her? What three things stood out right away? I was seeking their observations; on the ear tip – how it would indicate she had been neutered, how that was typically done to ferals and her calico coat meant most probably that she was a female.

Cinnamon was not big on strangers picking her up. No cat is. But given the right approach, she would tolerate it. When Cinnamon did not like the approach, she curled opossum like onto herself. No claws, no hisses, just pretend I am not here. Play possum.

Julie, was not so welcoming to new arrivals. When we were not around, there was evidence she pushed her feelings of not welcoming onto Cinnamon. One time I came home and Julie had a claw stuck in the front of nose. Cinnamon clearly did not take certain offensives lying down, she would defend herself. We would hear the occasional hiss and that mostly worked but then again, that went only so far. I used to separate them when I went out of town into different rooms for the pet sitter and I thought for their own welfare. The last time I did it, Cinnamon, who was not a complainer, meowed loudly and at great length when I came home. I never did it again.

I have to get through the rest of these days, to mourn the loss of this cat in my life while caring for my other animals, while doing my work, while taking care of all the minutiae of every day that needs attention. I am beset by all the things I should have done, could have done for Cinnamon.   The tears come, not always with notice. It's always worse when someone is nice to me. Why is that? And tell her too, I love her and am sorry for how we have failed her and how much I miss her in this life.

<u>References</u>

- Pierce, J. (2012). *The Last Walk: reflections on our pets at the end of their lives*. Chicago: The University of Chicago Press

DOGS

When Olivier brought home Daisy, as a surprise three-month-old puppy, I was of so many minds in response. "**Not** welcome" and "**SO** cute" and "a little funny looking" sat right there together in my head. A dog!? Who was going to take care of a dog!?! Don't get me wrong, In the past I had volunteered to walk shelter dogs, pined for the ones I became attached to but would not take home, loved all my childhood and my sister's current dogs. But we had cats, easier to care for, don't need walking three to four times a day, can leave them overnight, cats. Then, there was the very first thing I said to Olivier about Daisy: "You have just changed our lives forever." And they were.

Not just us, dogs changed all our lives. Dogs, our proverbial best friends, in our conquering of the natural world, without them, we would have gone slower, longer, or not at all. Such partners, giving so much and asking so little in return. Cultivating the planet, herding flocks, keeping watch, heralding danger, exploring new continents. Dogs, who carried our burdens, kept us safe, hunted better than we could and left the catch to us, shepherded flocks -who listened more to us than we have ever listened to them.

Dogs, using scent and pheromones in ways we will never know, to "see" a world and richness for us and sharing finds we could never discover alone. Dogs, who recognizing our human reliance on the visual, our own weakness in sound and scent, in deference, or so they might communicate more easily with us, developed direct affiliative eye contact, inner eye brow raises, gaze alterations. Dogs, who have watched closely our every move to know if they are safe around us or learn what we would like from them, giving everything and asking only for the leftovers shared in companionship. All with, no score keeping, no grudges, no punishment, no sulking. And in their careful consideration of us, dogs who honed those special, just for, only about us bonds. No dog is the same. Those that picked us, that chemistry that defined the connection, the stuff you don't get to quantify and qualify, that you know you know when you have had that dog in your life.

If nature selects for the most and best in usefulness, then we have only to look at dogs and know that would be love.

CANINE BEHAVIOR, COMMUNICATION, AND BODY LANGUAGE

At a pet store I go to, the manager asked for advice about communicating with his dog. He was concerned that his dog was not willing to listen to him and felt that the dog was being downright rude when he would speak to him. I asked why he believed this was happening. He explained that when he would speak to the dog, the dog turned his head and looked away. This look away, meant to the owner that his dog was being dismissive of him and not interested in what he had to say. Now, such behavior might be true for a human but it is not true for a dog. While the dog may have and probably was, reacting to the tone of his owner's voice, the turned head here is more to diffuse tension, increase distance and avoid conflict.

When a dog looks away from somebody, this is stress related behavior that either might seek to communicate appeasement or be self-soothing. So, in fact the dog was communicating to his owner both an effort at pacification and relaying that the situation he found himself in was stressful while hoping to make things calmer by removing visual pressure and looking away. There is more than one way to increase distance. If the owner and the dog were understanding each other in this scenario, the owner might have either lowered or changed his tone of voice along with adjusting his body posture, perhaps in relaxing or turning sideways to alleviate the stress. In response, the dog might glance back at the owner with a more relaxed face, maybe even followed by an open mouth and perhaps a little tail wag. Problem solved. But we are different species and translation can take some effort.

So, how to then decipher what is going on with such, at times complicated, inter-species communication? Scientists studying the behaviors of socialized wolves and dogs, found that dogs originated "communicative face/eye contact with the human earlier and maintained it for longer periods of time" (Miklósi, Kubinyi, et al. 2003) when compared to wolves on problem solving tasks. Dogs look at us more than wolves do and with purpose. You know when your dog rolls the tennis ball under the couch and gets you to remove it by first staring at you, and then under the couch, and then at you, and then under the couch, etc.? That gaze alternation to get your attention focused is communicative and part of a problem-solving task. In the study in question, the researchers used food in a sealed container. These authors propose that one of the selection pressures in the domestication of dogs was a predisposition for behaviors that relied on originating and sustaining contact with the human face and the eye and:

> "the corresponding behavior in dogs provides the foundation on which developmentally…complex communicative interactions can emerge between man and dog." (Miklósi, Kubinyi, et al. 2003.)

It is often said, that intelligence for any species, is best ascertained and measured in how well an animal uses their abilities to navigate their environment. Where domestication by humans is part of the context of the environment, perhaps the cleverer of the species is the one who figured out the other one first. A dog's dominant sense is that of smell, yet the dog uses vision and gaze in communicating with humans who rely on the visual. Pretty smart.

Researchers delving into just what our dogs are paying attention to in our presence, found, to some pet owners' utter lack of surprise, the object of attention is us.

Not only are dogs able to read our faces and follow our gestures but they are also very aware of what we are focused on and when. A separate study by Christine Schwab and Ludwig Huber on attentional states, looked at dogs in their homes with their owners. The dog owners asked their dogs to remain lying on the floor. How long the dog cooperated was measured with food in the room and without during the time the owner: a) watched the dog, b) read a book, c) looked at television, d) turned away from the dog, and e) left the room. Most dog owners already know that the most deferential dog was the one being watched and the most independent dog was the one who found him or herself alone in the room. Or was it "Lie down" might have meant just that, if the owner was in the room? The jury might then still be out on whether the dog was supposed to understand the command as being an ongoing one. Our studies show great compliance on the part of our companion animals. If literal interpretations are allowed, the compliance factor increases exponentially. Smart dogs or smarter dogs, you decide.

We are a visual species and we rely on this as our primary mode of communication and information gathering. Dogs depend on olfaction, and next hearing, for their main sensory inputs and information exchanges. In the co-evolutionary journey of persons and canines, the canines are the ones who can be said to have perfected the multi modal in adding the visual to their repertoire with us.

Scientists further report that over the course of years and years of domestication, the process may have contributed to the development of social skills in dogs enabling them to be able to interpret human behavior in social and communicative contexts. Basically, this means that dogs can read our facial expressions, our body language, and to react to both. The reciprocal still needs some work, how well do we understand dogs? Fox kits bred for their penchant for moving toward humans without fear and aggression are found to be just as proficient as same age puppies in following the social cues of humans. What may have started out as a simple business arrangement seems to have developed into so much more than that, at least on the part of the dogs.

Humans, primates that we are, rely on sight as our main sensory process. We are visually complex. How we measure out eye contact may vary culturally but facial expressions made, registered, and responded to, are some of our most extensive forms of non-verbal communication. Visual gaze figures differently for non-human animals. It is more direct. At least with each other.

> "One of the most effective signals used by more dominant dogs is the direct stare. When two dogs first meet, subordinate individuals break eye contact earlier than dominant ones…Confusion can often arise if a dog continues to be stared at despite having already broken eye contact…dominant animal's intentions may be reinforced by more direct threatening signals, including baring of the teeth and snarling."-Bradshaw & Nott (1995).

Such an observation is even more significant, especially when we realize the extent this allowance has been modified on the dog's olfactory and aural end, to meet our species' visual needs. And how extensive and refined, the domestication process must have been and continues to be.

Differing theories exist as to the domestic dog's exact forerunner. Whether the dog came into human lives from an unknown wild canid, jackal, or grey wolf (with matrilineal genetic evidence weighing in the heaviest for the grey wolf) is still being researched. Where and when domestication took place is also in question, with most agreement that it happened up to 15,000 years ago. Another consensus, is that domestication of the dog was not just one event but several, with the last several hundred years, being ones of intense selection by humans for specific traits and breed diversity. Over the process of these many years of domestication, how did the domestic dog and humans come to be so deeply intertwined in each other's lives? And why has this foundational human animal bond forged over thousands of years of co-evolutionary processes reached such enduring value to this day? What is it about the dog?

Whatever wild canid or canids contributed to the heritage of the dog we know today; the ancestors of modern-day dogs were no doubt those canines able to best brave prolonged exposure to humans to go along with the beneficial food scraps and refuse they left behind. Tamer animals and humans entranced by them, struck implicit alliances leading to keeping, hunting, vermin control, pack carrying, shepherding, travel guides, companions and unwittingly, for the dogs, providing sustenance in leaner times. What may have started out in the most practical of ways, has resulted in some of the most intricate results. There is much written about civilization being owed to the human relationship with the horse, but it is truly the dog that is the most sustaining and meaningful.

Dogs may in fact come from a family of social carnivores but are also very much opportunistic scavengers that like humans have omnivorous diets. While comparison with the wolf is typically drawn upon for insights into dog behavior, there are major differences in these species. The social structure of the wolf is a family unit with one breeding pair that live together and can hunt as a pack or separately. The wolf family unit stays together. The feral dog, in a natural environment, will coalesce into group living situations of related or unrelated members of up to six individuals. These groups are fluid where membership can shift and change easily over time. Dogs scavenge independently and rarely hunt as a pack. Food is not a family affair for dogs. Wolves eat together with the hungriest, not the most dominant wolf, eating first. Dogs bark. A lot. Wolves have a range of vocalizations included in their vocal repertoire but barking does not figure into it. The second go to for humans in communication is vocal or sound, another sensory system to tune for dogs interacting with us. As Dog Expert Extraordinaire, Alexandra Horowitz, observes:

> "Given the relative scarcity of barking in wolves, some theorize that dogs have developed a more elaborate barking language precisely in order to communicate with humans."- Horowitz (2009).

Along with the visual, including body language, vocalizations are the most accessible forms of communication humans can relate to. But the significance of vocalizations (body language too) for each species can vary. The dog's vocalizing serves to get our attention, requests and alerts can follow. Similarly, the cat has developed more extensive meowing around humans to get us to "listen." A look at our own human verbosity and selecting for barking as a valued trait in dogs, supports that we may have contributed to an even more "verbal" dog.

What does appear to be consistent, is that our cats and dogs have increased vocalizing around human beings. Such adaptations of sorts to aid effective communication amongst species are consistent with certain universals for most species – such as vocalizations having characteristics of low pitch and guttural tones being more predictive of danger while higher frequency sounds can indicate pain or alarm and high pitch in short intervals can also signal play. Collective practicality; we all need to know and be able to react effectively to what danger sounds like. No matter which animal it comes from. For dogs, think of the interpretations given to whimpers, yelps, screams as compared to growls and low sustained barking. We innately understand those utterances. Barking has been highly selected for in dogs by humans for its value in guarding and alerting behaviors. Research on components of barking found:

> "Barks are loud, and different barks have varying numbers of harmonic components, depending on the context in which the bark is used. Stranger barks were the lowest in pitch and the harshest: they are nearly spat out. Isolation barks tended to be higher frequency and more variable: some ranged from loud to soft and back again, some went from high to low. Play barks, too are high frequency, but they happen more often one after the other than the isolation barks."- Horowitz (2009)

And where there is vocalization there is listening. Dogs have superior powers of detection when it comes to sound compared to humans. Adults, young children being the exception, hear sounds below 20,000 Hertz. Dogs, hear sounds with higher pitch or frequency – up to 65,000 Hertz. On the lower end of the scale, the dog will also be able to detect sounds too soft for human ears to detect. Eighteen muscles in the dog's ear keep ears mobile and able to orient to sound. The base of the ear is the most moveable and even in breeds with longer or cropped ears will display movement and in addition to functionality can indicate emotional states. Hearing is a hallmark of the dog's power in sensory processing but not the only one.

Like many other animals, including wolves and cats, dogs rely on olfaction and chemical signals to investigate environment, other individuals and to communicate. Chemosignals and scent are deposited through urine, feces, pheromones, and body odors. Unlike cats, dogs will mark over or "overmark" urine of other dogs and this behavior occurs in both males and females (Bradshaw & Nott 1995). The information carried in these leavings is processed uniquely through a nose long enough to contain nearly 300 million (breed dependent) olfactory receptors, thought to be about 40 times greater than humans. The design of the dog's nose aids in functionality, sniffing in odor molecules without an exhale transmitting to the dog brain with the part responsible for processing seven times greater than ours. Chemosignals, such as pheromones get different or finer tuning, including through the use of the Jacobsen's organ, located at the roof of the mouth, a feature present in dogs and other animals but missing in humans, and one where the full extent of the systems or purposes of this organ is not fully understood.

Olfactory communication is hard wired and enriching for the dog and one we put to use. Humans have harnessed the extraordinary scenting abilities of dogs to track and detect for our benefit. As utilitarian as we might find such abilities, we often fail to appreciate the significance of sniffing every single object possible on a dog walk and we need to be reminded to slow it down and let our dogs sniff it all in. Such richness in the odors surrounding every step of a dog's walk are not lost on a dog and should not be denied. Sniffing and elimination are what walks are for. Every single smell counts and every urine mark is to be added to.

As bright as you may have already believed or are coming to believe, a dog may be, they do not speak Mandarin, English, Spanish or French or any other human language. They do speak dog. To best understand canine communication, we need to look at their highly specialized sensory processes such as hearing and olfaction and pay close attention to the significance of non-verbal communication from and to the canine point of view. We also need to put all the pieces of the puzzle together and know that we might always be missing some of them.

Dogs are sending and exchanging information in ways that we cannot decipher with chemical signals left behind and perceived and the range of frequencies and sounds they hear. We can best hope to advance our canine lexicon by being mindful of the whole dog. Dogs are already ahead of us in figuring this out, consummate multitaskers when it comes to communication; our best friend who relies on his sense of smell as the primary source of taking in information, has mastered the art of visual exchanges when it comes to human interaction.

This canine evolutionary journey was of course aided by human selection pressures and is undermined by us as well. Selective breeding turned vanity breeding leaves a legacy of mutations we seek for usefulness in farming, hunting or sport and drops to what appeals for cute. Working dogs are most comfortable working and hard pressed to find comfort in constrained spaces or with little to do. Service dogs are better suited to tolerating long periods of forced down time. Hence the Alsatian and Malinois police dog and the Labrador guide dog. Breeding for looks may satisfy the whims of an indifferent and changeable public sentiment but the compressed airways and too small craniums of the Pugs, French Bulldogs and Cavalier King Charles Spaniels are a lifetime of health problems, distress and pain for these and many breeds. Line breeding or the mating of relatives to preserve pedigrees can have disastrous consequences for wellness and welfare. In October of 2023, the Norwegian Supreme Court handed down a verdict forbidding the continued breeding of Cavalier King Charles Spaniel, as such breed was too sick or inbred to continue proliferation. Unfortunately, such a finding is only the tip of the iceberg.

Still after all our eugenics and through all our devotion, dogs have remained our friends, no matter what. Some of what we owe them might be found in figuring out what they might be saying.

Dog and humans communicate differently. Be careful not to give human motivations to canine body language. At times the significance can be diametrically opposed. For instance, direct eye contact for humans can indicate honesty and friendliness while for a dog direct eye contact can be adversarial, yawning out of the context of fatigue is a sign of stress, so is lip-licking out of the context of being hungry. Another example can be found in species specific differences—humans are primates and approach each other frontally while dogs are ventral and approach from the side. Worth repeating and remembering on every dog walk, even how we parse information utilizes different senses; humans rely mainly on the visual and dogs rely mainly on scent.

Context matters. To understand the scenario, you need to understand the environment it is unfolding in. In addition to observing the whole dog, consider what is happening and what has happened around, or to the dog. Noises, odors, movements, people, animals, etc., all play into a setting and cue us as to possible history and motivations. What is the dog paying attention to? A growling dog with ears back and squared off posture is oriented towards and reacting to something—what prompted that response? Detect the stimulus and the response becomes clearer.

Canine body language is all about looking at dogs. Dogs offer a wide range of emotion and intent in how they present themselves. They are also congruent – how they feel is readily apparent in reading their body language. There is little subterfuge where dogs are concerned. Facial expression (eyes, ears and mouth), body positioning (hair and tail) and stiffness communicate what is going on with the dog and what the dog is saying about it. Tense or rigidly held frame or muscles generally indicate vigilance or emotional tension while relaxed or loosely held bodies suggest comfort.

Look at the whole dog. Always factor in that each and every dog is an individual, with individual differences. To know what your dog is saying you must listen to the individual including the whole sentence and not just the words- study all parts of the dog in its entirety. Pay attention to everything at once; ears, eyes, mouth, neck, body, tail position, posture, tension, coat, and hair. Each aspect of the body contributes to the message and the conversation.

Ears: Hearing is the second most important sense for a dog after olfaction. Dogs filter the environment through nose and ears. Ear position is rich in information, they can show emotional state and functionality. Ears that are pulled back just a bit, are curious and friendly ears, while upright and forward, are primed for aggression if warranted. A relaxed dog will hold their ears in a natural position (watch for what this looks like in your own, at ease dog, so you can see the difference when they are not relaxed). Ears that are held flat to the head or rotated to the sides indicate fear or deference.

Dogs will orient (or point) their ears to what they are interested in (this also helps to receive more sound from a source). A raised and swiveled ear indicates an increase in attention. All behavior and emotion are happening on a spectrum, corresponding in time with shifts in environment. Ears are a good gauge to watch for this.

Breed differences give us a range of ear shapes and conformations. Floppy or pointed, Basset Hound to Chihuahua, these variations affect not just our ability to decode what is being "said" but the ability of the corresponding dog as well. Considering differently shaped ears, ear position is easier to recognize in a dog with naturally upright ears compared with a dog with hanging ears. Dogs with hanging ears do move their ears but the movements may be harder to detect to an untrained eye. Look at the base of the ear, hanging or upright, this is where you see movement originate.

Cropping ears or otherwise surgically manipulating their shape to conform to outdated practices is unethical and inhumane. Such procedures are painful and the resulting cropped ear not only interferes with natural communications between dogs, it colors human perceptions of the "sort of" dog that has cropped ears. Despite the fact, the dog had nothing to do with it.

Eyes: Differently shaped eyes also vary according to the breed, which may be either round or almond shaped. Again, you want to know what relaxed and normal looks like in the dog as your basis for comparison. Taking eyes and ears together, tells us a lot about how the dog is possibly feeling and what they might be reacting to. Are the eyes squinting or are they wide open and staring? A frightened or aggressive dog may stare with open eyes. A hard stare out of the corner of the eyes with the white of the eye showing (whale eye) is usually followed by the dog freezing and readying an attack if his unheeded warnings, as in body language is not read and responded to. Relaxed dogs may glance sideways at people with a relaxed overall facial expression as well. Squinting may indicate pain or deference while blinking may indicate stress.

Mouth: A calm dog has a mouth that is closed or moderately open, the tongue may be seen and there may be the occasional pant. A wrinkled muzzle with more of the teeth exposed (usually with a hard eye and a growl) signals aggression. Baring of teeth can mean two different things and should be viewed carefully, so as not to confuse the two: a submissive grin (usually seen together with eyes that are partially closed or squinting, a head lower than the body, and vocalizing that sounds like crying) consists of lips pulled up vertically with front teeth exposed. A closed mouth with tight corners and/or lip licking (without food motivation) and/or or yawning (without being tired) indicate fear, stress, or deference. Increased panting (without excess exercise or heat) can indicate anxiety or stress.

Hair: Raised hair typically means arousal, which means being excited about something. What the dog is excited about can often be gained from context, what is going on in the moment. Humans may not always be able to sense this the way a dog might, the arousal might come from a sound or chemosignal we are not able to perceive. How hair is included in the dialogue can most easily be seen with the change in coat hair standing on end (piloerection) along the hackles or withers (think of the top of the dog behind the neck where the shoulder blades meet). Piloerection is an autonomic, automatic – happens without conscious decision making, response to certain stimuli of concern that serves to make the animal appear larger and is a ritualized display signal to increase distance and posit possible aggression.

With any display signal or distance increasing behavior, it is important to realize that the missive, information, request, being transmitted is one not to aggress. Warning, threats or appeals to influence greater distance or changes in environments are geared to not escalate or engage in conflict. Canine societies, like most in the animal kingdom, revolve around degrees of deference and retreat to keep them peaceable. All bark and no bite is more than just a saying for dogs, it's a necessary way of life. Fighting is best avoided along with its injuries, compromises, and possible fatalities. Constraints to this happen in closed spaces or on tethers with little ability to retreat, history of severe trauma, or in unavoidable needs to defend in the face of perceived danger.

Certain breeds have naturally stiffer coats and some, like the Rhodesian Ridgeback, have a permanently raised crest of hair, with this type of breed you need to take into account the natural look of the coat. Dogs and cats will also shed or shoot hair excessively from their coats in response to occasions of stress.

Tail: Dog tails come in a variety of shapes and sizes. Add to the variety, breed variances such as naturally tucked tails found in greyhounds or naturally bobbed tails found in Australian shepherd dogs or curled tails such as those found in a pug. Tail position in these and other breeds is further compromised as display and signaling due to artificial selection. Additionally, custom still finds docking or cutting of dog tails (and ears) to be routine with certain breeds. Docking is another cruel and inhumane process often done without pain relief and one which deprives the dog of the use of the tail for expression and communication.

Tail position has been studied in a variety of animals but we are most familiar with wagging, which may not mean what we think it does. A wagging tail does not always mean a happy dog. How the tail is being carried, how stiff or flagged and how fast the wagging is, determines much of what is being conveyed along with hair position. Once more, hair on end signals arousal. A loosely carried tail which wags gently from side to side is a relaxed dog (look for this when your dog is relaxed to recognize it). A tail that moves faster on a happy and loose dog signals happiness. A tail on a stiff dog that "flags" upwards and moves tightly is an aggressive signal. Scared or stressed dogs often tuck their tails under their bodies.

There is so very much more on tails to know about. We are starting to learn that tail movement expresses and transmits much more information than previously believed. Researchers in Bari, Italy conducted a 2007 study looking at the direction of tail movement as an indication of emotion. The dogs tested were found to wag their tails to the right in certain instances and to the left in others. Tail wagging to the right at a very high rate resulted from seeing an owner, a medium rate when seeing an unknown human and a low rate when seeing a cat. Tail wagging to the left was seen when the dog was left alone or shown an unfamiliar dog. And this is just one study...

Body Posture: Find baseline by starting with observations of your relaxed dog to see what this body posture looks like. Sleeping dogs are a good standard to start from. Make sure that sleeping dog is one that is not overly tired or stressed, either can show atypical presentation.

Relaxed dogs generally do not have tense musculature or uneven weight distribution when standing, body tension is loose and movements are fluid. Dogs at play are also relaxed in posture and even as movements may be faster or with exaggerated gestures used in play fighting, play bows or rolling over, they are inhibited from full effect to have minimal force.

Assertive body posture is more rigid and erect, you will see the weight of the dog evenly distributed or leaning a bit more forward onto the front legs in preparation to move forward, muscles are tenser and the head is more raised. Aggressive body posture is amplified from the assertive with even more forward leaning and overt threat displays often accompany this such as barking or growling.

The scared dog will tend to try and look smaller than they are by cringing and hovering close to the ground; the head will often be lower than the body. Similar body posture is seen in a dog showing deference or submission but here the head is usually raised in greeting and you may see a raised paw or weight leaning back on the rear legs.

Look now, what is your dog "saying"?

<u>References</u>

- Bradshaw, J.W.S., Nott, H.M.R. (1995). Social and communication behaviour of companion dogs. In Serpell, J. (Ed.) *The Domestic Dog: its evolution, behaviour and interactions with people* (pp. 116-130). Cambridge, New York, Melbourne, Madrid: Cambridge University Press

- Horowitz, A. (2009) *Inside Of A Dog*. New York: Scribner

- Miklósi, Á., Kubinyi, E., Topál, J., Gácsi, M., Virányi, Z., & Csányi, V. (2003). A simple reason for a big difference: wolves do not look back at humans, but dogs do. *Current biology, 13*(9), 763-766.

- Quaranta A, Siniscalchi M, Vallortigara G (2007). Asymmetric tail-wagging responses by dogs to different emotive stimuli. *Current Biology.* 20;17(6):R199-201.

- Schwab, Christine & Huber, Ludwig. (2006). Obey or not Obey? Dogs (*Canis familiaris*) Behave Differently in Response to Attentional States of Their Owners. *Journal of Comparative Psychology*, 120 (3), 169-175

BRINGING A NEW DOG HOME

In the behavior business you get to see your fair share of new dogs, especially those new dogs coming from far off places, to new planets, as far as they might be concerned. I have met, just to name a few; dogs from the suburbs of Florida, ranches of Texas, islands of the Caribbean, off the planes from Korea, from backyards they have never left, away from their Mom, from rescues where they have lost count of who came from where when, shelters where they doubt the stories given, pet shops and breeders where the story is even more doubtful, and more. All arriving to new places, possible new homes and to the strange, grey concrete streets, constant din, and clamor of their new city. For each and every one of these dogs, one thing is true, their world just turned inside out and upside down.

Bringing a new dog into your home can be both an exciting and stressful time for us as humans, it can also be a stressful time for the dog. As many plans and expectations as you may have for your new dog, know that your dog simply cannot share them. He or she does not speak English, read, or know your intentions. They cannot know that you plan to be the best of friends, to go for plenty of sniff-where-you-want walks every day, teach them to play fetch and find the right toys and games to play with. They don't realize your plans to watch TV with them on the couch, find the best dog beds or even sleep in the bed with you. Your new dog cannot trust that you will find the right dog food, the best puzzle feeders and snuffle mats or make sure they get to the vet when they need to.

Things are changing very fast for your new dog. Whatever place, routine or family, dog or human, whatever sort of home, your dog has been in and is familiar with, is now gone and their world will never be the same. What your dog does know, is the overwhelming change of what has happened and is around them right now- the differences in this new environment, with all its smells, sounds and sights to be taken in with the immediacy of the moment. All this strange new world is unknown and different, including your presence. There is a biological necessity for this new dog to seek safety, security, solace, and routine.

- Anticipate what your dog will need and have it around before you bring your dog home. Things like:

- High quality dog food, get the good stuff. Dogs in new environments will be stressed and less inclined to give into hunger. Find the right sort of food puzzles for dry food and wet food, as in those that encourage and allow for natural behaviors like sniffing out dry food in a snuffle mat and chewing out wet and dry in pliable, hollow chews, but not those slow food frustration inducing food bowls please.

- Water bowls, stainless steel is a good choice. Make sure to keep these full, clean, and free from filming which can result from not washing. Change water daily.

- A six-foot cloth or leather leash (no dangerous extend a leash).

- A Y shaped harness to attach the leash to, a leather or cloth flat or martingale collar. Please, for welfare and wellness' sake, skip those anatomy damaging and behavior dampening prong, choke, and head halters.

- And remember, toys -lots of them, especially chew toys, stuffed toys and do not forget the balls.

- A brush, and several dog beds for each room to round out the essentials.

- Bring your new arrival directly home. Do not stop and run any errands or introduce this dog to your friends or family on the way home. As much as you may want to show off your new dog, resist. Meeting yet even more new people and new places is stressful here. You do not want to overwhelm your new companion. For the first week or so, limit both trips and visitors to your home to allow the dog to process all these changes including you and your home environment at an easy pace. There is a lot happening, help by lessening some of it.

- Make sure and take time to allow the dog the time to sniff around while still on their leash as you explore with them the new space. Keep the leash loose and let the dog set the speed of these investigations. Remember to always engage the dog in this process—talk to your new dog in a soft, happy, and calm voice. Explain what is going on. While your dog may not understand every word you are saying, your tone of voice and accompanying body language will reassure that he or she is in a welcome and safe place.

Start your walk around the indoor areas of your home. Include the block, yard, and perimeter area around your home as well. Keeping the dog on leash, will keep the dog connected to you, helping to foster both emotional and physical security (all new dogs are capable of escaping from unfamiliar surroundings given an opportunity).

- Initially keep your dog walks as relief walks rather than recreation. Take your dog for shorter walks in your neighborhood for the first several weeks, save excursions and dog park visits for the second month when you and your dog will know each other and the area better.

- Allow for your new dog to move at his or her own pace. It is essential for the dog to have ample down time or quiet time with no expectations. Sleep and rest are important. Let them. Do not put pressure on the dog by assuming an immediate friendship, it is perfectly acceptable and natural for this dog to be hesitant around you for the days it will take to get more at ease in these new surroundings.

- Set up dog beds and toys in rooms where you spend time and where you want your dog to acclimate to. Avoid garages, basements or other areas of the home that separate the dog from home life. You have brought this dog home to be a part of your life and your family, this is the time to incorporate him or her into the comfort of you and your home. Be judicious with crating or dispense with it.

- Containing dogs in crates has advantages for shelter, kennel life or to be used carefully in housetraining for very young puppies. Ex pens or fencing off areas you want to confine dogs too can be better choices but always remember, dogs are social animals and do best when part of a group and when trained to learn those human rules to navigate their environment and our expectations (more on that below).

- In a world where we decide when, what and where they get to do most everything, establish schedules and routines to provide a needed sense of stability and structure and afford control for the dog in this environment they have little control over. A consistent and frequent on leash walk routine and set meal times afford your dog with anticipation and reliance on the life events they need the most. Schedules have an added benefit for humans to follow as well.

- Feed your dog's meals in puzzle feeders to engage their time in natural necessary and rewarding behaviors. Chewing and sniffing meals out of the right puzzle feeders stimulates the cognitive processes dogs use in their natural environments, soothes anxiety, and allows for dogs to be occupied with necessary natural behaviors. Dogs like chewing for extended periods, let them. Skip the feeders that roll or tip over, require doors to be opened (better suited for cats or babies), slow feeders or lick mats (save lick mats for bath time and geriatric dogs) and look for puzzle feeders to stuff with maximum gnawability and chewability for wet food and snuffle mats for dry food.

Dogs that have not been exposed to puzzle feeders may need to acclimate to them, showing the dog how you prepare the feeder cues them into what is inside it (a puppy feeder on the market fashioned of clear material capitalizes on this), amp the value of food going into the feeder by choosing the more delectable to engage interest.

- Do not be concerned if your new dog does not exhibit much interest in play initially. Not every dog got to play before they came to you. And not every dog feels at home enough to play on arriving in a new place. This can come once your dog feels more at ease and safe enough to play. Each dog is an individual, with their own timing and preferences for toys and games.

   Offer a variety of chew toys, stuffed toys, and squeaky toys to help identify which toys your dog displays interest in. Keep the play invitation on the low-key side for novice players, your happy camper voice is good to use with dogs but softer is better with newcomers. Don't be shy with showing your dog how to interact with toys. Keep your movements and voice on the neutral side but freely engage with toys. Remember, dogs, like people, are social learners and can benefit from learning just how to work some of the more creative dog toys out there.

All dogs are different and unique. Relationships take time to build, so no forcing allowed here. Do not expect all things right away. There is no set time table here, every dog is an individual and what takes one dog weeks to feel at home with may make take another dog months. The dog should never be yelled at, scolded, or scared—ignore all bad behavior and reward all good behavior with simple calm praise. Think of focusing on reinforcing what you would like your dog to be doing instead and how to accomplish that rather than punishing what you do not want.

Is your new dog, coming home to another original resident dog? So very much here to tease apart. Think neutral territories to start, which always help in introductions no matter who is being introduced. Start with outside the home as in street corner (not owned territories)

(See also the chapter here on "plays well with others" on dog interactions and cat and dog introductions.)

Linear walking is another worthwhile practice to try out if it can be done safely. This is when both dogs are on lead, one behind the other dog (you need to two people for this and to pace appropriately), so that what one dog may think of as unobtrusive but the other knows, butt sniff can get in before the muzzle, t sniff.

Some dogs and puppies will usually social roll on meeting a new dog. That drop and roll and show of belly is a show of deference that is hard to resist, no matter the species. Always remember, with any environment, resource distribution counts, people energy sets the tone, etc., etc. And always to remember, always, owner attention is a resource, just as much, if not more, valued than other resources. Greet, love, pay attention to original residents first.

- Stay neutral, relaxed, or positive. Remember that over 90% of communication is nonverbal and that dogs are masters at reading this nonverbal communication. Should you be tense, concerned or discouraged your new dog will "read" this as easily as he can "read" contentment and ease.

- Begin a conversation by acknowledging your new arrival and announcing your presence when you walk in a room—say hello and use the dog's name but leave it at that for the first few days. Again, do not expect instant rapport, this dog has not met you before, does not know you and needs to see how trustworthy you are. Your new dog will be watching your facial expression, body posture and tension and anticipating your movements, so go slowly and be the friend your dog would like to have eventually without scaring him or her off as they get to know you.

- If you simply cannot resist interacting with your new dog no matter what, take a book and sit on the floor parallel to your dog leaving about three feet of distance between you both and read out loud -to yourself. Pick something you are interested in (no horror stories!) and stick with it for a few pages. This will allow the dog to become more accustomed to your presence without that pressure of on-the-spot intimacy.

- Even with the best house training your home is new to this dog. There are no familiar smells, routines, or safe and comfortable places to be yet. Should there be a break in house training, know that this loss of control may be stress related. Avoid any punishment of reprimands as they will only serve to damage relationships and create more stress. Give it time.

- Set your radio to a classical music station and make sure to leave it on for your dog when you are not around. The soothing music coupled with the melodic voices of the announcers will add to the comfort of your new dog's new home.

- Utilize the positives in relationship building. Add to your positive and reassuring behavior effective communication strategies. Get up to speed on force free, science based, positive dog training methods and canine body language to know how to best communicate and to understand what your dog is "saying." Even if you decide to work with a trainer, make sure to read and avail yourself of the wealth of free resources online or in books by experts like Steve Mann, Pat Miller, Ian Dunbar, Victoria Stillwell, or Sophia Yin. Two of Dr. Dunbar's books (geared to puppies but just as solid for adult dogs) are free to download, see references in this chapter for that link.

For more specific behavior help after your dog has settled in, look for a well-qualified and formally educated dog trainer or training classes. Ask all the questions of trainers and classes, including where they learned about training, their schooling, how they keep up on the science, and most importantly how they handle matters when things go wrong. Ask open ended questions and wait for the answers instead of filling the silence or pauses. Hear "dominance," "pack leader," "balanced," "results," or the like and know these are red flags and not what you want to pay for or have forced upon your dog. Either for the trauma to your dog or to build a trusting and loving relationship.

Anything that seeks to have you be "alpha" (an outdated way of looking at dog behavior) or use the buzzword of "balanced" (which means punishment or force is part of training) are additional tip-offs for what to avoid. You are looking instead for positive reinforcement; science based, force free training and material here.

Good training is a dialogue, that engages us with the dog, offers a dog manners to navigate the human world along with structure, stimulation, and bonding activities for you and your dog. Some of the beautiful things about force free training and behavior are how it lets us be positive ourselves, lets us thank our dogs for listening, lets us pay careful attention to them and their environment, including us in it, lets us acknowledge what they just did as "Good!" Keeps us engaged with them with love. Works with people too.

No matter what the come-on's, remember that training and behavior methods NEVER come with guarantees. Avoid at all costs dominance based or aversive methods. Worth repeating here, no prongs, training or e or shock collars, choke chains or head halters. **Do not train with pain**. Your dog does need to know human expectations and the rules that go along with that but those rules should never contain fear or intimidation. The relationship you are working to build is one based on trust and love and not coercion and fear. Again, beware of any methods that take your dog from you and prevent you from being overseer and part of the process, send away programs, or camps where promises and guarantees are made at the cost of pain and subjugation.

Even as your relationship with your new dog progresses, always permit adjustment periods to changes in environment or development and allow for individual personalities, likes and dislikes. There are some New York City dogs that would prefer never to walk on sidewalk grates and as long as there is a way around the grate, why not allow for that? The point here is that what you are aiming for is an optimal relationship and not for a perfect dog (even if all dogs are perfect already).

I was recently asked how to help a new dog overcome a fear of stairs in a home where the dog had been in the home for less than 24 hours. The dog was settling in but was reluctant to climb a flight of stairs after an initial slip. The new owner was concerned that the dog needed to overcome this fear right away to fit in to her new surroundings. The answer is that while there are ways to work on getting your dog to be more comfortable on stairs, this is just not the time to ask for it. For now, it is certainly acceptable, if not, preferred to carry the willing dog or if possible, coax gently that dog up the stairs. Helping a fearful dog is never a bad thing, as Suzanne Clothier points out- we never, ever see a wild animal (think of a mother tiger or polar bear) forcing their babies to "suck it up."

What is key here, is creating a trusting relationship first off, a relationship in which it is OK to be afraid of new and scary slippery stairs. Much of what your new dog is afraid of is necessary caution in encountering a novel situation. Given time and a safe environment, one in which you are helping to create, the situation becomes familiar and the caution gives way.

References

Dunbar, I., *Before You Get Your Puppy*
https://files.cdn.thinkific.com/file_uploads/5433/attachments/829/875/5e2/BEFORE_You_Get_Your_Puppy_-_Dunbar_Academy.pdf, retrieved January 15, 2024

Dunbar, I., *After You Get Your Puppy*
https://files.cdn.thinkific.com/file_uploads/5433/attachments/90d/cd5/5c3/AFTER_You_Get_Your_Puppy_-_Dunbar_Academy.pdf retrieved January 15, 2024

NEW PUPPY ADVICE

Before I had my own dog, I volunteered at the closest animal shelter to me in NYC to walk dogs. I loved dogs but was not up to owning one. I had cats, they were less responsibility not needing daily and multiple walks. Volunteers were warned not to have favorites, to treat all dogs with equal favor. Sure. It just cannot be helped, animal care workers you know what I'm talking about - that falling in love with that one special dog you have that unquantifiable chemistry with. You try not to let the other dogs know, but it is there anyway.

My secret favorite was a cocker spaniel puppy called Angel. A sweet and gorgeous caramel colored boy with the longest softest ears to drag along city sidewalks. The fringe on those ears picked up all sorts of sediment, large and small. I once snuck Angel into a groomer on a walk and had just his ears washed.

He was a dream I thought but apparently not when adopted. Angel was consistently returned. While displaying no behaviors of concern at the shelter, he was reported on guarding his food in the new home. Resource guarding, even as it can be worked with, is rarely tolerated. No one likes being growled at and interventions need to be carefully applied and managed. Another volunteer ended up adopting Angel. She told me she was taking him for the very reason that she would not return him. And while I hoped Angel would have a good life with her, I missed him and I talked about it. A lot.

Such talk then led my husband to bringing home a three-month-old cocker puppy, as a surprise, from a couple who were rethinking why they had gotten a puppy in the first place. Some surprise, this puppy was not Angel and I still was not on the adoption trail. But now there was Daisy.

"You just changed our lives forever" I famously told my husband. No more last-minute drinks or dinners in town – the dog needed to be walked. No spur of the moment, let's go anywhere for the weekend, unless they took dogs. Hello La Qunita. No more leaving shoes on the floor. Good bye new booties now chew toys. Everything had a "what about the dog" component. My not over the moon response on seeing her meant he wanted to bring her back the next day, to find yet another home for her. I didn't let that happen. No, I told him we now had a dog. And different lives forever.

Having a new puppy in your life includes all the changes for humans in accommodating the new arrival along with that sweet, warm, funny puppy-hood. Changes that come together with all the steps in teaching your little one how to navigate the human world with you. Puppy behavior is so much more than puppy kisses and cuddles, it's seemingly never-ending housetraining until it sticks, teaching which behaviors are not OK and which you fall in love with, like baby needle teeth that bite and wanting to be with you all the time even when they can't because these baby dogs are infants, social and bonding to you.

All relationships, especially those with a puppy benefit from structure, reinforcement, and clear and consistent communication. Begin with schedules to set reliable schemes and assure expectations of when those necessary life events happen such as mealtimes, walks, and play time.

Puppies are infants and do not come into the world housebroken. They are still growing with developing anatomy. Babies cannot fully control bodily functions or know innately where we would like them to relieve themselves. Provide frequent, as in every hour, walks to eliminate and to learn where you would like them to eliminate. Use puppy pads as back-up, grass scented and targeted spots for where you want your puppy to go when they are not outside. Remember to reinforce/thank/praise for eliminating in the right places and not to scold for the misses.

Crating because? Avoid using crates as cages to contain puppies for long periods and to prevent them from interacting with their environment. If you do not begin to teach a puppy how to behave in your home outside of the crate, how can they learn? Think instead of an area of the home which has you in it or in sight (no garages or basements!) for a temporary long term confinement area instead of a crate while puppy is growing, such as in the living room or your office area. This puppy station doubles in learning how to turn off puppy energy too, enhance with comfortable bedding, chew toys, and even a classical music radio station with added soothing human voices, all can add to the mix of rest now, play later.

For use in house training, remember to keep crating to a maximum of an hour. This method of teaching your puppy where to eliminate, utilizes a dog's innate instinct not to foul their denning area. In the natural environment, the animal would be free to leave at will to eliminate. This cannot happen in a closed off crate. To build on this for house training, take puppies out every hour or two (about how long they can "hold" it) to allow for relief and train with positive reinforcement for the locations we would like puppy to choose for voiding on their own. And if you can't make it out of the house, puppy wee pads placed next to exit doors are fine for starting dogs off to house training or when trips outside are not practical, due to timing or immunity.

When using open- or closed-door crates, make sure they have extra comfy dog beds and toys to increase appeal and drape a portion of the crate with a sheet to cut down on visual stimulation, be careful with this, make sure there is adequate airflow.

Begin leash walking lessons indoors to acclimate puppy to the leash and equipment. Puppies instinctively closely follow us around when young and leash training is ideal at this age. Keep your dog at your side by engaging with them verbally. Try patting your hand against your leg as you walk to keep them at your side. Pace your walk so you're at the same speed and practice stops that are fun ones. Stand still and ask them to stay with you – no pulling or dragging on your part allowed. And no scolding, this is supposed to be fun for the both of you, so make sure to praise puppy for listening as you go. Make sure your leash is four-six feet long. You want the leash to transmit connection, not make it disappear. Do not use extending leashes so puppy can learn to walk with you and not pull ahead, reinforcing this behavior as you let up on the lead. No chain link either, which can be tough on human hands.

Avoid choke, prong or head halters which are painful, inhibit learning due to stress and can damage puppy's developing anatomy.

Choose Y shaped harnesses instead that attach in the front. Continue harness use into adulthood over collar attachments for their benefits, including reducing pulling.

Once your vet has cleared puppy for outside walking, keep new puppy walks to routine locations initially. After puppy is comfortable on a set walk and is eliminating and not just in awe of every leaf and sidewalk object, you can vary your walks to offer new environments.

When working on outside walks and housetraining, know it takes time. Five months is about when most dogs will start to get it right. And then there is the big, strange, and amazing outside world to get used and take notice of. Remember, everything outside is new for your puppy so wonder and exploration can take over instead of relaxing and taking the time to relieve. Give it space, they will get there. Mete out that reinforcement of praise and treats at the end of voiding so as not to distract or interrupt efforts.

We often ask so much more of this canine species than our human babies in this regard -who get to wear diapers for years and then still require extensive potty training! (More tips in the housebreaking chapter.) Remember that the best inspiration for elimination is a good sniff at where another dog has already voided so let them sniff!

Puppies can learn that biting is a game with humans and with teething coming on in general between four-seven months, if not effectively addressed, a game that puppy, much to their confusion will not be allowed to still enjoy when older. For the dog, the rules have not changed even when adult dog teeth are bigger and harder bites come with them.

Stop playing any bitey games with hands and feet now. Begin retraining human interaction in this regard by making sure your puppy has plenty of chew toys along with stuffed toys around. Offer one instead of body parts when puppy engages in play biting. No scolding, just quickly provide the toy and a "here take this" instead. If you have an agitated shark response and nothing but human parts are sought, feedback at the exact moment puppy begins mouthing is key. Holding still and not responding or a sharp intake of breath are stop signals that puppy will take note of.

Make sure to praise immediately where there is even a momentary pause from puppy in return to your feedback. Please know, very huge point here: You are praising the pause, the stop in mouthing and not the mouthing. This is an important distinction to be aware of. Your puppy knows they stopped, even if it is for a second. Make sure you notice and acknowledge the behavior you do want by praising that stopping. Next, redirecting your puppy with a one-word redirection, and offering a toy to bite or chase when she is about to bite or is biting will also help your now better listener, if done consistently and repeatedly.

- Frequent and regular play sessions with other puppies and friendly amenable dogs, will develop much needed puppy socialization skills and help to develop bite inhibition with others. When it comes to dogs, puppies learn the rules of dog friendly behavior from each other and older dogs. Dogs signal requests to play with barks for visual attention and play bows, rump raised in the air and legs outstretched or bounced down and up mean "do you want to play?" Loose wiggly bodies and soft mouthing let each other know: it's all a game.

  When behavior is not wanted or welcome dogs let each other know. Too intense, too hard, or too rough, another puppy or dog may stiffen, yelp or snap in response. If the response goes unheard and puppy keeps going, play is interrupted; a good lesson to learn that playing nice is the only way to go.

- Interacting with others on a walk, dog parks, puppy kindergarten or puppy training classes are necessary and important for puppy's socialization around other dogs and people. Walking a puppy is a virtual magnet for humans and dogs alike. Make sure to supervise these interactions by letting people know the right ways to interact with your puppy. Not everyone knows this and you are the guardian for your dog.

- Pay attention to possible meet and greets by looking for friendly body language before the dogs are up to and next to each other. For canine exchanges, watch to make sure leashes are loose enough to avoid straining and the dogs are compatible. Decide if you want to go for the interaction ahead of time, rather than debating while your puppy strains at the leash in frustration. First, ask the other dog owner if their dog wants to say hi and monitor for response from dog and owner. A willing response from both and note that cross gender is typically your safest bet, can be a go ahead. Not every dog will welcome puppy eagerness. Move off quickly if there is any concern

- Jumping is a natural behavior to garner attention and how dogs will naturally interact with each other. This is fine if the other dogs think it is. Not necessarily a plus for humans. To teach puppies (works with dogs too) not to jump: Greet puppy by name to acknowledge them when you enter a room they are already in or when they enter one you are in. Then ask for a sit. Acknowledge/praise/reward the sit as soon as it happens. Or store toys close to a door or next to their bed and ask them to retrieve one.

- Keep in mind, the most important part of carrying out any requested behavior for any puppy or dog is the beginning of it. Ask with enthusiasm and repeat the millisecond, the response is noted. Always, always praise the fulfilled response, as in marking the performance with "Good Sit!" and follow with a treat and another exclamation. "Thank you" is always a good one. Pets and praise can sub in for treats after initial training or if you don't have one on hand. Make sure the dog has heard you and has time to process the request first. Give them a minute. Literally, the original and actual dog whisperer, Paul Owens, says we should give our dogs TWO minutes to process requests. Start with that minute.

- Be careful not to raise your hands when asking for a sit. This can "invite" a dog to jump up to them by your actions, exactly what you are not asking for with your words. You can also stand if seated if you see the puppy getting ready to jump and ask for that sit. The movement can focus the dog and pause the forward movement. Whether standing or sitting, the instant all four feet are on the floor turn and praise. This must be repeated over and over.

- If jumping is already established remember unlearning the behavior will take more one trial, especially one which she took so long to learn and practice.

Puppies need sleep, a lot of it, for growth and development. Make sure every room has resting spaces for napping or longer sleeps. And make sure those spots are conducive to rest, cut down on ambient noise and provide the best doggy beds.

Toys and chews are needed, they allow for naturally satisfying dog behaviors to be indulged. Developing a chew habit of "legal," as in what you want them to chew on, toys is the very best way to keep your puppy busy with meaningful and necessary dog behaviors. For puppy teething, think multiple chew toys and puzzle feeders to add to puppy toy chest and to help ease the pain of new teeth. Consider time and activity budgets here. You cannot always play with or pay attention to your puppy. Giving them a large variety of toys to snuggle, bite, chew and play with at all times, provides different outlets.

Ditch the food bowls for the right sort of puzzle feeders. No slow food bowls that frustrate. Pick the right puzzle feeder that you can fill with kibble mixed with wet food that offers your baby dog the opportunity to methodically chew their meal out, a soothing and time-consuming way to spend time. Snuffle mats are another good choice for dry food and treats. These mats work on maximizing the dog's superior sense of smell to locate each piece of kibble or treat and turn snack time or meal time into time well spent on a rewarding multi-sensory activity fest.

Learn the "why's and how's" of puppy behavior and training so you can provide it yourself and understand what trainers and behaviorists are doing right and when they are doing it wrong. There are excellent and well researched training websites, videos, books, and manuals available by good and compassionate trainers and educators. (See references below for links to two free puppy training/behavior books by veterinarian and dog behavior expert, Ian Dunbar.) Along with the good, there is also material using force and punishment, that is dangerous to apply for your dog's welfare and for your personal connection with them. Make sure to take the time to learn the difference.

With force free and science based training classes, learning in a group setting affords social and observational learning and utilizes the model rival method for humans and non-humans. Investigate trainers and settings with great caution and care. Look for formal credentials and ask which methods are practiced and observe. This is an unregulated industry and buzz words and methods utilizing "results," "balance," and "alpha," and "pack leader," etc. training are based on human aggression and punishment and to be avoided to create a safe and trusting relationship with your puppy.

Study after study shows the benefit of using force free, positive methods and the fallout from force and punishment. Your puppy's gentle nature will be impacted seriously by aversive corrections. *Be wary of any environment training your puppy out of sight. Avoid fear and displacement behaviors by focusing always and only on force free, positive training methods.*

We are quick to correct behaviors but often miss the opportunities to reinforce all the good stuff that our dogs and puppies are already doing. This is another component of learning and training often overlooked and one easily done when focused on.

Offer praise constantly for all the good, quiet, or wanted behavior the puppy is doing even when not asked. Tell them what you like when you see it. This so called, "put it on command" is just labeling the behavior you are observing to mark it or acknowledge it and acknowledge/thank them for it. Say "Good Quiet" or "Good Girl" for being just that, "Good Sit" when they are sitting, "Good with me!!" when they walk nicely by your side, etc., etc., etc. Your relationship with your puppy begins with you. Make it a good one.

References

- Dunbar, I., *Before You Get Your Puppy*
https://files.cdn.thinkific.com/file_uploads/5433/attachments/829/875/5e2/BEFORE_You_Get_Your_Puppy_-_Dunbar_Academy.pdf, retrieved January 15, 2024

- Dunbar, I., *After You Get Your Puppy*
https://files.cdn.thinkific.com/file_uploads/5433/attachments/90d/cd
5/5c3/AFTER_You_Get_Your_Puppy_-_Dunbar_Academy.pdf
retrieved January 15, 2024

HOW TO TRAIN A DOG

The other night, as I was sitting in the car finishing a conversation on the phone, as we do these days, a man and his dog passed by not noticing my presence. Things were not going well from the sound of it. Frustration and consternation were apparent in the man's tone, cadence, and pressured insistence on repeating the dog's name over and over in mounting exasperation. After every third name repetition, the volume and tone would increase in angered "Come on!!"

Looking out the car window, I could see the dog, a puppy, sitting on the ground, exploring, apparently fascinated by the grass, the leaves, the wind, anything but taking care of the business of housetraining or learning the mechanics of being on lead. Focusing on how to walk on a leash or how not to be necessarily amazed by the wonders of the new and exciting outside world do not come easily to dogs not used to new situations or environments and certainly not to puppies who are easily entranced and distracted. Not wanting to embarrass anyone, always cautious with the possibility of unseen repercussions with unwanted attention, I said nothing.

The next day I saw the pair again. This time I was not in the car. I asked to pet the puppy, still distracted but overjoyed with pets !!!! After telling the little dog how fantabulous he was, I remarked to the man, on how world class adorable this little guy was, such a great puppy to have! And then on how overwhelmed with processing all the sights, sounds, smells of this unknown universe is for puppies. How just about every single thing, is to be investigated, discovered, and amazed at. Who has time to pee? "I never thought about it that way" the man said to me.

Dogs do not come into our lives knowing what the rules are for living with humans. Instead, they come perfectly and ably equipped to be dogs. To sniff out the world around them, dig and paw at things to further examine, mouth to inspect and play, jump and butt sniff in greeting, all very necessary dog things to do but not necessarily welcome by humans. Until we tell them otherwise. Teaching your dog manners for a human dominated world requires understanding the species you are working with.

The dog has many canine behaviors, preferences and disinclinations that are perfectly typical, acceptable, and necessary in a dog's world. Take jumping up on people, this is dog greeting behavior, an absolutely fine way to say hello for a dog, so too is making the most of the opportunity to grab a bite from the table or the garbage. And pulling is only natural when you are more excited about getting somewhere with four legs which move you faster than people with two legs. These and other natural behaviors are dogs being dogs and not for trying to be in control of their humans.

Dogs being dogs also seemingly includes their affinity towards and attempts to please us. They have adapted gaze alternations and inner eye brow raises to match our own. Such developments and devotion goes a long way towards our believing they might have all the answers for knowing what is right and wrong. They don't. Still, a dog's behavioral needs and how dogs experience the world from their own super powers of hearing and smell to greeting and other social behaviors can be markedly different than our own human experiences and behaviors.

How then, to make this work? How to teach the rules dogs need to be part of human families? How to get a dog not to go ballistic at the doorbell ringing? Not to jump up on those guests coming through the door? Or smell everyone in those inconvenient places? How about keeping a dog from the opportunistic food grabs off the table, counter, or garbage? Or how about walking at your side without pulling to get ahead? Or all the other things, dogs may not come to you knowing about but you would like them to?

Where do you begin? You have already heard all this stuff about how you are supposed to be the one in charge of the dog, the leader, the alpha in the pack. But do you really want to stare down a puppy? Or knee a dog in the chest who is overjoyed to see you? And does all that even make sense to a dog anyway?

It is good to keep in mind that much of the natural behaviors we now see in dogs are behaviors that we have selected for. Think barking to alert us to dangers, or the chasing and stalking seen in working dogs, for instance. While dogs are a social species to begin with, we have also actively recruited (as has been the case with all species initially domesticated), for amiability towards humans. This agreeableness creates a receptive audience to learn from us.

That dogs like us is a wondrous thing, truly it is. Their apparent nonstop eagerness to please, lends itself to making us believe they understand everything we ask of them. Would that it would be so! If only we could just explain to them what we want and then have them do exactly that. But dogs, no matter how much they like us, no matter how co-evolved, do not come to us speaking human language. We teach our infants and children words, next phrases, next sentences and then letters that make up these words, phrases, and sentences, all the while expecting our unschooled dogs, at every age, from every place, to know those human words, phrases, and sentences along with all manners of things they have never been taught or might be able to learn.

How to then teach this dog that never went to school? We know all animals learn. Learning theory tells us this happens in different ways. There is through association; where we rely on the significance of paired events – for instance every day at five am, the dog gets fed. Five in the morning becomes significant for the dog, now that the association of breakfast is paired with it and food is expected. Or operant conditioning; a specific behavior is rewarded, prompting its repetition – you spend the day at the office and get paid for it. You go back each day because payday comes each week. Or through observation or socially; a mother cat brings back live prey to kittens to teach them exactly how to enact the killing bite to dispense with them. Learning also happens through insight; inferences based on thought and past experiences.

Juveniles learn from parental caregivers and conspecifics the basics needed for survival, how to find food, shelter, affiliates. Most dogs we meet have had parental contact and learning interrupted in being removed from their mothers at earlier than optimal stages. Even as maternal caretaking and litter mate socialization has been discontinued, the puppy or dog is innately equipped for survival skills.

Dogs are opportunistic scavengers so little there needs to be taught, aside from what not to eat. The younger the dog, the easier the learning, but learning happens throughout life.

Capitalizing on associative and operant conditioning is the training we are mostly doing with our dogs. Much of the teaching delivered to dogs is through associating actions and concepts with reinforcers so they are repeated or deterred. Rules of nature follow in all behavior that is rewarded will repeat and those behaviors that are punished will be avoided.

Punishment is an effective teacher and one which has serious impacts. Association with the punisher creates fear of its presence and of the dangerous and painful things to be avoided. Punishers can be habituated to, so their magnitude needs to be increased to be effective. Additionally, the stress of punishment can effect displacement behaviors or negative changes in the body. The use of reinforcers that are not forceful, can be argued to be more effective in learning and indisputably in welfare.

To fine tune this with dogs, in a nutshell: we remove punishment, manage the environment, allow for natural behaviors, show them what we want, make sure they know how to do what it is we are asking for, when they get it and even when they almost get it, make sure to mark that they did it either verbally or acoustically, next reinforce it and make sure to make the process, this learning, a meaningful and pleasant one.

Environmental management is necessary to set everyone up for success. Things like, no more uncovered trash cans in the kitchen, food left out in dog reach and knowing where the best place is to begin to train with the least distractions for the both of you.

This same process can be tailored to different species, cats, horses, rats, rabbits, etc., the process of learning is universal while the difference is in the species. Different species have different natural behaviors and value different reinforcers. Lettuce may be a fine reward for a bunny but of little interest to a puppy. We need to know the natural behavior of each species to train them. And just as rewards matter, so does approach. For instance, that happy and excited voice most dogs will respond to is a deterrent to cats. Soft, elevator voices work better with cats.

Learning canine manners in a human world is to a dog's advantage. Something they pick up on when sit becomes an automatic sort of doggy please. Any size dog can benefit. Little dogs can be in greater need of structure just because they are so little and oftentimes their size excuses the need for training. We can ignore the pulling, little jumps, smaller barks, "cute" growls, and snarls because of size. Chances are, you will pass at least one or two whirlwinds of smaller dog and owner on a city street. All that barking (and jumping or pawing) gets the attention the dog is asking for at the moment. The behavior serves them well- being so much smaller than everyone else around them can be scary. And that size means it can be more easily ignored and not as difficult for the humans around them. Bigger dogs may not get so much slack, we don't often see the bigger dog exploding down the street, pulling and barking, with their owner along for the ride. These dogs get some sort of training in responding to human direction in most areas where people live in close quarters.

Let's take a look at how this works with barking. Asking a dog to stop barking denies this now natural behavior and need to notify humans and others in the environment of changes detected. Ignore the alert and the dog continues as the alert has not been effective. A sort of "Did you hear me???," "The stranger is still at the door!," "The other dog is still across the street," The cat is still on the fence," "I am still home and not sure when you are ever coming back!!!," the list goes on and on. For this sort of barking issue, instead of reacting to the bark with your own barking (what might yelling sound like to a dog, hmmm?), first acknowledge the bark.

Use the dog's name and one or two other words to signal an affirmative on the alert being received, such as "Fido, yes I know." Or "Thank you Fido." Now ask for "quiet" or "shush." You can begin to train these commands by offering a high value treat directly in front of a barking dog's nose. The barking will stop, even if it is for a microsecond, because the dog will sniff the treat.

Know that, barking and sniffing cannot happen simultaneously, so offering the treat at that moment rewards the quiet behavior you want (it does not reward the barking). This is what you want to mark or acknowledge. The very instant the pause ensues and barking stops, even for that micro second, offer the treat and say "good quiet" or "good shush" in a happy and excited voice. This noting that exact moment of response, no matter how fleeting, is critical and allows the dog to begin to learn this behavior, in this case: stopping barking, goes along with that word said at the same time. Good to remember the cue, especially it's followed by praise, a treat and, best of all, their happy person.

First ask clearly: Remember, the dog's name is not the request. Get the dog's attention that a request is coming by saying their name and then ask for the behavior. Too often when we work with our dogs, we repeat their name over and over again, in an effort to get them to do something. "Fido! Fido! Fido" never translates to "Fido. Sit." Once you have Fido's attention, ask for what you want and wait for the response. This may not happen instantaneously. Especially when first learning, they need to process what you are asking for. As much as you may believe that your dog knows what you would like him to do, he cannot read minds and needs clear direction to formulate a response and reinforcement.

Next, once you see the response go ahead and let the dog know that what they are doing is exactly what you are asking for, this is where marking the behavior verbally or acoustically comes in. If you asked for a sit and your dog is sitting, say "Good Sit!" repeatedly rather than "Sit! Sit! Sit!" repeatedly. Be careful not to ask again and again for a behavior when it has already been offered. We may be happy with the response but we are hopelessly confusing the request for the response for our dogs. This is a common training error, so thinking ahead on what to do or not to do, helps everybody with clear communication.

Lastly, reinforce it meaningfully both time wise and value wise for the dog. Real life rewards, marked and offered at the moment the behavior happens, are the best insurance that the behaviors will reoccur on request. With dogs, our timing needs to be spot on. We are trying to link the behavior offered and the request in a world where many, many things are happening at the same time.

We have mere seconds to associate the requested behavior and the request. As soon as you see the behavior requested being offered, say something to note and reinforce it. "Good Stay!!" Adding the "Good" as praise to the marker is another layer of positive, useful in teaching all species.

Breaking things down clearly for the dog and for you makes for successful training. Follow that marker/acknowledgement with a reward for added reinforcement. Reinforcing the behaviors you are asking for with good salient-to-your-dog rewards like high value food treats, toys, loving praise, petting and play is the most effective way to teach requests and proof responses to them.

You can amp the reinforcement value for the performed behavior by offering multiple treats in rapid succession. Keep praise always on offer. Once you have a solid response, proof as you go along, always offer a "Good (<u>fill in the dog's name</u>)!" as both marking/acknowledge the behavior and for positive reinforcement.

Verbal praise or smaller sized treats as opposed to bigger ones in training sessions or on the street are good to have as back-up. Smaller bites are easier to handle and the dog will enjoy a little bite just as much as the larger one they seem to inhale. With less to chew on, the dog will be readier sooner for what is coming up next. For dogs that tune out treats in certain environments, look to squeaky toys and tugs as possible reinforcers.

In training and in life, keep acknowledgement, gratitude and treats in your repertoire. Offering a reward on an intermittent basis has the greatest effect, but can be the most frustrating. Not to know when the next recompense will be offered can be stressful. Behaviors may continue to be performed but waiting for when reinforcement will arrive is frustrating so keep them coming. Never forget that praise and life rewards, as in going out for a walk are reinforcing and you can never run out of kind words.

Thank you. Once a behavior has a recognized name, a "thank you" can be just as reinforcing as a "Good (<u>fill in the behavior</u>)!" Another benefit of using "thank you" is the energy shift in how we think about it when we say it. Your dog is offering a behavior at your request not independently. It is beyond helpful to how we view training and our dogs, to recognize that the behavior we are seeing, is not their idea but one willingly given in deference to our asking for it. We all want recognition and thanks for our efforts. Thank you is just as important to dogs as it is to people.

Use your words. When your dog performs the behavior requested, mark it so it can be associated with the request. Some people like using a clicker to mark behaviors. For others a clicker is cumbersome, an additional thing to do in the instant you need to identify the desired behavior. Find what works for you and use it. I can say "Good Sit!" to mark sitting way faster than I can click. That click has always appeared delayed to me, no matter who was doing it. Again and again, with training timing is everything.

We are working with dogs, who will always be able to process more information much faster than we can. Just think of how much more they can smell and hear or even see than compared to humans to begin with. We may be familiar with the dog's super powers of scenting and hearing but that extends to what is seen as well. Dogs have the ability to process visual information up to 80 cycles per second, compared to our 60 cycles per second. Add to the speed of all this processing, the speed required to pair a request (and a symbol when you add in hand signals), a response and the need to impart significance to it. There is a reason you keep hearing that training is a mechanical skill. The more you practice and focus on coordination and timing the better your skill level will become.

Along with verbally marking those behaviors asked for that you want repeated, use your words to build vocabulary. Dogs quickly associate and observe the labels we use every day for the things that are important to them. Walks, dog park, beach, treat, ball, play, Mom, Dad, etc., etc., etc., all have very clear meaning and words we end up spelling out or even miming when they figure out the spelling. I like to channel exuberant energy with greeting behavior into getting a special toy. To know which toy, I would hold up the toy and give it a name. Following John Pilley's method for training the famous Border Collie, Chaser, you call the dog's name for attention with the toy in hand and repeat the name of the toy, do this four or five times. Repeat several times over the course of several days. Name of toy learned. Try it.

Remark and acknowledge the good behaviors, especially the ones you did not ask for. We are so very, very quick to correct or redirect, to call out the unwanted behaviors we see. Take the same speed to praise and thank the dog for all the "good" behaviors that just happen without being asked for. Sitting or lying quietly, walking easily at your side, and more can all be noticed, acknowledged, and thanked. Giving a name to those behaviors and reinforcing for them (verbal praise and thanks counts) is called "putting it on command," perhaps now better termed "putting it on request." This approach to training is often overlooked and in doing so, we leave much on the table. There may be thought that this is somehow a less effective approach in training but all the criteria exist for associative learning along with dialogue, confidence and relationship building.

Allow for response time. The dog is learning. Permit some time for them to figure it out or to switch gears if they are otherwise engaged. They need to think about what you have asked for and process the request into a response. A dog that is standing needs time to move into a sit or a down, while larger and older dogs may need even more time. The need for perfect timing is on the human side of the equation here not the dog who is already working on it.

Work up to more distracting environments. You may get quicker responses and consistent results each time you ask for a behavior in your kitchen and living room with requisite praise and food treats and radio silence in the park. This is perfectly natural. You do not have squirrels, other dogs and the like at home to compete for your dog's focus. Expect that an environment that has much to interest a dog will demand divided attention and up your engagement level, rewards, and more manageable areas of the environment to practice training in.

To address, work on those behaviors outdoors that you already have a solid response to indoors. Once you see the response outdoors, praise handsomely and/or treat repeatedly. It is so much better to have your dog perform one sit behavior outside than have you ask repeatedly for another behavior that they have not mastered and that is not forthcoming.

Keep in mind that going backwards to what was last working is the way to go forwards. If your dog knows how to sit outside or touch outside but will not go into a down when asked, request the sit or the touch. Sometimes, asking for the comfortable behaviors can lead right into the more difficult ones next up.

Short, frequent training sessions are key and have been found to be the most effective in training and reducing anxiety. Training is an opportunity to dialogue and focus on our dogs. We have not enrolled them in military school. Such a dialogue should be force free and add to the relationship not detract from it. Some of the benefits of working with force free training and behavior is how it encourages positivity in actions, consideration, and care for environmental impacts, encourages and allows us to engage with our pets without force, focusing on the good we see, lovingly.

Keeping training drills on the short side, keeps them fun for both parties. Do minimal repetitions when asking for a behavior, minimal as in three to five and not more than that. Over repetition is stressful and counteracts the added benefits of training. Incorporate games into training exercises, with frequent breaks for down time, sniffing around and or play.

Even highly schooled bomb sniffing dogs practice in short training segments every day to keep sharp and keep it fun. The pet care technician program I taught was right next to the New York Stock Exchange. Elite explosive sniffing dogs work with handlers in this area. Field exercises with my students often included observing these canine human teams working vehicle check points approaching the exchange. We were fortunate in being able to speak with some of the handlers and interact with the dogs, hearing about ongoing training for these 'sniffer" dogs. Target scents are hidden 15 to 20 times on each dog's shift to keep the dog sharp and yes, the dogs get rewarded for finding the targets.

Do not initially expect your dog to know what you would like them to do, show them. For example: "Sit" is great for focusing your dog, becomes a canine "please" (you'll see), works on impulse control and keeps them in one spot. Yet, your dog does not know that the word "sit" means put your butt on the floor.

Sit: With a standing dog, work on teaching sit by getting your dog's attention first. Call their name to get the focus on you first. Next, saying "sit" use a treat to lure your dog into position. Do this by holding the treat closely up to their nose and slowly pull it back over the head, keeping the treat in the same close to the head position. As the dog moves his head to follow the treat you are holding over his nose his hind quarters will drop on to the floor in a sitting position. As soon as the dog is sitting mark the behavior with a "Good Sit!" while giving the treat at the same time. If the dog jumps up, you are holding the treat too high and luring the dog into a jump as he attempts to follow the treat.

This happens all the time when training sit, the human pulls their hand up as the dog tracks the treat. Most of the time we are not even aware we are doing it. You may also lure a dog into a jump unintentionally if you jerk your hand upwards once the dog sits rather than releasing the treat. Treating the dog in these scenarios is fine since the dog is following the lure, just remember to only mark the requested behavior with the word "Sit" once you get it.

Avoid manually placing your dog into a requested position. This is confusing for all dogs and pushing or putting pressure on anybody is uncomfortable and painful for those dogs with hip and back issues. Luring your dog into place with a food treat gets the dog to accomplish the movement on his own individually. Giving a behavior independent of being physically manipulated into it, is necessary to practice, proof, and repeat it independently.

Always remember the leash is not a directional tool, you are. Tugging on a leash to get a dog to move does not teach the dog what you want. Pulling on a collar, head halter (best not used at all) or harness can be painful, uncomfortable, and risks damaging anatomical structures in the head, neck, shoulders, and upper abdomen. We often excuse our own pulling as a response to being pulled by the dog. Such force in return, can in fact increase pulling as the dog has to brace against the force we are now exerting to stay upright and not be lifted off their front legs. Being aware of what is in the environment your dog may want to get to for sniffing and or eliminating and communicating with your dog along the way of requesting a direction or behavior is always better.

Start with their name for attention, paired with a request using either a word or a gesture pointing the way to go. Talk to your dog, your voice engages them with you and your presence. No one will mind except for your dog. You want to train loose leash walking while allowing for necessary sniffing on walks. Leave the phone in your pocket and take the time to look around and at your dog while your dog gets to sniff around.

When asking for "come" it is vital to remember that this is a request best trained in a happy and excited tone and never a punishing one. Never use this request to place your dog in an area of confinement such as a closed room, crate, or kennel. For those times you do need to confine a dog, use instead a specific request as in "go to your crate."

To train "come" you need to be an excited, happy force to return to, without punishment. Seriously, you want your dog to want to come to you and not be afraid of what you will do when he gets there. Clapping your hands softly, call the dog's name and ask for "come," repeatedly all in that jolly training voice. Do this either while dropping down to a crouch or take a few steps backwards (make sure you can do this safely) away from the dog. You can also pat the floor alongside of you as a focus on a place or target.

The very most Important part of any request, especially a recall request, is the beginning and the gain in momentum. As soon as the dog begins to approach, encourage, and reinforce mightily repeating "Good Come!!!" and offer a reward that microsecond your dog arrives.

Once the recall is solid, add in asking for a sit once a dog has returned to you by luring with your treat reward. It is also is a nice way to keep him with you. Release your dog with a "Go Play!" immediately as a further reward and to further emphasize all the wonderful things that happen when you learn and perform the "Come" behavior. Repeat. With a partner, you can call and release the dog to each other. Two or three (not more) repetitions of this are good to practice at the dog park or beach. Where there is hesitancy in a response, you may be too far away. Close the distance between the dog and you making the recall distance shorter and ask again. Is the dog still hesitant? Get closer.

Teaching counterpart commands back-to-back is particularly effective in training such as "off" and "take it" and "come" and "go play." The requests reinforce each other. Remember to reward enthusiastically for a response and to avoid reprimands especially once a behavior is performed. If a dog hesitates on a recall but responds to a repeated request never scold the dog for not coming quickly enough. Worth repeating here, the first part of the recall is where you need to build forward motion, so employ happy energy, sufficient distance, and repeated requests. No scolding. Scolding only serves to punish the dog for finally listening to the request. Praise and treats are in order instead. And more practice and more releases.

Never forget that lectures work (sometimes) with people and never with dogs, as much as you would like to go into all the reasons why barking at or lunging at other dogs is wrong, your dog simply cannot follow the argument. All that your dog will know is they stopped barking or lunging way back before you started on the lecture making the lecture an unrelated event. Way better to pair your short correction with the lunge.

As to corrections, try for an "Off!" or a "Now!" or a sharp syllable here in lieu of a "No!!!" which tends to be too heavily loaded with negative sentiment. Always mark the response to the correction with recognition and/or "thank you's."

Correction requires redirection too. The volume of your tone serves as a correction and the single syllable keeps it from being overly punishing. When the dog responds with the requested behavior, reinforce and reward to keep this behavior occurring on request. Always, always follow with a redirection – what to do next.

To change a behavior, you need to offer an alternate behavior that is just as good if not better than the one you are seeking to change. Give them something else to do instead, hopefully something just as good from the dog point of view.

Be proactive in seeing when a scenario is ripe for a break from the behavior being requested and a redirection is in order. If the doorbell rang but the guest has not yet entered your home, the goal is asking for/getting the quiet from the onset. Step one. Next give your dog something else to do—redirection—such as "sit" or "go get your ball" or "go to your bed." Or even better, have a food stuffed chew toy to occupy your dog as the something else to do. Make sure to reward, mark and acknowledge that redirection and performance too.

Train for the needed and the fun next. Teaching your dog how to roll over or shake paws or play dead can be a fun and bonding experience for the both of you. So, if you can make it fun and are both willing to learn how, go ahead. The basics that you want to incorporate for safety and structure are those requests that will keep your dog with you, away from something or coming back to you. Work on fundamentals first like "sit," "stay," "leave it," or "off," and a recall command such as "come," crucial to keep your dog coming back to you whether at home, the park or on the street corner.

Buy at least one good dog training manual and read it. This is necessary for you and your dog even if you have taken or are considering training classes. Research the best material out there on dog behavior, body language and those offering positive reinforcement and force free methods and read ahead. Even without the classes, books are a must to get you on the right track training wise. *Easy Peasey Doggy Squeezy* by Steve Mann is first-rate or *How To Teach A New Dog Old Tricks* by Dr. Ian Dunbar is fabulous for all dogs (Dunbar gives great insight into behavior, the only thing to leave out are the outdated leash pops) and *Do Over Dogs: Give Your Dog a Second Chance for a First Class Life* by Pat Miller is excellent for rehomed dogs or those in need of remedial work.

It is tremendously helpful to study methods and techniques for training your dog by learning from people who are highly qualified experts, people who teach dogs without harming them or forcing them or using aversive strategies to compel them to perform. There is a wealth of freely accessible information to be found online by searching for the right names and buzz words. Professionals such as Dr. Ian Dunbar, Dr. Sophia Yin, Dr. Patricia McConnell, Steve Mann, Suzanne Clothier, Karen Pryor, Pat Miller, and Victoria Stillwell are just some of the authorities you can find on the web, have written books worth reading, appeared on television shows, hosted webinars and seminars, all of which offer valuable instruction on dog training and learning theory. For enlisting professionals to train or modify behavior, good to know that only one trade organization mandates members not use force – The Pet Professional Guild, their members can be found on the PPG website online grouped by region.

Unfortunately, there are also way too many examples of how not to train your dog as well. In our unregulated industry and in our own sometimes-misguided quest to get likes and viral hits, we find too many numerous examples of misguided dog training rampant on social media, in countless books, articles, websites, videos and television shows. Well-meaning dog owners new and old do the right thing and read up and research what is best for their dogs and often run into the wrong information and apply it. With the best intentions and looking for guidance, an owner may be told here to scruff or shake a dog into submission, roll dogs over, pin them down, knee them in the chest, grab their muzzles, strap electric shock collars on them and shock them, restrain their movement and constrict their breathing with the painful pressure of choke chains, prong collars and head halters and more all in the name of "training." This is compulsion, force and torture not training.

Some of these dominance-based trainers call what they do "balanced" or "results based." It is certainly true that you can force any animal to do anything with a strong enough aversive. When the motivation is to avoid punishment, you do get "results." You also get fall-out, stress and trauma along with those results. Anyone can say that alternating force with reinforcement is a sort of balance. Trainers who teach with punishment and force and the clients who retain them, no doubt believe their methods are the most effective ones.

Culturally, we may view avoiding punishment as an indulgence, a "spare the rod, spoil the child" approach. Add to this, beliefs that what kindness and considerations people are entitled to should not be extended to dogs. Other beliefs may center on a view, that dogs or certain breeds of dogs, merit being trained with a heavy hand. Or there are those thoughts, that this is how we have always done it, no matter the species. Extend the thinking to horses, for instance, who have been traditionally trained with force and aversives, despite knowing the behaviors and nature of horses, and knowing force free training can be an even more humane and effective way to work with them, this continues. Movements exist to change training methods in this arena but are a long way from being mainstreamed.

What may not have been considered thoughtfully enough, is the whole picture. Along with the results you do get when training with force; research, and observation show, you also get an animal that is mostly afraid of the handler. Or if you are extraordinarily able with timing, you get an animal that is afraid of the handler and the punishment the handler gives, who is stressed even by the appearance of the handler and training and performing requested behaviors.

Watch compulsion training, all that lip licking, yawning, blinking and looking away is appeasement or "please stop" there is no such thing for a dog as "calm submissive." Along with "results" you will also need to increase your punishment to get the same response as the effect is short lived as the animal habituates to the pain or accepts the pain as the price to pay for a behavior they feel the need to perform. And then there will be those displacement behaviors or anxieties which will start to appear in other areas like breaks in house training or displays of fearful aggression.

The America" Veterinary Society of Animal Behavior (AVSAB) notes that:

> "Even when punishment seems mild, in order to be effective, it often must elicit a strong fear response, and this fear response can generalize to things that sound or look similar to the punishment. Punishment has also been shown to elicit aggressive behavior in many species of animals. Thus, using punishment can put the person administering it or any person near the animal at risk of being bitten or attacked."

As to the dominance theory those result based trainers usually espouse, the AVSAB says:

> "recommends that veterinarians not refer clients to trainers or behavior consultants who coach and advocate dominance hierarchy theory and the subsequent confrontational training that follows from it."

Countless scientific studies support the effectiveness of positive dog training compared to compulsion training or the use of aversive training methods. Along with the AVSAB, myriad organizations which promote animal welfare, "positive" trainers and behaviorists support this position. Yet compulsion dog training continues and oftentimes the participants and viewers are unaware that what they are doing and watching is cruel, inhumane, and just plain wrong. You get to do it the right way.

References

- AVSAB Position Statement The Use of Punishment for Behavior Modification in Animals. http://avsabonline.org/uploads/position statements/Combined Punishment Statements 1-25-13. Pdf Retrieved December 1, 2013

- AVSAB Position Statement on the Use of Dominance Theory in Behavior Modification of Animals. http://avsabonline.org/uploads/position statements/dominance statement.pdf Retrieved December 1, 2013

TO CLICK OR NOT TO CLICK, THE HOW-TO TRAIN QUESTION

I am a believer in using my words. This counts for training too, with verbal markers instead of clickers. Seems offering a robotic signal in response to something the animal is just learning or not sure of, to begin with, is not fair or kind, especially when that animal is cooperating on a supposedly mutual goal or learning progression. Feedback best follows on more than one level. I also find clickers an awkward accessory to coordinate and manipulate in training. Using visual observation and acoustic markers, as in with eyes and voice, are way easier for me and make for a more elegant and fluid delivery for the dog.

In training and behavior modification, marking specific behavior and the timing of reward delivery are crucial and need to happen in less than mere seconds (studies show that more than three seconds are a missed opportunity). In practice, it soon becomes apparent that being able to successfully time marking the requested behavior at the exact moment it happens, is a skill that requires experience, attention, and practice. Once that skill is achieved, for myself and for most of the trainers I have observed, it is way more precise when marked verbally, where vision and utterance can be simultaneous compared to the extra time needed to manipulate a device or clicker. And, despite theoretical differences, surely (Clever Hans aside), we know that it is nearly impossible with the range of communication in body language and emotion to ever be neutral around an animal much less uninformative. Verbal markers punctuate the points we are making with the rest of our bodies.

But where does this all leave us with the best way to train a dog? Apart from when or if to use a clicker, what about the rest of it? Do you use positive reinforcement? Or be a pack leader? Or, ask your dog to imitate you? Seriously, social learning is the latest thing.

Ask three different people and chances are you get three different answers. Ask three different dog trainers and it probably gets even more confusing. What if you don't want your dog to work only for food? Isn't that bribery? Why should anyone work for free anyway? Don't pack leaders make sense to wolves and not dogs? And just how is a dog supposed to imitate you anyway?

Debates on the ideal method to teach a dog those manners for a human world range from which of the latest theories and techniques to use, with the largest divide falling on whether to employ force or not.

There is, of course, science on what works best. There is a lot of science, the latest research continues to show force free, positive training methods to be the most effective and humane. Despite or in spite of, the evidence, force, and punishment are still chosen by some for immediate, if not lasting, "results" and "balance." Or because dogs- at least to certain minds, being "just dogs" do not merit more considerate training styles. How force fits into learning is important to distinguish in any discussion on training or behavior modification of any kind. Again and again, science will tell us, the most effective methods are force free. Yes, punishment can produce immediate results. It will also destroy trust, give rise to fearful and displacement behaviors, and need to be increased in intensity to prevent habituation.

Those drawn to humane methods may find even that formula, may not narrow it down much. The latest protocols in positive dog training are each one supposedly better than the next in terms of effectiveness, whether it be using words as opposed to clicks, or silence, or body language or eye contact or none at all.

Who endorses what protocol can be even more important. Users respond to trainers and behaviorists more on personal magnetism than scientific credibility. As scientist and behavior expert, James Serpell says: "We're much more impressed with charismatic media figures than scientists who are thoughtful and methodical."

The business of dog training is booming for those wanting to find services, thinking to be, or developing a dog training business, or simply wanting to be part of an "in" crowd of the method and trainer of the moment. Cesar Milan, and other compulsion-based trainers' much disputed dominance theories and inhumane applications may have set dog training back 50 years but there is no denying their ardent supporters embracing force-based training techniques in group or private classes, board and trains, webinars, seminars, or their own respective media successes. No denying that a good deal of people are buying into what they are selling.

And the merch: The gadgets to use, and buy. Each method comes with the "best and only," equipment - whether clickers, collars, halters, harnesses, leashes, shirts, wraps, brushes, buckets, automatic treat dispensers, remote sensor, toys and more. Settle on what sort of training you should be doing and the devil in the details of who to follow and what you should be using to do it, guarantees complications. It can feel as if only with the latest and greatest gear, can we do the best job at training. But gadgets and personas do not train dogs, people do. Or do they?

Most dog trainers have definite feelings on methods used, including how to mark requested behaviors. Details, technique, and finesse matter. Hugely popular amongst trainers is "clicker training" - manipulating a device to make a sound to "mark" the moment when a dog performs a requested behavior followed by a reward to reinforce it. The popularity of using a clicker to train is on the rise with videos, webinars, online training academies and even conventions devoted to their use. It was reported that more than 1400 participants attended "Clicker Expo" in 2016, a number that grew steadily over the years, even into the Pandemic, when the event went virtual. In 2020, Clicker Expo had over 70 presentations and workshops to choose from.

Clickers or other event markers of behavior are believed to function as a link, marking an event acoustically with the reinforcement to follow ("bridging stimulus" or "secondary reinforcer"). To fully appreciate how the device fits into dog training, it is useful to look at the origin of acoustic signals in laboratory studies: scientific research looking at stimulus and response in associative learning relied on automatic signals to mark an event or to signal when a reward would arrive to reinforce a desired behavior. These experiments were designed specifically to eliminate any human interaction to muddy the waters of what an animal was responding to. A neutral signal was needed.

Animal trainers working out of sight, distance or in different modalities with their subjects, also found the devices useful. Dolphins may be hard pressed to hear our voices underwater but whistles and clickers carry. Clicker trainer founder, Karen Pryor was a former marine mammal trainer. However, out of the laboratories and marine parks and next to our dogs, in the very real world of how dogs learn with owners, trainers and environments, we find little, if anything, that is neutral or robotic.

Surprisingly, or not, depending on experiences, force free trainers can reserve force free for dogs and not for humans. We can reserve our feeling of empathy for the animals we work with and skip the people. As tempting as it can be to pass over the person in consideration, we need to consider how this thinking affects not just our interpersonal relationships but the short shrift we are giving to the attention paid by the animal subjects and clients in front of us. Schisms in how we train whether with emotion in tone or to click or not to click and how, can be a not so positive divide in the world of positive dog training. We can all appreciate that training, good welfare-based training, is necessary but we are divided on how to use a verbal marker or accepting that these markers are multipurpose in marking an event, informing, and communicating.

Clicker confusion for the animals can be a concern as well. In most group dog training classes, training is often done without sufficient spacing to make each click distinctive for each animal. The dogs' vigilance in watching every move we make before handing over a treat, makes up for this. In treat dispensing we perform multiple preliminary movements before the treat is dispensed. Where is the actual marker then? If we add in imprecise timing to the mix, and focus only on sound, an animal is more likely to associate the most recent click in time no matter who it is coming from. The discussion is further complicated when noting that studies on clicker training routinely examine one individual animal with one trainer.

When it comes to how animals learn, there is no one way. All animals learn in a variety of ways though trial and error, by association, through insight and socially. Most of conventional dog training whether utilizing force or reward relies on associative learning. Dogs are also wonderful social learners, learning ably from both other dogs and humans. Part and parcel of that social and associative learning is how the way we interact with dogs impacts on how well they learn and perform for us.

When it comes to relationships, a study by Drs. Jamesion, Baxter and Murray examining the nature of interactions with working dogs and their handlers, found that rates of accurate results in detection dogs rose with familiar handlers and positive relationships. Who we are with matters. Connections with our animals are so significant and prized that setting up a rival for our attention is a particularly effective training method using social learning. Alex, the famous parrot, was taught via the "Model/ Rival Method" wherein a human rival for his owner's attention would perform/model a requested behavior for Alex to perform.

Social or observational learning can also be used to teach an "imitation rule" for dogs to follow with our own selves as the model. Scientists Claudia Fugazza and Adam Miklosi published their "Do as I do" method of training in a 2015 paper. The technique, consists of first gaining a dog's attention using eye contact and spoken language. Next the trainer demonstrating the behavior requested. Followed by giving a "Do It!" command (repeating these steps as needed). And then marking/thanking/informing with praise or food or petting as a reward was found to be:

> "more efficient than shaping/clicker training for teaching dogs complex object- related tasks and goal directed sequences of actions." (Fugazza, Miklosi. 2015)

Just who is doing the teaching counts too. In a more recent 2018 study, the researchers showed that puppies learn most ably from humans, as opposed to their mothers, who make eye contact and speak to the dogs before demonstrating a new behavior to copy and from watching unknown puppies perform behaviors. No matter the species, watching someone new is apparently more interesting than listening to mom.

If you live with a dog, you already know just how much our dogs learn through simple observation. We are their study subjects. There are good reasons we are all spelling out words like "walk," "car," "dog park," and "beach" around the dog. And if you have more than one dog at home, you have constant evidence of what they are learning from each other. The dogs have learned not just rules and requests, they have also learned what certain words represent. What about what hundreds and hundreds of words represent?

Then there's Chaser, the famed border collie who John Pilley taught over 1,0000 words and who captured media attention. Deservedly so, Pilley and Reid's paper on Chaser outlined his method of training:

> "We taught Chaser one or two proper-noun names per day. Several trainers taught Chaser using the same procedures. All trainers were consistent in their use of the correct proper nouns because the name of each object was written on the object. Most of the training took place in our home and front and back yards. Each time we gave Chaser a new name to learn, we held and pointed to the object to be associated with the name and always said, "Chaser, this is . Pop hide . Chaser find ." (Pilley & Reid, 2011.)

I spoke with Dr. Pilley in a phone interview, shortly after the paper came out. Pilley told me Chaser's appetite for learning was inexhaustible, at times more than her human teachers could respond to. When I asked if she also felt that way about the repeated trials to support the research paper, he conceded that she might have not been as enthusiastic about those but wanted always to please her people. Dogs.

So, if we can just show dogs what we want, why continue to train them any other way? We may have never given it much thought, be the most comfortable doing something the way we have always done it, the way everyone else is doing it or the way the latest celebrity trainer says we should be doing it. We can also believe that because we are working with dogs as opposed to humans, different dictates apply.

How does the science rate efficacy of clicker over verbal over food? When it comes to comparing associative learning methods, recent studies have shown no difference between clicker training, using a verbal marker or training with food alone and even better results with social learning methods. Research has also raised significant questions about how we use verbal markers to communicate and how the lack of a relationship between the dog and an often, unknown trainer paired with the dog used in experiments can impact study results.

These questions beg us to consider taking mainstream dog training out of the perhaps antiquated world of operant conditioning and learning by association and relying more on the way dogs and people are already learning more effectively -socially. So, why is it taking us so long to get with the program? Why are there more people going to conventions on the latest in something like clicker training and zero conventions on social learning? Where is the hook?

When asked what the benefits of clicker training are, hundreds of respondents familiar with the method completed an online survey listing their belief that it was an added incentive for dogs being trained along with enhancing performance. They also referred to the clicker as a form of communication.

The survey results on pros of clicker training, published in 2018, by researchers at La Trobe University showed 586 surveys were completed with 92.3% of the respondents being female and 6.3 % being male. One of many data points showing an overwhelming majority of women are using force free methods or completing surveys and attending workshops. Is this an indication that culturally, force is preferred in male dominated industries such as military and police? Their training methods with humans and dogs would support the question. This demographic speaks volumes as to gender preferences for training styles. Positive dog training appears to resound more with women as opposed to men.

It doesn't have to be a clicker; it came out in the answers that survey takers were not necessarily attached to their clickers. The practice is more about the acknowledgement/marking/grabbing the behavior as it were rather than the mode used to do it. When asked how to characterize clicker training, participants defined it more as a method, rather than an attachment to a mechanism, agreeing:

> "clicker training refers to training that uses a mechanical clicker, but many also included training that uses a verbal marker" with "'verbal "yes," verbal "good," whistle, mouth

click, finger snap, as being equally effective."' (Feng, Howell, 2018.)

Even with the consensus that markers can range in acoustics from the spoken to the click, there remains no small disagreement amongst trainers or scientists about which is better or why. Including the hundreds of trainers in the survey study, markers are thought to be more than just a signal for an event by offering feedback to the dog.

To test whether words, sounds or no marker would make a difference in which was used, a 2016 study by Chiandetti and Avella compared training by unfamiliar to the dog trainers with, a clicker and a treat, a spoken "Bravo" and a treat, and a treat only. For this experiment, dogs were trained to open a bread box and to generalize the task to an object that had the same function but appeared different. The researchers found no significant differences in any of the treatments. They noted:

> "learning seems to be independent from the type of sound anticipating the food reward, and even more strikingly, it seems to be equivalent either with or without the clicker sound or the word "Bravo." (Chiandetti & Avella, 2016.)

Even tone of voice counts. It's all about the neutral. Right? Well, that just might depend. In the study, experimenters held to saying the word "Bravo" in a flat and uniform tone. But how might these results be different if the experimenters spoke with the sort of inflection and emotion that we use in everyday training life? Especially when we share by manner and tone how pleased we are (or not) with a response. The scientists concluded on this very point, with a thought worthy of much elaboration and consideration (emphasis added):

> "The fact that dogs pay high attention to other human cues besides the rewarding ones, as for instance those communicative signals shown with ostensive communication, makes social forms of learning more effective than learning based on clicker training, as recently demonstrated by Fugazza and Miklósi. **In this sense, we can expect that an enthusiastic regulation of the trainer's tone of voice might**

modulate the efficacy of the learning. Indeed, dogs (and horses) respond congruently to verbal commands as human infants do. In our experiment, the word "Bravo" was pronounced in a neutral and consistent way across trials, thus resembling more the automatic click-clack of the clicker than an enthusiastic trainer. **A further investigation should consider the possibility that the melodic contour of the trainer's voice could improve the dogs' learning.**" (Chiandetti & Avella, 2016.)

What then might be the advantage of clicker use? Proponents of clicker training, claim multiple benefits of the method, including those dogs that are clicker trained, are trained faster and acquire complex behaviors more efficiently. While, those who eschew clickers, maintain that clicker trained dogs are more excitable and impulsive.

Testing such assertions, yet another study, done later in 2018, by Feng, Hodgens, Woodhead, et al., compared dog owners training their own dogs using either clicker training (clicker plus food) or food only training and found no specific advantages or drawbacks with either method in terms of dog owner relationship or impacts on the dog's impulsivity or problem training skills. Still and all, certain pros and cons of clicker training surfaced. Owners were followed over a six-week training course and while they did report difficulty with the clicker method initially, they also reported that it had a benefit when teaching a behavior where the dog was not in eye contact with the owner (nose touching a cone). Makes sense, if the dog cannot see you, a distinct and precise sound can certainly help. The authors surmised that the:

> "study provides the first evidence that clicker training may make certain tricks less challenging to train, but also that it may not produce benefits as greatly as previously reported." (Feng, Hodgens, 2018.)

Here, we can be reminded of the reasoning behind the original application of clickers or automatic sound use in lab and distance training, with distance being paramount. Of further interest, would be how verbal markers would compare here.

Clickers in training may not outweigh other markers or make for a superior training method but again, this does not mean that aspects of its use do not have advantages for training in certain environments or with certain species or disadvantages. Clickers, acting as a secondary reinforcer, are often used to signal the imminent arrival of a food reward which can keep an animal performing a behavior, even when the reward does not follow.

Studies find animals continue to offer cued behaviors for an extended period of time without being rewarded. These tests, to determine how long before a behavior is "extinguished," are surely stressful and frustrating for the animal subject to them. Conversely, when the signal predicting the reinforcer is faithfully followed by a reward, the animal can experience a sense of control in the pattern. That association can lend comfort to taxing environments.

Pairing a neutral reliable signal to predict rewards can enhance welfare in certain situations. Captive and domestic animals deprived of choice and control, are routinely stressed by changes in schedule and routine and shelter animals even more so. Cats in shelters have been found to exhibit signs of extreme stress in their reluctance to eat or interact socially and in increased hiding behaviors. Cats do not do well in new or changed environments and even less so in shelters, being extremely territorial and most comfortable in their own home territory. Deprived of familiar environments and slow to acclimate to new situations and people, certain cats may take comfort in the neutral aspect of a device announcing a reward, as opposed to an unknown and untrusted human.

Research done in 2017 by Kogan, Kolus, et al. looked at how training shelter cats with clickers (verbal markers were included in the definition of clicker training), might reduce stress and increase welfare along with adoptability. The cats were successfully clicker trained to offer a variety of behaviors such as target, spin, sit and high five. The cats aced it. Such favorable results prompted the authors to remark:

> "this type of training allows for predictable interactions thereby increasing an animal's sense of control and the predictability of it's environment, and as a result, their well-being and welfare." (Kogan, Kolus, et al. 2017.)

In the end, clicker training's popularity begins and ends with the people using it. As popular as it may be, it is not what is teaching our dogs. Dogs are excellent learners whether we use a clicker or a verbal marker, food alone or show them what we would like them to do. Dogs learn from us all the time, whether we spell the words out or not.

There is a story told in dog training circles of a trainer demonstration on how to teach a dog not to pull on leash. The trainer demonstrates a "red light, green light" technique in which the moment the dog begins to pull the trainer stops moving and the very second the dog stops pulling, the trainer moves forward. This exercise is one the most challenging to perform for people because the timing must be exquisitely coordinated to effectively teach by association.

Precisely stopping with moving forward when the leash is being pulled creates an association of pulling and not moving ahead, while instantaneously moving forward when the pulling stops, creates an association of loose leash and forward motion. Remember, there are one or two seconds to pair these events and teach this. Applying timing and executing a response to the red lights or green lights are hardest for the people doing the exercise and not the dog. Try it.

The story goes that the trainer demonstrating the technique worked holding a leash with several knots in its length. He liked how the knots gave him a better grip on the leash and for no other reason. After explaining how timing, skill and response worked to teach a dog not to pull, the trainer was asked only one question by his audience: "Where could they buy that special leash to stop the pulling?"

It's not the leash or the clicker but how we say the words we use, the training, teaching, and communicating, that stops the pulling.

References:

- Chiandetti, C., Avella, S., Fongaro, E., Cerri, F., (2016) Can clicker training facilitate conditioning in dogs? *Applied Animal Behaviour Science*, 184, 109-116.

- Feng, L.C., Howell, T.J., Bennett, P.C., (2018) Practices and perceptions of clicker use in dog training: A survey-based investigation of dog owners and industry professionals. *Journal of Veterinary Behavior*, 23, 1-9.

- Feng, L.C., Hodgens, N. H., Woodhead, J.K., Howell, T.K., Bennett, P.C., (2018) Is clicker training (clicker + food) better than food-only training for novice companion dogs and their owners? *Applied Animal Behaviour Science*, 204, 81-93.

- Fugazza, C., Miklosi, A., (2015) Social learning in dog training: the effectiveness of the Do as I do method compared to shaping/clicker training. *Applied Animal Behaviour Science*. 171, 146-151.

- Fugazza, C., Moesta, A., Pogany, A., Miklosi, A., (2018) Social learning from conspecifics and humans in dog puppies. *Scientific Reports*, 8, 9257

- Jamesion, T.J., Baxter, G.S., Murray, P.J., (2018) You Are Not My Handler! Impact of Changing Handlers on Dogs' Behaviours and Detection Performance. *Animals*. 98(10).

- Kogan, L., Kolus, C., Schoenfeld-Tacher, R. (2017) Assessment of Clicker Training for Shelter Cats. *Animals*, 7, 73.

- Pilley, J. W., & Reid, A. K. (2011). Border collie comprehends object names as verbal referents. *Behavioural Processes, 86(2), 184–195*. doi:10.1016/j.beproc.2010.11.007

- Serpell, J. (2017) In *Malignant Behavior: The Cesar Millan Effect*. Sandgrain Films. https://vimeo.com/243498663 retrieved 11/5/2023

HOUSETRAINING 101 FOR PUPPIES AND DOGS

Back when I was housetraining my Daisy, I slept with her leashed on the floor next to me. The other end of the leash was wrapped around my hand or wrist. The idea being, when the dog would move in the night to seek relief, you would be woken and would take them out directly to eliminate. It worked. Lucky for us, we lived at the time in a house with a backyard so we did not have far to go.

Variations on the strategy for daytime use include, attaching the leash to a belt loop with a larger puppy or dog or holding the small dog in your lap or arms, especially after returning from a walk where they have not eliminated and noting all the precursor movement behaviors of needing to eliminate. All variations require taking the dog out immediately when they give signs (see below) they need to go. All housetraining requires frequent walks and enthusiastic reinforcement when the dog does eliminate where you would like them to. That is the nutshell version.

Not using the outdoor facilities is a frequent puppy happening with all the sidetracking that comes along with their fascination for the outside world or from a puppy being exclusively trained to use puppy pads indoors. And then there is what happens when breaks in house training happen.

First, make sure to review and put in the place the basics of house training:

- As early as possible on awakening, the puppy or dog should be taken to the preferred area (outside) to eliminate. Bear in mind that no matter our own preferences or schedules, elimination is a biological necessity. Limiting this for human convenience can have severe wellness and welfare consequences for animals. This may mean an earlier walk than we might like or training to use a wee wee pad in the home if a walk is not possible.

- For dogs not sleeping in the same room or even when kept in the same room, a remote camera can be helpful in determining when the dog wakes to adjust timing or routine walk time.

- Walks should be given within 15-20 minutes after eating.

- Allow for maximum sniffing on walks. Sniffing is relaxing and picking up another dog's pee mail is inspiring. Dogs are known to "overmark" or urinate over another dog's urine. Let them.

- Walks should be given after playtime. Exercise can prompt elimination.

- Four routine walks a day are ideal, first thing in the morning, midday, late afternoon and before bed. After eating, first thing in the morning, last thing at night, after excitement, are some of the times puppies really need to go. Five months is when puppy bladder and rectums are most fully developed to control eliminations and when a regular schedule of routine walks typically takes hold for full house training.

- At every instance of elimination in a preferred location, as in outside or wee pad inside, the dog should be rewarded with always effusive praise and even better, praise and treats but always, always, praise. Time the praise and treats to the microsecond the dog has finished and not as they start. We want to associate the reward with completing the behavior and not for just beginning the pee, so they do not learn to "ration" out the elimination for rewards. (I have seen some smart dogs who eliminate in segments and wait for rewards before completion.)

- Breaks in housetraining can stem most commonly from lack of an appropriate place to eliminate, preference for available places, anxiety, fear, or storm/noise phobia. Additionally, check in with your vet to make sure there are no related health concerns.

- Never, ever, ever punish unacceptable elimination. This creates fear, distrust and teaches the dog to eliminate in a different spot, not to hold what their body is insisting be voided. Think training for what you want the dog to do rather than punishing for what you do not want them to do. In this scenario, you want the dog to indicate to you the desire to go outside for relief, to perform that behavior there, not in the house, and for you to reinforce the desired behavior.

- Review any crating protocols, including advantages and disadvantages of that practice. Let's start with taking a closer look at crating and go from there. Done correctly, crating works to house train due to the dog's natural avoidance of soiling a denning area. "Correctly" here means taking puppy outside to eliminate every hour for puppies under 4 months old and every two hours over four months old. Those hourly rules are devised after how long an infant dog can naturally and reasonably hold on to not eliminating.

Keeping dogs confined to crates for longer periods can backfire on the house-training front as the confinement is too long, health is affected adversely, and or the dog eliminates in the crate. Dogs innately do not want to foul a den but dens do not have cage doors, they have exits that dogs can use at will to leave when they need to eliminate.

Crating also does little to train dogs to be comfortable or versed in living in our rooms with us. A more effective method to limit available floor area, for both safety, socialization, and housetraining concerns, is creating a puppy proof confinement area where you are both close by puppy and spending time around that area. This can be done, either with portable fencing (ex-pen) in an open space as in your living room or in a smaller room without fencing (such as bedroom or office).

A well-appointed puppy proof area can also help with settling down, naps and stationing to work on toys, and puzzle feeders for breakfast, snacks, and dinners. Make it puppy proof by removing objects of concern but make sure it has a comfortable bed, water, plenty of toys including good chew toys and wee pads placed farthest from water and puppy bed. It also important to be reminded that puppies are not physically mature enough to hold urine and feces for very long.

When house-training either from confinement area (or even open-door crates), taking puppies out **every hour** to the locations (either in the house or outside) to where you would like them to eliminate and waiting for this to happen and then praising and treating the very second after full elimination is the recipe for success.

After an outside walk with no elimination, keep the leash on the dog or the puppy and pay attention to any squirming, circling, fidgeting that typically signal an oncoming elimination. No time to head downstairs? Walk over to pee pads and just wait with puppy. Limit talking so as not to distract by your presence or trying to figure out what you have to say and let the puppy focus on taking care of business. Make sure to hold the praise-like-crazy happy camper voices or baby talk (we all do it) for the second they have completed the entire go.

Keeping walks and other significant happenings on a reliable and consistent schedule affords that elusive sense of control over events in any dog's life. All dogs need to eliminate when they need to but most do not have the freedom to go out and do it on their own. Dogs depend on us for when to go and where. A set routine provides relief breaks that can be anticipated and depended on. And for when necessity takes over, dogs' petitioning, whether it is squirming, barking, or standing at the door, for extra relief breaks needed to be listened to. Stomach upset can happen to anyone.

Routines are also helpful for humans doing the walking and housetraining. Much of the stress and lapses in housetraining during the pandemic were linked to lapses and breaks in our own waking, getting ready for work, feed and take the dog out for a walk every morning, afternoon, and night routines.

Established housetraining can have set backs. It happens. Those breaks in our routines affect our pets and there is stress in that. Other changes can also affect dogs, new places, new people, new events. Start again with the basics. Always, going back to foundations is a good step. Next, for help with remedial house training, first breathe (in, out, repeat often). Your pet is amazingly skilled in reading you and your body language, if you are upset, uptight, angry, or frustrated this will come through loud and clear. No matter how upset that little face looks at you in response, it is not because they understand they have done something wrong, more they would like you, so much, not to be mad at them, please.

Again, and again and again: Do **not**, no matter what, **ever,** ever scold or push your dog's face into the mess or the floor during an elimination. Ever. You can gently and quietly, pick them up if their size allows when you see them relieving themselves. This will stop the action as the muscles tense in response. If size does not permit, leash them quietly and gently and go outside anyway.

Scolding and dismay often leaves dogs under the impression that they will get in trouble for eliminating while in proximity to you or in locations you have caught them at in the past. This can lead to finding new and often hard to find places to relieve themselves instead. You want to turn this around. You can do this by, again, removing all punishment. Next, by keeping your dog in their halter and leash and attached to you. This way you will know when they need to go because you will be aware of their every movement. You can tie the leash to your belt or loop it around your wrist (and don't forget the sleeping with the leashed dog next to you). Once you do see those intention movements of sniffing, squirming, circling, etc., let your reaction be a positive one. Be matter of fact and go directly out with them. Even if they are done with the act. This helps to work on creating the association of where elimination is preferred. Go outside and either walk with or place the puppy on the ground and take the time to allow them to sniff around, which may after a few good and relaxing sniffs inspire a finish. Praise and treat.

Most dogs will "tell" you they need to go outside to eliminate by standing in front of the outside door. This usually comes without any sound effects. You can train a need to go out cue as well. Start this by praise and modeling a behavior to indicate needing to go out, such as ringing a bell hung on an exit door as you both leave for a relief walk or to let the dog with a yard (supervised please to capture the praiseworthy moments) outside. Praise the bell ring with a cue for this, "good out (your dog's name here)," etc.

Pee pads are not a dirty word. Put them down to help with house training, "in case," or continue to leave pads in the house but place them as close to the exit doors as possible, front and yard doors (and in the "secret" spots the dog is now using). The first hurdle the puppy owner may face with using the pads is getting the dogs to target the pads to begin with. To understand and address what is happening when your pet missing the mark, consider the circumstances:

- Are pads and paper changed frequently? Change pads and paper as soon as they are soiled or as quickly as possible to ensure your pet will have a clean surface to use. And clean the spots you don't want them to use, even harder. Pets are drawn to the scent of protein deposits waste leaves behind, this means those areas you do not want your pet to be using are still calling to them. Most cleaning products will not eliminate those scents even when we can no longer detect them. A good enzymatic cleaner is key in removing all traces of relevant odors that most other cleaners leave behind. Avoid using bleach or other caustic chemical cleaners that are irritants to skin and sensitive mucus membranes.

- Where is the pet relieving themselves? If the elimination is happening close to your exit door, your pet may be trying to get as close to outside as possible. Move the pads or paper next to the door.

- Is the elimination close to the pad or paper? Provide a larger surface area. If elimination is close or in proximity of the pad or paper your pet may simply need more than one pad or paper. Some dogs will position themselves to relieve themselves and move forward or backward before, after or during the process. Your pet may be starting on the proper surface and run out of area before he or she is done. An extra pad or more paper can help with this.

- What is your response when you see a miss or when you see your dog use the pad? Did we already say no punishment?? Puppies will shred and tear pads for the plain fun of it. They will also do this for the attention it brings them. Make sure your puppy has plenty of toys to play with and pads are not the only objects to interact with. Do not chase a puppy tussling with a pad. This turns the pad tussle or chase into a game. Find another toy or treat to trade for the pad instead. Hold the treat directly in front of the dog's nose, too far away and it has less smell and less value in proximity. As soon as the dog drops the pad, offer the treat, take the pad, and thank them.

- Has the pet been trained to use an alternative? Covering the top of the pad with several layers of newspaper for the pet that has been paper trained as a puppy can help. The dog that has been paper trained will more likely target the newspaper which will soak through to the pad underneath.

Dogs instinctively prefer a soft surface to eliminate on, in the natural world that surface would be grass covered. Manufacturers of training pads scent these pads with grass to attract this natural preference. The question can be asked, that without being trained to eliminate on actual grass or outside initially, is the puppy is learning instead to eliminate on a pad that smells faintly of grass inside and not the grass and the outside environment itself?

Once training pad use has been accomplished, i.e., no scolding for not using, careful placement, in both the places the puppy is already eliminating on or setting up a large enough confinement area along with bed, toys, food and pads, an adequate distance away from all that so that the puppy will use them. Your dog is now puppy pad trained.

Good news? Well, such an accomplishment has two sides to it. If your puppy has been effectively trained to only use puppy training pads, they may very well be reluctant to eliminate outside of the house thanks to successful puppy pad training.

I have worked with an older puppy who will wait as long as canine possible before eliminating outside. I also have a much older dog who will hold out on outside walks. These are not dogs who have decided they prefer one location or one surface over another, these are dogs who have been trained where to go. This is learned behavior.

Such behavior was taught for good reason initially. Health concerns and fragile or puppy immune systems can limit contact with potentially hazardous pathogens in certain environments or those carried by other puppies or dogs. These concerns prompt keeping puppies or compromised dogs inside and on pads for relief breaks. From the physical side of wellbeing, mission is accomplished. From the training to toilet outside, mission not happening. A more ultimately successful overall house-training program includes training for your puppy to relieve himself outside as well as inside.

New puppy owners are often given advice from their veterinarians to restrict their dogs' interactions with other dogs and with the outside environment -which includes walks- until fully vaccinated. While limiting a puppy's exposure to possible agents of infection is a good thing on the avoiding possible infection on the one hand, it is highly problematic on the other. What is also being limited is socialization and learning, along with adding being exposed to the outside environment of new and varied sounds, moving bicycles, scooters, cars, different people, different dogs, and eliminating outside the house and not in it. Some veterinarians recognize this and suggest carrying the puppy to a toileting location. (You've seen those people carrying those puppies on the sidewalk, that's them.) This can limit exposure and permits the dog to become comfortable and familiar with the outside environment, especially grass patches, tree pits, sidewalks, etc., as areas for relief.

The younger the puppy, the greater the beneficial impact of socialization and learning. Between the ages of around 4 to 14 weeks, puppies are more readily open to learning about their environment, and that varied and new outside physical world with all those other people, animals, and behaviors in it we expect them to learn for navigating the human world. Not allowing puppies to benefit from this "sensitive period" for learning can have negative lifelong consequences if not managed and addressed. Lack of socialization carries over into adulthood. We can often see such marginalized socialization expressed in frequent alarm barking, tense body posture and language, frequently accompanied by anxious and fearful behaviors around new environments, people (especially children and men) and dogs.

Behavior problems are the main reason that pets are relinquished to shelters, so early socialization is vital. It is equally important to have the discussion on how to alleviate health concerns with your veterinarian while working towards safe socialization by being mindful of the dogs and environment you do interact with. Things that might be considered may cover if the puppies and dogs you might encounter look healthy? Have they been vaccinated? Is the surface area targeted free from debris and garbage?

Once there is an all clear to bring your puppy and dog outside, reversing puppy pad training may be needed. And, yes, there are steps for that too:

First: no scolding or punishment no matter what. You want to teach your dog what to do instead.

Second: at home start saying "good go pee" (or whatever other term you prefer) the very instant your puppy urinates or defecates on their pad, so they can learn the label and the request. Remember timing is everything in training so pay attention to that instant of the completed go to reinforce.

Move pads closer to exit doors and group several pads together to increase surface area to target.

Third: put your dog on a strict walking schedule with the morning walk as early as possible, before 8 am with walks every 4 hours after that. The schedule should be consistent and regular - the same every day. Time to say it again, schedules have the added benefit of offering a sense of control over events, especially important ones. Those schedules and routines, especially tacked on to our own, are also easier for owners to keep up with. Pairing meals and walking on schedule lessens stress and adds structure. Walks after meals are best. Dogs need to eliminate after eating. Don't miss this window.

Do not leave food down before the dog's morning meal. Give the morning meal, with whatever added treats in it that will heighten appeal, right before walking. A 15-minute gap between food and a walk is the longest period to allow.

When outside allow for sniffing as much as possible. Especially inspiring are the sniffs where other dogs have eliminated. Dogs naturally "overmark" to add their urine to where another dog has left theirs. Be aware of what is happening with your dog on the walk. Make sure there is no phone to distract you. Engage with your puppy when they are not busy sniffing by talking to the puppy and asking for an occasional sit, stay, touch, or look, etc. Practicing this at a corner is an added safety bonus. Ask for "good go pee" several times in a happy, not over loud, voice next to areas where it looks like other dogs have peed and puppy sniffs.

When there is elimination is outside, wait until puppy has completed the act and then praise like it's the 4th of July. All house training, including outside efforts call for owners to greatly reward the corresponding efforts -"Good pee outside!!!" (You know you've seen those people too). Do not return directly home, signaling that a walk ends as soon as your dog goes and the fun is over. Walk another half a block or more, and then return.

Should your dog keep to puppy pad training and consistently wait until returning home to use the pads, go back to basic house training. That part of house training where you keep the dog on leash attached to you so that as soon as you see signs of an impending bathroom break, pacing, looking around, wiggling, whatever intention movement your dog uses- go outside immediately and take that walk again.

Be Patient. Remember your dog is being a really good dog, doing what he was trained to do. The tricky part is learning something new.

No pee pad issue but still not there on house training?

Continue to walk your puppy frequently and on as close to a schedule as possible. Worth repeating, schedules are beneficial for our pets and provide a sense of control where we are the ones deciding when and where they get bathroom breaks. Remember, if you walk your dog and nothing happens do not release the dog when you get home. Keep them either attached to you on the leash or on your lap. Pay attention to the signals that they need to go (squirming, fidgeting, etc.) and proceed directly outside.

When your puppy does go, have a party for him, you always have with you the greatest reinforcer, praise from you to him. Tell him the very second, he has finished voiding, what a fabulous little dog he is.

"What a good wee outside (<u>your dog's name</u>)!!!!" The great thing with all that loving praise is the timing can be instantaneous, no fumbling for treats. And when you start with the praise and a treat follows, that's an added party favor. Whatever your dog likes best that you can do outside together, do.

Make sure your walk is long enough and paced enough, for several long pees, even female dogs will urinate several times if given the opportunity. A voided bladder is a good thing especially when house training. A good sniff at everything they would like to sniff around the block is a good distance for exercise, smells, and the distraction or two for excitement. And if your dog is distracted by squirrels or birds at a distance, use the distraction to your advantage, let the dog focus on a few (not catch), to expend energy (another "motivator"), and then pick puppy up and walk where there are less squirrel distractions and they can focus on business.

Always, always make time for sniffing. During your frequent routine or as needed walks, try and find locations where other dogs may have frequented. Sidewalks surrounding and in front of your house will work better than the familiar backyard. Trash bags and hydrants on city sidewalks have great sniff appeal. Dogs are the most interested in new smells and especially the smells of other dogs' urine and will leave their own pee mail to add to prior missives from others. It's a collective effort.

HOW TO STOP BARKING AND KEEP A QUIET DOG BUSY

Living in New York City, as I do, means for most city dwellers, apartment life. Home includes next door, shared walls, and, neighbors. The suburbs, too have their share of town homes and apartment complexes, appealing to residents with affordability, amenities, and location. Living in multiple dwellings, also translates to party walls and common areas. All conducive to smelling what's for dinner next door as you walk past, hearing other people's disagreements, late night TV or music streams, when the walls are thin or the volume is loud enough. Such can be part of the smell and sounds of a city, apartment complex or your upstairs neighbor, what comes with how we live and share space with each other. You learn to let the mostly, occasional disturbances go to keep the peace. Tolerance is key. We may hesitate to ruffle feathers over a birthday party where everyone sounds like they are having way too much fun but draw the line at dog barking. People have much more to say about a dog barking. Often with good reason.

Dog barking, especially prolonged barking can raise more than just noise grievances from neighbors, it can raise concern for the barking dog's welfare. Notes slipped under doors, complaints to coop, condo boards, and landlords can cause owners to search with greater urgency for solutions for the dog and the neighbors. Add to this, the anxiety of the transplanted dog owner and dog, formerly living in a detached home, now an apartment dweller, trying to adapt to constrained spaces and other people's input, welcome or not. Reassurance is in order to know that complaints and adjustments are only problems, when they can't be fixed.

Excessive barking can be worked with, if we understand the individual, the need, and the message communicated. Experience and applying the research, has shown over and over, that behavior modification can be most successful given careful, thoughtful approach and the magic ingredients, consistency, and time. To begin to address the change here, the underlying emotional state, motivation and need of the animal needs to be considered and where necessary, responded to. That dog is barking for a reason, what is it? Barking is a way to transmit information for dogs, signaling a range of alerts and also may express and be part of restlessness, anxiety, frustration or boredom.

Barking is a natural behavior for dogs. Other canids, like wolves, do not bark nearly as frequently as domestic dogs do, especially as they mature. Increases in barking behavior is something we either have selected for in dogs or one that has co-evolved with us. Humans rely primarily on the visual and verbal when it comes to sensory processing. Sounds both get our attention and are a significant part of how we communicate with each other. There is a distinct value we put on the alert/alarm barking of the watch or working dog. So, no fair blaming them for what we have asked them to do in the first place.

Functionally, barking can serve three broad purposes: alert, alarm, and solicitation. People arriving back home at the end of the day or the UPS driver passing by can elicit a bark as in: "did you see that/I see you!?/look at me so we can play!" An unknown person at your door or a new dog in the yard next door, can be cause for a red flag being raised or "Stranger/Danger!" That bark alternating with gaze fixed on you and then an object out of reach, transmits a request for help or "My ball is under the couch and I need you to get it please!" And then there is the plaintive, as in sounds of despair, that signal a whole other alert as in: "Are you ever, ever coming home??"

Acknowledging the purpose of the barking by letting your dog know they are heard, either by telling the dog you did see the cat across the street, heard the doorbell too, will get the ball and thanking them for the alert, then redirecting after the fact by asking for another behavior can help the barker to know their communication has been received.

Barking as an expression of an inner state is also an alert and equally as noteworthy. For barking where the underlying emotional state is negative, changing the environment to change emotion is necessary or technically, eliciting a "conditioned emotional response." Counter conditioning is often required to add in or linking a negative with a positive. Which can look like- Scary garbage truck pulls up outside the house, you toss dried salmon away from the source of the sound but still close enough to just hear the truck and build up to reducing distance over a period of time. Enjoyment of salmon changes feelings of anxiety to pleasurable feelings of relishing treats and associating anticipation of treats with future arrivals of garbage trucks.

All this is part of why letting our barking dogs know their communications to us are effective is so important. We all know that every situation and individual is unique, guidelines and protocols are that as well, guides to put interventions in place and gauge response and adjust as you go:

Acknowledge, direct and redirect: Every dog will naturally alert to a change in their environment, whether the change is the ring of a bell or the presence of a new person entering a room. How they respond to new people varies from vigilance, mild curiosity or alert or alarm barking. Always allow for an initial bark or two. No one wants to be told to shut up right off. Especially when they are trying to tell you something. Thank them with one or two words for letting you know. Thanking them for the alert, is more than a polite response, it can shift our own energy in the moment to more empathetic as opposed to reactive.

With excited barking in response to a familiar individual entering a room, it can help for those individuals to reassure the dog through greeting them first off when entering the home, a room, or an area where they are. A simple "Hi Spot" or "Rover we're home", will help take the pressure off them to inform you how excited they are and to and to let the world know about it. Follow the greeting on a return home by asking your dog to get a favorite toy or keep something they can mouth in play next to the door and offer that to them to play with.

The next step in refocusing your dog's energy is asking her to get her favorite toy. This is also a great redirection for a jumping-so excited-you're-home-don't- know-what-to- do-dog and allows all that energy to be channeled to finding something special, taking hold of it and bringing it to you. Of course, you will be visibly thrilled that the toy has been retrieved. Share that acknowledgment too – "So good you found (fill in the name of her favorite toy)!!" Now redirect yet again to a yet another thing to do - a place to go with that special toy.

Teaching "Quiet" Strange noises and newcomers frequently set off alarm barking. Acknowledge the bark by telling the dog in four or five words that you heard/noticed it too. For example, "Thanks, I heard that too." Only after your response to the barks do you want to work on the next part, directing and redirecting. A good start on this is teaching "quiet."

As with any teaching any new behavior, it is more effective to begin the learning in a low distraction environment to begin. Teach quiet when barking is more alert than alarm to proof it.

An alarmed dog is too aroused to learn, in that scenario, more effective to end the barking is distance or change in environment (more on this below). For alert barking, keep the acknowledgement phrase timed to after those one or two alert barks to let your dog know you have heard the message. Next, hold a treat right in front of the nose or an attractive toy they are keen on. As soon as there is a micro second of a pause to sniff or regard the toy or treat, the barking will stop. Even for a second or two of that pause, sniff, or regard, it will stop. The dog simply cannot change focus to sniff a treat or regard the toy presented to them and bark at the same time. Immediately say "Good quiet!" or "Good hush!" or whatever request you want to always use for this and offer the treat or toy at the very same time.

That reward is important - remember you are marking and rewarding the quiet and not the barking. That pause in barking or treat taking may be momentary. You may have to repeat this again and again and again. This can take time to learn and only when that time is spent with patient repetition. It is highly unlikely to change an established pattern of behavior alerting with one trial of learning something else.

For this and all behavior change, we are looking at a process and not an event, one we can best accomplish with consistency and attention. Such a process takes time. Keep in mind, that barking is alerting us to a major concern for the dog and if we do not give the attention/response, the dog probably feels they must continue barking. Yelling at a barking dog mostly creates more barking as we are pretty much communicating to the dog that we are concerned too. Again, rewarding a dog for not barking or listening to our request for "Quiet" is **not** rewarding the barking it is rewarding the response.

Panic Barking. For a highly reactive dog in a stranger/danger full on panic barking attack, "quiet" may not be enough. With this sort of barking being a distance increasing behavior, granting distance is often necessary. In a closed space, throwing treats behind where the dog is standing can get them to retreat to a safer space. In some scenarios, safety concerns may call for barriers or other rooms for the dog to be separated. In an outside environment, distance from the source of alarm is what the dog is asking for or for the alarm to stop. What needs to be determined is what is the "safe" distance of registering alarm without being triggered by it, to determine threshold and work from to begin modifying behavior.

Behavior Modification. Make sure the dog has a clean bill of health and any indications of pain or discomfort are accounted for first. When it comes to modifying this level of reactivity, you need to be two steps ahead of when those triggers (other dogs!!!, scary people!!!) present. While a dog is under threshold (the amount of time with or next to the bad thing which is not stressful or not so bad), you can begin to employ some desensitizing (increasing exposure through minimal increases in duration or decreasing distance to the bad thing), counter-conditioning (creating good associations with the bad thing), response blocking (preventing visual) and redirection (circling, changing direction, etc.). These are all good strategies to have in your tool box. With desensitizing and counterconditioning having the most potential towards long term change when thoughtfully applied.

Basically, you are looking to see the very first sign that your dog has noticed the trigger or stimulus they are reacting to. The further away from the trigger, the better to start this. At first registration the dog should not be overly reactive beyond alerting, whether that is head turned towards, ears pricked, momentary stilling. The dog may have heightened vigilance but still be comfortable. Distance counts. So does seeing or being aware of the trigger, if not, there is nothing to desensitize or counter condition to. At the exact moment the trigger registers and you see calm behavior, call the dog by name, praise them, and treat the calm behavior. Treat taking is tricky here, because for the stressed dog, treats are not worth focusing on. This is where using their name and praising them comes in as reinforcement instead. Do not push how close you can get to a trigger.

Put a good foundation down of reinforcing calm registration over days and increase distance and duration in literal baby steps. And make those increases separately. Where advance warning is not happening as in, the trigger turns the corner and is directly in front of you, change your own course to avoid it.

Response blocking, as in standing in front of a dog to prevent them from seeing or moving forward, is good to know about and use in certain circumstances but can be frustration inducing as the consistent strategy. Response blocking does not address the motivation of the behavior, is hard to employ calmly for the human, or change the emotional state of the dog quickly enough in the moment. Taking this process closer to home can be a bit of challenge when the unexpected knock comes at the door or when living in an apartment building where controlling the environment is easier said than done because we cannot manage the triggers completely. But the more work done on desensitizing and counterconditioning, under threshold, at a distance or outside of the building, the closer we can get to generalizing this to when it is happening over threshold.

Again, this is why you want to work on modifying the tone and frequency of barking by targeting all occasions of barking as well as addressing the environment and overall underlying emotional state of the dog. You will have the most success initially with the alert barking for outside movements/noises as opposed to higher distraction events like actual people arriving. Building on improving response from low distraction to high distraction is the best way to proof behavior.

Know Body Language Know how your dog is feeling from the barks and from their body language. Determine the distance at which your dog is comfortable with the trigger and where you start to see behavior change that indicates the need to alert or when they are approaching threshold and from there, meltdown.

Signs of stress, in order of severity:

8. Stiffening up, staring
7. Lying down, leg up
6. Standing crouched, tail tucked under

5. Creeping, ears back
4. Walking away
3. Turning body away, sitting, pawing
2. Turning head away
1. Yawning, blinking, nose licking (*least frightened and threatened*)

When you see a sign of stress, make note of whether it passes as in one lip lick or progresses in intensity? Do you need to give the dog more distance from the trigger or reassurance? Does that work? Never forget that going backward to what was last working is the way to go forward.

Short, fun, frequent training sessions help with barking associated with anxiety and reactivity. Without structure, we leave dogs with the need for a lot pressure to be vigilant and constantly in need of asking for things or alerting us to them. Much of how these needs or alerts is communicated is with barking and other behaviors. Good, force-free training also gives our dogs dedicated time together, confidence, boundaries and creates a stronger relationship, one where they are more tuned in to what we are asking of them and they are responding to it.

Keeping training sessions on the shorter size, as in less than five minutes, and avoiding repeatedly asking for the same behaviors, keeps these exercises something to look forward to and not military drills to be avoided by both trainer and trainee.

First know that training is so much more than teaching tricks and commands, it is about finding a clear and consistent way to communicate with another species, one who has no extensive "language" to understand our explanations and whose idea of life rewards is at times very inconsistent with ours. Because of this, we are limited to an almost pantomime if you will, one where we must identify without words, what stimulates a response (from the dog's point of view not ours) and "condition" that response, we "associate" behaviors and rewards (meaningful for a dog) and we reinforce - repeat so we both know that's what we're looking for. For instance, your dog already sits when asked. Think about this- you call their name for attention "Lady," ask for the behavior "Sit," she does, you say "Good Sit!" and reward. That request, mark and reinforcer are what you may be initiating but the translating to responses are all on the part of the dog.

We are quick to correct but not so quick to validate and miss amazing training opportunities because of that. Praise is part of training and a reward that should be freely given for being "quiet" or "good," etc. Putting a behavior "on request" is one of the most reinforcing, easiest, and effective ways of training if we remember to do it. Mark the good stuff they are already doing without being asked! If your dog is sitting with you, lying down, or waiting patiently you can label the behavior they are doing at the moment and praise: "Good quiet (your dog's name here), good quiet!" The trick is to time the acknowledgement and praise to when you see them doing the behavior you want to reinforce.

Remember, that your dog's name is not the command or the praise. Label the behavior, remark enthusiastically on it. We do not nearly praise or thank enough at all.

Work also with your dog on what she knows already to make it stronger for the both of you. Dogs process information visually faster than we do. As humans we need to work on our timing of asking, labeling, and rewarding so our pets know what we want.

Schedule and Routine are significant for reducing frustration and anxiety. The routine of schedules and the patterns and exchanges we put in place with training are ways of providing the dog with control over the unpredictability of a world where the dog must make everything happen by "asking" for it.

Knowing when an event will occur, eliminates the very need for a soliciting bark to remind us, request, or alert us, that it is time for breakfast, dinner, treats, dinner, or a stuffed chew toy. Nothing is ever guaranteed, but I have seen this single intervention reduce barking in dogs of every size. Keeping schedules and routines gives a sense of control over the life events dogs are the most concerned about, meals, walks, playtime, and interactions with owners. Those very same schedules and routines, once in place can help us to keep to them as well. Repeating patterns of behavior keeps them ongoing in all species.

Make sure walks include ample opportunity for sniffing it all in. Allowing our dogs to sniff everything on a walk, exercises their cognitive processes in how they sift, order, and perceive information about the world. A dog's primary sense is olfactory not visual. They need to sniff. Smelling it all in is the best exercise there is for their brains and more enervating than a run. Satisfying this behavior also lessens reactivity and stress.

Give your dog more dog things to do in their day: Feeding their diets in a puzzle feeder for both breakfast and dinner meals will prolong the satisfaction of eating and chewing out a good meal. Avoid frustrating slow feeder bowls and make sure that puzzle feeder allows for prolonged and satisfying chewing instead. Sniff mats for sniffing out dry food or treats are also good choices.

Go back to stuffed toys, chews, and puzzle toys full of treats and leave these around to engage with. Dogs need chews and toys to engage with. You can't be everywhere. Think enriching activities to feed the mind. Try new and different toys to leave for them and keep one by the door (a special chew or toy they only get when you leave). Puzzle toys that you have buried for a day or two in your dirty clothes hamper (gives them the best smell of you) and that you have showed how to play with, are a great addition to squeaky and stuffed toys.

Stationing or teaching your dog to her "spot" or "place" is frequently recommended to target barking. Offering this to a dog that has frantically alerted to a change in the environment as a command has the feel of putting them in a corner with their back to the room. Do not ask a stressed dog or barking dog to go directly to their spot. Better to use as a redirection after you have acknowledged the bark and redirected with a request to get a toy and then take it to their spot. That sounds and is way better.

Start training go to your spot by throwing treats on the intended spot you want your dog to use. A good place for a spot is a comfortable bed that is far enough away from the door or window that reaction is lessened. Just throw treats on to the bed to begin. When the dog goes on to the bed, as in places at least two feet (four is better) on it, mark that and reinforce simultaneously by saying "Good Spot!!" Do this two to three times a day for three days. Always remember to mark/reinforce the requested behavior when it is given. Even or if you have run out of treats, thankful praise is always available.

Next, with a treat ready before you ask, try saying "Go to your Spot!" If the dog goes to the bed, mark/reinforce they have done as asked with "Good Spot!" while immediately tossing that treat at the same time. Your dog may look at you and wait for the throw. Go closer to the bed and ask again. Wait. Ask and wait again if no response. Get even closer to the bed and ask again. Wait. Still no response? Ask for something your dog does know, like a sit and reward that. Stop the session for the day and try again a day later.

Supercharge positive associations with those beds you want your dog to use for stationing by feeding chews and puzzle toys on them. Leave a really cool stuffed toy there. Keep stationing practice short, sweet, and frequent like all training sessions. And don't forget to acknowledge that first bark!

Classical music has been proven to soothe dogs and cats, leave the radio on for the music and the soothing voices of the announcers for calming effects and the company of those voices when you are not around or even when you are.

Teaching alternate behaviors that are just as good as the ones you are seeking to replace is the magic sauce in training for different responses. If barking is arousing and communicates a need what is just as good? Providing needs – see schedules and routine – and loading the routine and living environment with activities to indulge natural behaviors and places full of positive associations and comfort, as in puzzle feeders, chew toys and that their bed is the best place to chill.

HELPING YOUR DOG NOT TO PULL

Frania, I watched your video and only wish that my 4-year-old, black, solidly built 37-pound Cocker Spaniel was as docile on leash. I think he has some field spaniel in his bloodline as he's bigger than any other Cocker's I've seen. He's stubborn, meanders, pulls, doubles back and will spend 15 minutes smelling a blade of grass. If I'm not paying attention, he'll pull my arm out of the socket if a squirrel runs by !! :-) Other than that, he's affectionate, playful and a good dog without a mean bone in his body. Any ideas? -Barry

Nobody wants a dog to pull Barry. Nobody. When I posted videos on dog walking and helping your dog not to pull on leash, they received a lot of views, comments, and questions. Good thing for our dogs and for us, we all want to do this the right way. There are a lot of suggestions out there on how to get to easy walking and stop a dog from pulling but which ways are the most effective? And what would your dog like you to know about pulling?

Do not be deceived, Daisy is very much the "ever-eager-to-experience-the-world-now" spaniel dog as well. We have just been working on how to train dog manners to human liking longer. Your spaniel is interested in spaniel dog preoccupations, how very dog like.

Dogs pull mostly because they have got four legs and can get to where they want to go way before we do. Getting to taking a sniff around and chasing squirrels are where they want to go. Dog expert, Dr. Ian Dunbar once made an excellent point about pulling that I think works here. Dunbar talks about making an agreement with his dogs;

they could sniff to their heart's desire, just as long as they did not pull him to do it. Fabulous, allowing for a companion dog's natural behaviors and desires along with consideration for ours in the same sentence.

Sniffing is so beyond important to your dog. Just think of how much they have traded off in their co-evolutionary processes with us, adding in eye-to-eye visual communication, how we mainly process information, as opposed to how they mainly process information, through smell. That blade of grass means the world in those 15 minutes to your spaniel. There is so very much to know from a smell, who was here last and when, what they ate, how they were feeling, and so very much more we have yet to uncover but which the dog already knows. And the squirrel? Absolute unparalleled fun to chase after a prey animal, not *The Wall Street Journal* or the Internet but then again, those appeal mostly to humans.

On to my ideas on easier walking for the both of you: work on being more in tune with what you want from each other and when you want it. This is training. Communicating to your dog what you expect from them in a way they can understand, tempered with humane consideration which allows for their wants and needs.

First and foremost, know that pulling back on your dog's forward movement or in general, is one sure way to get your dog to pull back. They must pull against you just to stay upright and keep all four feet on the ground. This is the concept of "oppositional reflex" at work. You can see it when you see a dog straining to go ahead while they are held back from a collar or a back clip harness. Observe as the pulling intensifies on either end as the dog is restrained and their front legs lift off the ground. The dog needs to increase pulling even harder forward to maintain equilibrium. We can see this with our own bodies, if we are pushed against from the front or the back, we move in opposition to the force to stay upright. More on why using a Y shaped front clip harness is best for this reason and others, follows.

What about all that excited pulling forward to meet another dog? You know that dance of hesitation you can do as you consider: Is the dog friendly? Boy or girl? Should you let them say hi? No doubt, all the while you ponder the what-to-do part, your dog is straining vigorously to go forward, pulling on their collar, harness, leash, while you pull backwards against them as you both work to maintain balance. Any stress, hesitancy, or trepidation you are feeling in advance or in response, is travelling right down the leash adding to the frenzy. Far better in this scenario to decide ahead of time. Choose either yes or no to allowing for greetings as you see and size up the oncoming dog and owner. Once decided, let the greeting happen straight away with a loose leash or if not, keep moving.

Know that cross gender encounters are typically smoother encounters. Having a female dog, when it was a go for me, I would ask three questions as we approached before stopping:

"Can they say hi?"

"Boy or Girl?"

"Friendly?"

You always want to start with asking a handler if the other dog is OK with the interaction. Other dogs may have their own reasons to avoid encounters as may their handlers. Determining gender counts, and while for some males it may be obvious, shaggy coats and postures may not be revealing. Having a handler weigh in on mood and temperament of the dog to be greeted is valuable information in a possible upcoming interaction. Depending on the responses you hear in turn, you may have good reason for not wanting this to be a meet and greet.

If I thought it better or the answer was a no to greet the other dog, I would smile in response, say nothing and step around my dog so I was the buffer between both dogs and just keep walking. If the answer was an affirmative on all counts, I would lower the leash so the dog had space to maneuver and not have to strain against a tense leash creating more anxiety and pulling while always monitoring the encounter closely. Whether you say hi or keep going, pulling is not needed by your dog.

Please avoid equipment that uses force or pain to control a dog. Shock collars (also called, "training" or "e" collars), choke collars, prong collars and head halters all fit here. These collars and halters work to regulate a dog from pulling by electric shock, constricting the neck, applying direct pressure with small painful points or apply uncomfortable pressure to the highly sensitive eye and muzzle area and severely inhibit natural behaviors. Small or flat faced dogs, with even more sensitive and fragile anatomy, should never, ever be walked without a harness.

Prove this to yourself- watch the next dog choking against their collar or pawing at or with subdued, cowed demeanor on a head halter (for more on head halters please see Suzanne Clothier's excellent piece) or the dog pulling on a prong or choke because the pain and inevitable physical damage being done to the neck and back is now the price of being able to go forward.

It makes so much more sense ethically and effectively to stop the pulling as a handler from your technique on the other end of the leash. Skip the extend-a-leashes please, they can be dangerous and will in fact, encourage pulling as the dog learns to pull and is reinforced by the release of the lead in return. Do put your dog in a Y shaped halter with a front clip attachment instead. A good, well fitted, "no pull" kind of harness can be a wonderful thing as they guide your dog's movements below the sternum without putting pressure on and causing damage to neck, trachea and back. Make sure to try on several of these harnesses to find the right fit and avoid strapping that is cut too high under the front legs or too low over the front legs, both make easy walking painful or restricted.

To develop and work on an easy-walking-no-pulling technique, start walking your dog on a 6-foot leash in the house. Inside your home, the dog will be better able to key in on your cues and movements without outside distractions. Consider the dog's motivation and attention in this, talk to them, to both engage their focus and show your interest in them. After several calm and easy walking sessions you can also try attaching a lead to your belt at home so the dog is attached to you and you can both get used to moving well together. The lead should be long enough to allow the dog to lie down comfortably next to you and short enough so that you are aware of the dog's movements.

Remember to always set your dog up for success by asking for what you want first and allowing for a response. For instance, if you want to get up from your desk, let your dog know where you are both going. Say "Rover, let's go get a drink of water" and wait for Rover to respond. Rover may not know what each word means but he probably knows "let's go" and he definitely knows what your body language is for "let's go."

OK, so now that you have worked on paying better attention to each other and walking at home, try it outside. Relief comes first, if your dog needs to eliminate, let them. Pay attention to body posture – slowing, sniffing, circling, squatting, are all signals a dog has to go. Now. Let that happen without correction.

Once a relief pee or poo is out of the way, start practicing your red light green light technique (more on this below) and be mindful to allow for more needed relief stops on the walk. Cue the dog on walks too. Sniffing that blade of grass forever again? Time to move? Cue with a name and request first. That leash is not a joy stick and not to be jerked. Ask for forward movement, turn and take that step forward and allow your dog to follow.

Ask for your dog to stay next to you, either when practicing walking inside or out by using a verbal cue, like "with me." As your dog walks alongside you, remember to mark and reinforce that behavior by adding "good" to the request, so your request phrase of "with me" is now a marker or secondary reinforcer with "Good with me!" Or more simply, a way to tell your dog that what they are doing now, has a name, and that you are thankful they are doing that named behavior you asked them to do.

Please, do use treats (the small easy to eat on the go kind) as you go along to keep the exercise meaningful and rewarding for the dog. With a mid-size dog, keeping a treat at your outside thigh and level with your dog's head will keep them positively glued to you, remember to give the treat every few steps). For smaller dogs, bend down to offer that treat every few steps. Remember to mark or acknowledge the behaviors that are happening and that you would like to be repeated by saying "good with me!" or other cue every time you treat. Very excited dogs may not respond to treats outside, that's fine. For some dogs, a special squeaking toy or tug toy works. Or commendations. You always have your praise as reward. Everyone wants to be acknowledged and thanked, including your dog.

Always keep your training fun and less than three to five minutes. Short and frequent training sessions are proven to be the most effective as opposed to long and boring. Not to mention, humans are more likely to do it if you break it down into easier to accomplish segments time wise. Adding training sessions to routines and schedules can also keep them happening. Asking and thanking for one or two behaviors before a walk can tune in the focus to the handler.

Practice the red light green light technique. This is the exercise where you stop moving the very micro second your dog pulls and you move forward the micro second the dog pauses or stills in response. Sounds simple but every human struggles with this when it comes to timing.

With the red light green light exercise, just like cars at a traffic light, there is no pausing between beats. The dogs pulls, you stop on a dime. The dog pauses/stills you move instantaneously forward at the exact same time.

What people frequently get hung up on, is the moving forward once the dog pauses. We pause too, whether in observation or deliberation, when we need to be stepping forward in the pause or when the pulling stops. That moving forward when the pulling stops is meant to be saying to the dog, "you stopped pulling so we get to go ahead." If we do not move when the pulling stops, the whole exercise is totally confusing for the dog. They pull, you stop, they stop pulling, you are still stopped. See? You need to move when the pulling stops.

This is an excellent exercise to stop pulling but needs to be done with exact timing. It is also a great exercise to develop focus and attention on the dog for the human doing the training, a benefit to your training expertise once you get it down. And you will get this down if you practice and pay close attention to timing. Make sure to add plenty of verbal praise (treats are always good if they will take them) at a green light. And always end all exercises on a good note, as in a successful response, treats or a "thank you!"

Allow for easy walking by letting your dog sniff along the way. When you approach what is uber attractive and prompts more than sniffing, try and see it before your dog does. Keep proactive and set him up for success, by either releasing them from the training session, whether it is to say hi to a friend or passing the pet store with a phrase that redirects the movement "let's say hi to Rover", "let's go pet store" or whatever other word you use to say "all done with training for now" before the pulling temptation to get there right now presents itself.

Do let them sniff that grass forever- it is so much what the walk is for. When the squirrel runs by, give them the way not to run after it by seeing it before they do and call their name to get their attention and move off in the opposite direction. Pace can be useful here, as in quickening your walk with a jolly camper "let's go!" voice or circling to move forward in a different direction than the distraction. All good to know, and only if you can do this in a smooth and fluid movement without jerking or pulling on him.

Try also offering an alternative behavior like a few moments of tug or a special squeaky toy that your dog likes (only offered specially for these occasions). Those alternatives in the place of squirrels and the like, need to be just "as good as" and treats do not often make the cut. The squeak of a special toy, a ball, tug toy or what appeals to the dog you know are key to offer here instead. Keep making sure you are always talking to the dog. Again, this will ensure that you are engaged in what they are doing and will keep them tuned into and aware of you. And don't worry, no one will care that you are talking to your dog, except your dog.

All this is to say, that this is about training. which takes much time, communication, consideration, practice, and effort on our part to get those four-footed dog bodies which go much faster than our own to move with us.

LET THEM SNIFF- GETTING THE MOST OUT OF YOUR DOG WALK

The house and land we lived in on Florida's Gulf coast was sprawling. Enough property that you could easily be out of earshot of another person. This was before the days when we all walked around with a cell phone in one hand. If I was looking for my husband, it was easiest to call Daisy and ask her to find him. When asked she would look up at me as if in acknowledgement and then drop her nose to the ground and get to work. The sniffing and tracking she would do followed every step in the path he took to wherever we located him. If he had first went to the back garden and then around the grapefruit tree and then the garage, that was how her tracking followed along.

I never stopped delighting in asking him if he first been in the garden and then around the tree before the garage or letting him tell me his stops and comparing them to Daisy's tracking. She followed every step of his in the order he took it. Whatever she "saw" in the scents she uncovered in her search were full of information, of time and presence, of flavors and shadows we can only begin to imagine.

Every dog shares sensory processing that makes olfaction their primary sense, the way they take in information about their world. Humans process the world through sight and are primarily visual. Dogs process the world through smell and sound. This universal aspect for dogs and the disconnect with humans can leave dogs deprived of opportunities to perform the most basic and necessary of behaviors when outdoors with us. Mostly we are not letting our dogs take that sniff around nearly enough or at times, at all. I ask clients if a dog gets to sniff as much as they would like on a walk and always ask for walks that allow for even more sniffs. And this one intervention, one of the easiest to apply, can yield some of the biggest benefits.

A dog's sensitivity to odors is 10,000 to 100,000 times more sensitive than our own. Keeping your dog from sniffing it all in when outside, with all there is to smell out there, is like you crossing the street practically blindfolded. Dogs are built for "smell-o-vision." From an olfactory cortex (that part of the brain that scrutinizes odors) which is 40 times greater than ours, to a hundred times more nerves than our own linking the brain to the area of the dog's nose that detects odors. It is this very highly specialized nose that we are depriving of all this incredibly informative odor when we prevent our dogs from sniffing around on a walk. Being able to use their principal sense is intrinsically rewarding for dogs, it's how their made. Watch a dog getting the chance to smell to see how absorbed, focused and content they appear.

The dog walk is more than just an opportunity for elimination. All dogs, city dog or country dog, need the time and opportunity to sniff around as much as possible to properly "see" what is happening in their surroundings on a walk. Sniffing is taking inventory on the world and its happenings. For those dogs relegated to puppy pads and suburban dogs given backyard access only, please know every dog, no matter who or where, needs a walk every day. A walk they get to sniff on. As John Bradshaw writes in DOG SENSE:

> "Smells are very important to dogs, much more than they are to us. Dogs don't just use odor to decide what to eat or not: It's their primary way of identifying people, places and other dogs. Smell is their dominant sense, the one they use in preference to all their other senses, whenever they can." (Bradshaw, 2011)

Getting the chance to properly sniff is more than just a luxury in perception. Peter L. Borchelt, an animal behaviorist practicing in metropolitan New York, made the observation, that being allowed to sniff was in fact, an almost athletic activity for your dog. At a lecture I attended, Dr. Borchelt said that a good walk (again, backyards only don't count here), with sufficient sniffing opportunities is a chance to allow your active, apartment or house bound hound a proper chance to expend the right amount of energy to return home properly tired from canine efforts. A good sniff around the block can make a difference. This canine expenditure in cognition, and perception and olfaction is a true workout. Forget the weight bearing saddle bags, the timed runs, and never force a dog to use a treadmill. Let them sniff instead.

Make sure to vary your routes on walks to allow your dog the chance to discover new smells left behind. Discovering the latest or newest information, or reading the "pee mail" from other dogs is important and cannot be beat for encouraging elimination in return. Please include those walks for dogs with their own backyards to sniff around – they know most of the smells in their garden already, give them more to explore.

Sniffing around can also lessen stress and give a dog a greater sense of optimism. Studies tell us that allowing shelter dogs olfactory enrichment (things to sniff they might like) lessens those behaviors associated with stress. We talk about the benefits of positive training but activities dogs find innately rewarding can be even more reinforcing. A 2019 study comparing dogs practicing those activities where they got to use their noses to find scents or food independently ("nosework") with dogs practicing physical exercises keeping in step with their owner ("heelwork") in which both groups were given food rewards for completing the exercises, found the nosework dogs outperforming the heelwork dogs on a cognitive bias test designed to measure optimism. A possible indication, that those dogs that get to sniff are happier dogs as well. The researchers also noted that nosework offers a chance for these canine companions at our side to work autonomously and exercise choice and control over their actions while indulging in a naturally satisfying activity.

Take a look at how your dog walking stacks up on your next walk:

- Are we ready for the walk to finish before it starts? Running to the tree in front of the building and that's it? Losing patience, in a rush, need to finish to get to work or start dinner? Are we yanking a dog without any signal or warning from a lamppost or street sign just as it gets interesting for them? Or are we not even looking, wrapped in our phones? Is our attention elsewhere and not noticed that our dog has started to sniff and yanked them away before they can complete a breath in? Do we recognize that sniffing and circling to find the right spot to go and pull them away before they can do that too?

- Do we think dogs are people and a good butt sniff means something other than getting to know the dog being sniffed? Do we think that sniffing the poop or pee mail is gustatory rather than simply olfactory? It is not. Sniffing another dog's urine is about investigating the complex richness of the chemical signals left behind to find out who was there last, when, and what happened and other dog news we have yet to uncover but the dog knows if we let them.

- Our pace, inattention, biases, and the equipment we put on dogs during a walk can affect welfare and prevent sniffing. Does the leash we are holding offer enough slack that your dog can easily move forward towards an appealing to them, scent and or, drop their head to sniff easily? Or even worse, are we putting choke chains, prong collars and head halters on our dogs to inhibit any sniffing along with other natural behaviors? While dogs may pull and choke against prongs and chokes and theoretically avoid the constraint by not pulling, the head halter offers no relief, placed high on the muzzle around the sensitive eye area often resulting in visible drooping and slinking and needs to be added to what not to use on our dogs.

The good news is that making time for sniffing, including finding, and exploring new and varied routes to sniff along, not only allows a dog to adequately "see" the world, sniffing is exercise for your dog's brain that feels good, lessens stress, and makes them happy. Increasing this exercise in olfactory processing will increase welfare, help lessen anxiety for dogs and that makes their humans happier too. Digesting all that information is a lot of work. It is as equally important for the dog to have something mentally satisfying to do as it is for the dog to have something physically satisfying to do.

Try it out, go for a "sniffari" on your next walk and the next and the next, etc. and see how much calmer and happier your dog will start to appear. Letting your dog sniff his or her way around the block, opens up a whole new world for you both.

References

- Bradshaw, J. (2011) *Dog Sense* (p. 242) New York: Basic Books

- Duranton, C., Horowitz, A. (2019). Let me sniff! Nosework includes positive judgment bias in pet dogs. *Applied Animal Behaviour Science*, (211) 61-66

HOW NOT TO TRAIN A DOG AS SEEN ON TV

I have thought a lot about this chapter, about the benefits of preaching to the choir in the "what not to do with dogs" chorus we all sing in. No matter how much we might know, still want to learn, we, all of us, still need to learn how to have that conversation with those other choirs singing different songs in different churches.

Reasoning someone out of a position they have reasoned themselves into rarely works. And those communication strategies to echo someone else's thoughts to get your own in can come across as insincere. New information, where there is interest or perceived benefit in that information, can change beliefs and from that, actions, but that change comes from the individual, at their own discretion. Human behavior change to effect how we work with animals is having a moment and that is a good thing. One hard and fast takeaway, is knowing that when talking to those with opposing beliefs, finding common notes, threads, or consensus where it exists, and going from there can be the most successful in starting and continuing a dialogue, and from that, perhaps, change happens.

Before formally studying animal behavior or learning theory, and back when the "Dog Whisperer" was a huge hit, I went to hear the star of the series speak. It was an engaging presentation and he was a charismatic presence, well received by his audience. Likeability goes a long way to engaging viewers until they take a closer look at the methods used. Eventually, backlash over the use of aversive and forceful methods seen resulted in the networks canceling the series. For a while. They are back. With company. Netflix's recent reality dog training show highlighted a training outfit that prominently featured shock collars as part of their training packages. Choke, prongs and slip leads are standard equipment shown in use on the show. Social media overflows with self-proclaimed dog magician gurus who flood, force, and push scared and overstressed and excited dogs into submission. And the likes are exponentially viral.

As loud as certain voices may be in opposition, compulsion training appeals to dog owners and to a culture where it more than sells. The buy-in from audience and owners is real. As much as we can decry the trainers for their work with punishment, the force and cruelty in "balanced" or "results based" training, along with the pain and coercion that drives it, is the very simple fact that these trainers, this training, is successful because their clients support the methods that are being used. Whether because their belief follows the reasoning that this is how dogs learn best or that this is how dogs best deserve to learn. Such thinking can reach to the upper levels of a society, no matter how civilized.

Two of President Biden's dogs, consecutively, made headlines biting security personnel. Avid concern followed each string of incidents. Including a degree of concerned trainers and behaviorists offering guidance with positive science-based methods (this writer among them). The prongs and choke collars the German Shepherds were outfitted with were also noted, along with the history of military training the Biden family dogs had experienced. Not one of the experts weighing in was publicly taken up on their offer of force free training or behavior modification. Both dogs ended up leaving the White House to go onto whatever remedial methods of training or homes were offered, neither of which was disclosed.

A deeper dive into a look back at how compulsion training played out in one reality scenario is seen in an older 2013 episode of one Bravo's Real Housewives series. The shows featured the housewives embroiled in the usual dramas. Avid viewers tune in to watch how those dramas play out. But what about those cast members who have no control over their surroundings or the script? Witness the training showcased with one of the Beverly Hills housewife's, Kingsley (a one-year-old pit bull puppy at the time of the show).

Kim, the California housewife, acknowledges in the segment that she needs to train this big puppy. Good intentions. Where this seems to fall apart is in the training methods employed by the trainer Kim has hired. We see the back of the trainer kicking what appears to be the dog as a method of control, advocating Kim and her son to "thunk" or hit the dog with her legs and to grab on to the tender scruff of the dog's neck to force the dog to sit. There is much made of how this puppy should not be treated tenderly and the relationship with "Mommy" that he has forged is disparaged. We see no teaching of Kingsley as to what we might want from him. None. No teaching of "sit" or "off" or "quiet'." Instead, rough manhandling is promoted which teaches mostly to fear the handler. Studies bear this out extensively. Punishment creates distrust and defensive aggression as the animal seeks to protect himself. Teaching any animal, human or non-human with positive reinforcement is proven more successful, not to mention kinder and more humane.

Kim, seems to question these methods. In her Bravo blog of November 12th of that year, she writes: "The trainer told me he's very spoiled! You think??" No Kim, your doggy is a puppy and a high energy, playful puppy who likes to play and do dog like things like bark, and chew and bite and jump up. He needs to learn how to be redirected in a positive manner. Kingsley needs to be taught how to channel those behaviors to coexist in your human family. Positive dog training without aversive strategies is the way to go here.

Kim writes on that when she spoke to the trainer:

> "he told me that Kingsley is not going to like the authority of a trainer and what he represents and that he could become aggressive with him." (Richards, 2013)

Absent anthropomorphism, the reason for such thinking is not explained. Dogs are highly domesticated to work well with any human that handles them with sensitivity, consideration, and respect. Typically, if a dog reacts aggressively to anyone, including a trainer, there is an issue with the approach and handling. A dog can become defensively aggressive when he is fearful and is protecting himself. Kim goes on to write:

> "He also said that Kingsley should have a choke collar and leash on when he arrives, because Kingsley may sense him on the other side of the door!" (Richards, 2013)

Restraining an animal with a powerfully aversive method such as a chain which chokes off their air supply is compulsion training, which is not only cruel but which will work counter to what you are trying to teach. This sort of pulling and pain inducing will hurt and frustrate the dog, add to that the owner's high energy and you have a recipe for extreme arousal. One way attack dogs are trained to attack is by keeping them on lead and just within reach of whatever it is they are trying to get to. Way better here to counter-condition and desensitize Kingsley to strangers and teach him to sit off to the side and offer treats for quiet behavior. This trains "sit quietly while I open the door" not "opening the door means you get choked."

Sometime after the show aired, there was an update on Kingsley. There were media reports of the dog biting, lawsuits and of him being "given up."

Massive followings are hard to come by in this business. Only a handful of professionals make a real living in paid appearances. Securing a place in the higher echelons can depend more on reputation, timing, and charisma than insights into the science and its applications. The number rises for consultations and training classes but getting a toehold is often more about connecting and connections than qualifications.

Absent mandated education, which would set standards for a degree of knowledge of foundational concepts of learning theory, welfare, body language, natural history, etc., for dog trainers or behaviorists, advantage comes in not just being thought to have "the" methodology but in being "the" person to implement it. Such paradoxical thinking can add to professional trainers and behaviorists forming elitist groups. The criterion for in groups may vary from group to group but as all exclusivity does, it depends on the superiority of edging out others.

Standing can also give a bigger megaphone to get messages across. Messages that often have less to do about the subject of dog training and more about the personalities themselves. What, if anything, force free dog trainers should be doing, need to be saying, to those who use force in training has fueled controversy. As social media continues to expand and our insatiable appetites for likes and attention spans grow in opposite directions, the popularity of training methods and their divides become more pronounced and devolve into all manner of keyboard fighter tactics. Substance and objectivity can quickly get lost in the thread, replaced by the why-else-do-it, satisfaction of character assassination as opposed to debating the merits of approach.

Countless scientific studies support the effectiveness of positive dog training compared to the use of aversive training methods. The AVMA position on dominance-based training methods is that they are cruel and inhumane. Animal welfare organizations and countless trainers and behaviorists support this position. Yet force and coercion in dog training continues, including the use of aversive equipment. Magnetism of individuals takes precedence over scientific principals and ethics. Prong collars, choke chains and shock collars are all featured in plain sight in the dogs living with the featured players in other televised reality shows, either as part of the story line or as the main plot featuring the supposed quick fixes. Displacement behaviors and fearful effects after the fact, are not followed.

Karen Overall's foundational text on clinical behavioral medicine outlines the protocol for behavior modification starting with first managing the environment and second removing all punishment. Conversely, we also know that we pride ourselves for strength, believe that someone must be in charge, might can make right and empathy can be for bleeding hearts. Science works in both directions, also telling us that punishment shows immediate results. Initially that is. Punishment needs to be increased to avoid habituation, will elicit fear towards the dispenser and has been shown to prompt displacement behaviors. The question then becomes one of ethics and morality vs short term effectiveness. For all of us.

When I was teaching, before any other course work, the first topics and exercises I assigned were to promote empathy with other species. It has to start there and it needs to include all animals. Trainers discussing the controversy in conflicting methods are often heard referring to their own personal journeys in crossing over from the use of force to force free. They get there not by another's approval or by accepting that what they are doing is OK but by the realization that it is not. And that punishment and force frequently makes matters far worse. That far worse is often the very reason that most clients come to a force free trainer after using alternate methods.

The dedication and passion we have fueled by compassion for animals can side step the very animals we need to start with, ourselves. We need to treat each other with the very kindness and respect we want for companion animals, including standing up to those who would harm them. From there we can have a conversation. Cards on the table. Force is inhumane, unethical, immoral, counterproductive for the results we seek and not something we can agree on. It's not how to train a dog.

<u>References</u>

- AVSAB Position Statement The Use of Punishment for Behavior Modification in Animals. http://avsabonline.org/uploads/position statements/Combined Punishment Statements 1-25-13. Pdf Retrieved December 1, 2013

- AVSAB Position Statement on the Use of Dominance Theory in Behavior Modification of Animals. http://avsabonline.org/uploads/position statements/dominance statement.pdf Retrieved December 1, 2013

- Richards, K. (2013) Kim: Lisa's Faint Looks Like a Fakey. https://www.bravotv.com/the-real-housewives-of-beverly-hills/season-4/blogs/kim-richards/kim-lisas-faint-looked-like-a-fakey. Retrieved November 6, 2023

DOG BITE PREVENTION

When I was a little girl, I would try and visit my grandmother's house. Try, because two houses before the one she lived in, was a German Shepherd Dog. That dog was as tall as I was and when I tried to pass their house, would stand in the middle of the street, directly in front of me, squared off, face to face, and bark non-stop. That was it.

Whether the motivation was protecting territory or spooked by children, I never thought about, I was young and not up on dog behavior at the time. What I did know was, those barks were meant to keep me from moving past that dog. I stood in place and cried and cried. After some time, I would turn around and go back home. I never told anyone about it. The dog never did more than square off and bark repeatedly. I never made it to visit my grandmother when the dog was around. And I never got bitten.

In the United Sates, the second week of April is National Dog Bite Prevention Week, dedicated to increasing knowledge to prevent dog bites. That one week is good to focus on dogs and bite prevention, even so, dog bites can happen any every day and every week in the year.   Learning about and managing dog bite prevention is accomplished through better understanding of canine behavior, the frequency of dog bites, who is most targeted, under what circumstances, the why and when a dog is likely to bite.

Behavior issues are in the top reasons given for pet surrender at animal shelters. Aggression, more specifically is the behavior of concern, also one of the top reasons, clients consult an animal professional. My own dog clients, fall mainly into one of two categories; training and aggression. The pandemic has increased socialization and anxiety cases but aggression and training still lead. This is where the applied in applied animal behavior comes in – how do the theories and experimental findings come out in real life and which interventions and individuals would they make more sense with?

There is much to go into applying those interventions and the hardest part of it can be getting it to stick. With the people. Every case, needs care to consider the science, individual, circumstances, history, and dynamics. Then there is the work in not just making changes but in the amount of time we devote to doing them in real life. Most of the interventions call for change in how **we** interact with our dogs, changes in our behavior to effect changes in the dogs' behavior. Once the medical has been ruled out, research tells us the biggest difference can be made when removing punishment. This sounds simple but punishment has an appeal in its immediate effect of stopping behavior in the moment (it also damages relationships and has serious fallout in displacement behavior and increased defensive aggression). When we take it away, we need to have alternate strategies of desensitization and counter conditioning and be committed to implementing them. Some of the most dedicated clients I have met to working through interventions, are the ones with biting dogs.

There are over 43 million households owning at least one dog in the US. Over 66% of those households consider their dogs members of the family according to the Humane Society. And any one of those dogs can bite. In our dog loving society approximately 4.5 million dog bites happen each year according to The Centers for Disease Control and Prevention ("CDC") with children between the ages of five to nine, being the most at risk. Two thirds of the children bitten are boys. And men are more likely to be bitten than are women.

The pandemic, has also affected the statistics. Compression and social isolation no doubt factors, whether momentary or a trend, the population being bitten has reportedly shifted recently in France. A study surveying French oral and facial surgeons collected data from June 2020 to March 2021 of 87 patients treated. The findings included evidence of more bites in girls than boys. Bites to all children were mostly to, in order of occurrence, cheeks, lips and chin and eyelids. Data also showed severe bites had increased in 2020 over the prior two years and that biting happened more frequently while petting or playing. How petting or playing is constituted as defined in the study is not clear, nor is the presence of adults during the bite event.

The numbers are impressive and as sources such as The National Canine Research Council and the ASPCA, reminds us, need to be put in perspective. Children are twelve times more likely to be hit by a car than be bitten by a dog or seven times more likely to be hurt by a sharp object than be bitten by a dog or one and half times more likely to be injured by a bicycle than be bitten by a dog. Further, with 75 million pet dogs in the United States in 2007-2008 and 400,000 bite injuries to children the overall bite rate is extremely low.

A popular misconception is that stray dogs are doing the biting. Fact is, children are more likely to be bitten by a dog they know, and be bitten in their own home. The CDC notes that the likelihood increases with the greater number of dogs in the home. What is most important to note is that most, if not all, dog bites can be avoided. Dog bite prevention begins with education for both children and adults about canine body language, appropriate handling, training and how dogs communicate with humans along with necessary and adequate supervision of children around dogs. While much education is targeted towards children in these efforts, unless the supervising adult is equally proficient in understanding why dogs bite, and how humans should act around dogs, little can be reinforced to benefit our health, safety, and our dogs' welfare as well.

Paramount to understanding dogs is learning that dog body language is not human body language. Canine body language has its own unique vocabulary to communicate emotions and intent. Take greeting behaviors; while humans may value eye contact, hand shaking or hugging as friendly salutations, it's different for dogs. For them, a sidelong glance, lowered head, assuming a "t" end to position to another dog, followed by a muzzle and good butt sniff are a more appropriate way to say hello. Add in that a jubilant dog welcome consists of jumping, not a welcome human behavior. For the fearful or aggressive dog there are a whole other set of distance increasing behaviors, starting with looking away, lip licking and yawning that communicate discomfort and deference for dogs and taken out of context in human parallels, do not translate. Identifying what a dog is saying is part of what keeps everyone comfortable and safe.

How then are we doing when listening to what dogs are "saying"? A 2010 study in the *Journal of Nursing, Social Studies and Public Health* found that children correctly identify a dog's emotions, on average 17% of the time. Not an impressive statistic. Recognizing fearfulness in dogs was accurately reported by 41% of girls and 29% of boys studied. Adults may not be doing a much better job at reading their dogs than children. Dog expert, Stanley Coren's viral post on not hugging your dog looked at a random sample of 250 photos posted on the internet of people hugging dogs, in 81.6% of the photos, dogs displayed at least one sign of discomfort such as turning the head away, lip licking, yawning, lowered ears or "whale eye" -where the white of the eye is visible at the corners or rim. Coren points out that the uploaders of the photos are most probably posting proof of their happy and close relationships with their dogs. The dogs would most probably not agree, but how close are those relationships in fact and how happy is the dog?

Dogs, like most animals, manage conflict with highly ritualized display signals that signal a request for distance and/or demonstrate or promote deference. Physical altercations are to be avoided and appeasement is signaled to maintain peaceful resolutions.

Minor signs of stress such as lip licking or yawning out of the context of being hungry or tired, and or looking away, are strong indicators that distance is requested and that what is happening to the dog is not what they would like to be happening. Signs of increasing stress such as whale or round eye, flattened ears, creeping, trying to leave, tucked tail, weight held towards the back of the body, rigid body tension and raised hackles are definite calls for more space. All of which should be respected, that these signs are often not heeded and dogs tolerate a good deal of unwelcome contact is both a testament to a dog's magnanimous nature than our own lack of consideration.

We need not to forget that biting is part of canine natural behavior. Dogs bite. Dogs may bite for any number of reasons- fear, pain, protectiveness, lack of control over an aversive situation, lack of socialization, lack of or inappropriate training or handling, and of severe stress and desperation when the warning signals they are giving are not being acknowledged.

Veterinarian and animal behaviorist, Ian Dunbar, developed a bite categorization system that looks at the immediate history, antecedent and current setting of the event; i.e., the object of the bite, type of bite delivered and environment. Understanding dog biting in this scheme is accessible contextually to how domesticated dogs live with people:

1) What is the target of the bite? Is it a Person? An Animal? A Thing?

2) What is the severity of the bite? Mouthing with no teeth? Puncture wound (1-4 bites)? A tear in one direction for less than three seconds? A multiple bite attack or severe mutilation?

3) What are the circumstances surrounding the bite? Trespassing? If so, is this in the home? Or private with public access, as in a fenced yard? Or is the dog leashed in public with the owner unaware of biting inclination?   Or is the dog leashed in public with the owner aware of biting inclination? Or is the dog roaming free?

Dunbar's system makes more rational sense of how and why a dog might be biting and how and why, it might be justifiable from the dog's point of view. An intruder breaking into a home where a dog lives, is more likely to be bitten justifiably, we might believe, from the resident dog protecting family and territory from a stranger as opposed to the dog at liberty taking bites along the way. We might also classify as justifiable, a quick, soft, single bite on another dog versus a person in play or a nip to guard or protect a sore spot or a litter of puppies no matter who the recipient. We can raise the bar at biting where meals or a favorite toy is concerned but these things count to some dogs just as much as sore spots and puppies.

Justifiable from any viewpoint does not necessarily equal acceptable in every scenario. For dogs to live successfully with humans, they need to have what we call "bite inhibition." Bite inhibition means just that. The dog must independently hold back and regulate the force of the bite on humans and other dogs.

Bite inhibition is learned in development, as a puppy from the mother and from litter mates. When a puppy bites too hard, the mother offers a maternal correction and litter mates offer feedback either vocally or both with body language. When the biting is extreme or the feedback unheeded, the feedback escalates to withdrawing from interactions. In these exchanges, puppies learn what acceptable social contact is and what it is not. We can also see this sort of feedback in the dog park or at puppy play groups, when an offended dog yelps and/or retreats from what is not a welcome interaction or what is not fun for them. The dog who bit too hard often gets a second chance at playing nice but not a third.

In play or in passing, getting away from an antagonistic individual is always a safer and healthier strategy for both parties. Ignoring threats and warning signals or pressing forward can increase the signs given in intensity. For example, when deference shifts to defense, a dog may raise the hair around the shoulder blades, indicating arousal and shift weight to lean forward into a "ready-to-go" stance and bark repeatedly. This often is still more of a threat than a promise. Standing squared off like this, indicates the dog is ready to move forward and the barking is attention getting and warning intent. But this dog is mostly looking for more space from another. Such moves can be, depending on context, seen with conflict behaviors, like paw raises, that switch in or happen concurrently.

Canine body language communicates affect, intent, and influence and in doing so, can illustrate what a stressed or happy dog looks like. Developing our consistent use of an objective lexicon to describe what a dog is doing enables us to observe the dog's behavior without infusing it with human goals and emotions that might not apply and color our responses such as: "mean," "nasty," "guilty," etc. To such ends and to help shelter professionals qualify behavior, assessment tools like the ASPCA's SAFER include being versed in a visual glossary for what terms like "whale eye," "open mouth" look like and in what contexts.

It is necessary to note that breed and conformation differences have an impact here. "Ears back" on a cocker spaniel or basset hound with long floppy ears look quite different than "ears back" on a pit bull or Doberman with upright, cropped ears. Even so, both dogs will move the base of the ear back. Focusing on the base of the ear when observing this aspect on the dog is more good practice in paying attention to overall body language to inform emotional state.

While describing what stress can look like is helpful, seeing it is even more helpful in learning to recognize it. Reliable videos are an invaluable tool in demonstrating stress signals and canine body language. Elemental Media's "In the Company of Dogs" , Maddie's Institute "Canine Body Language" and AnimalBehaviorist.Us "Stress Signals in Dogs" are some of the good ones to begin with. Interactive web sites designed specifically for children, but just as valuable for adults, that incorporate teaching canine body language with games, videos, and photos such as _doggonesafe.com_ are excellent for, again, parents and children. The more comfortable we are with understanding what our dogs are saying the more likely we can respond to it.

Dogs never "just" bite. Whatever the stimuli, trigger or stressor that sets off the reaction for the biting dog, it is preceded by any number of neurological and behavioral changes prior to biting happening. Even where those warning signals have been ignored in the animal's history, they happen, even if some are skipped or if they are happening at a vastly accelerated pace.

Those times when biting happens without apparent warning are often seen when a dog bites out of protection, when surprised by an unexpected hug or kiss on the face or when a dog has a history where warning signals are continually ignored.

Worth the emphasis of repeating, most often dogs avoid biting in their efforts to signal the need for distance or deference, with a suite of warning signals, starting with lip licking, or yawning out of context of being hungry or tired, to looking away, ears pulled back, whale eye, stiffening, tail tucking, creeping, freezing, lunging, wrinkled muzzle with bared canines (open mouth displays can be positive ones, depending on how many teeth are evident and how much gum is showing), weight shifts, raised hackles, alarm barking, snapping, growling, etc. (see also the chapter on the "Canine Ladder of Aggression" in this book.) And then there are "conflict behaviors" indicating the animal is conflicted in emotion and response. Paying attention to and keeping those signals in place, is one of the reasons you will hear that trainer proverb "never train out a growl," as in never take that warning option off the table so the dog has no choice but to go to the next step.

Proper adult supervision around children and dogs cannot be stressed enough. Children need to learn how to treat dogs humanely and adults who know how to do that need to be doing the teaching. We also need to know our children, a 2012 study found that less shy children were more likely to take greater risks around a dog they did not know, even with a handler present. Actively preventing dog bites with all dogs, especially those that may be anxious, stressed, or defensive, requires observing the dog and surrounding environment first: Dogs that display fear, who are growling or barking should not be approached or stared at directly as they are asking for more distance. Listening to this request takes the pressure off the dog, not listening can result in snapping or biting. We have heard repeatedly, with good reason, that dogs that are eating, chewing on toys, sleeping, and caring for puppies should not be approached as biting can happen in these scenarios. Tethered dogs, with no control over their space or ability to retreat, are more likely to bite when approached, as are dogs that are behind a fence or in a car, which may bite to protect territory.

How to respond to the dog that is past threshold, excited, reactive or in proximity, includes the well-publicized, strategy with any stressed dog, including an aggressive dog, the "stop, drop and roll" technique:

- Stop all movement, this decreases any perceived threatening behavior. Be still.

- Drop our eyes, eye contact can also be perceived as threatening.

- Roll our bodies to the side. Turning sideways to the dog as opposed to facing a dog frontally is perceived as less aggressive for dogs and many other animals. Back away slowly once the dog quiets. Never run.

Any number of variations exist on this method, with all approaching the same goal, from wrapping arms around the body (stops movement), reciting nursery rhymes (lessens fearful behavior from humans) and dropping to the ground turtle fashion to protect the face and mid-section in the case of an actual physical attack. Barrier protection and decoys are worth mentioning and if the user can skillfully emply them without exacerbating the situation. Open umbrellas and bicycles can be effective barriers, throwing objects or treats past the dog -never, ever at them – can gain distance.

Remember, the first rule with all stressed or scared dogs is to stop what we are doing that is stressing them. Identifying the stressor or stimuli that is evoking the response needs to be matched with, either eliminating it which helps to remove or reduce the intensity of the trigger or perceived danger to the dog, or providing distance from it, and or reducing the time exposed to it.

Modifying behavior with dogs as puppies is one effective way to develop bite inhibition. When interacting with puppies, humans can help to develop bite inhibition. Puppies have plenty of "legal" things they are encouraged to bite, as in toys and puzzle feeders. Puppy mouthing is also perfectly acceptable to allow to a point. Feedback can come from humans too. Once the puppy bites too hard, which means you feel teeth, the human can gasp sharply or extort one short syllable. Not "NO!" which can put too much negative energy in its use and delivery. "Hey" or "Enough" are not as emotionally loaded.

That short retort, delivered once needs to be timed exactly to the biting. The startled puppy will be able to pair a short, well-timed yelp or gasp or "Hey" (this the correction or feedback) to biting too hard. As soon as the puppy stops, to the second (again, timing is everything here), the bite, thanking them or praise should follow. Always acknowledge and praise or thank a puppy when there is a pause or the biting stops. *It is important to remember, you are marking and reinforcing the stopping and not the biting.* Keeping a stuffed toy or tug toy nearby and at the ready to quickly offer when puppy starts mouthing can also help to get the puppy to chew on that instead of humans.

Then there is the adult biting dog. A grown dog that has diminished bite inhibition has not benefited from or been exposed to the suitable socialization or interactions with other dogs or humans. It is also possible, that somewhere there has been a deficiency in the development process or the warning/signaling/pre-intention movements before the bite was trained out of the dog or ignored to the extent that the dog skips the growling or freezing or whatever else comes next on the ladder of aggression.

Severely traumatized or painful dogs may bite and there may be substantial challenges in rehabilitation. *Nowhere in the course of this work, is any advice being intended nor should any reader not consult with an appropriately qualified, formally educated professional in the field when working with aggressive behavior.* Caution needs to be carefully exercised, research shows the use of any aggression or force typically exacerbates aggression and not lessens it.

The first step in modifying behavior is management. Manage the environment and situations the dog is in to prevent or avoid scenarios where biting might occur. Safety is a primary concern here for all parties. Highly stressed, reactive, biting dogs can benefit from basket (the cage muzzle that allows the dog to pant and open their mouth as needed in the basket) muzzle training both to limit handler concern and biting incidents.

Basket muzzles are not all bad, but how we look at them can be. A few years ago, I travelled to Vienna. A beautiful city and one with plenty of dogs and handlers on the street. Most of the dogs I saw had muzzles. "Must have an aggression problem," was my first thought. Until I boarded the tram and saw the signs requiring all dogs on trams wear muzzles. My whole perception of muzzles in Vienna changed, for the dogs there, the muzzle was a happy sight to behold, signaling a trip with their person.

Removing punishment is the second step, and this includes what the dog, themselves, might perceive as punishing. The use of aversives can be dangerous with the biting dog. Defensive, fearful, or stressed dogs may perceive any harshness, force, or punishment as something else to protect against and the biting can actually get worse.

Behavior modification follows by working with the dog through creating appropriate positive, force free associations towards triggers or counterconditioning and careful and gradual desensitizing to fearful stimuli. Such work is usually best accomplished through carefully applying the appropriate interventions and working with a well-qualified professional using force free methodology.

Missing where a dog's threshold is, going too fast or too intensely, can add to the history of trauma and intensify the dog's reaction. When a dog is asking for something or somebody to stop or go away, it is an immediate request. Distance or cessation from the trigger needs to be introduced directly as does intensity in the experience and reduction in time exposed to it.

How successful the outcome might be, will relate to the time given the process, how effectively the behavior modification is delivered, how skilled the practitioner is, how closely the handler/owner works in tandem and the degree of the lessened bite inhibition.

Desensitizing and counterconditioning relies on gauging a neutral response to the trigger for the dog and increasing time, intensity, and frequency of exposure to a trigger or stimuli- but not all at once. Learn where the individual dog's level is for first reacting. That level is one you want to stay under. The idea is to expose the dog to the stimuli: the other dog, the person, the bus, the bicycle, whatever the trigger is, while they are still not reacting to it but they have noticed it. The first registration of that scary or alarming thing, is the interface of border of the flight zone or envelope of space, the dog feels safe to start from.

There is a distance that the dog can register or see the stimuli or trigger but beyond that initial registration, a reaction is not elicited. For instance, the dog alerts to someone on the corner with a cowboy hat and sunglasses, the dog may still momentarily or lift their head or point their ears or any other sign that indicates they are aware of the trigger but there is no escalation beyond that. Acknowledge the dog by calling their name for attention, praise them, and treat on registration. Repeat several times keeping distance consistent.

Work on gradually, as in baby steps gradually, closing that distance. You may find that as soon as the dog notices the trigger, they will look to you for praise or a treat. Give it to them. At that point, the dog is under threshold. Concurrently (timing is always everything) acknowledging the alert by labeling the behavior ("Good Noticing Rover"), followed by rewarding the alert either with praise or treats, provides social support and positive associations to the trigger. Reinforcing the dog's neutral response under threshold can set a foundation to build on to incrementally decrease distance. Can a step forward be taken and can that be reinforced? That is your distance/intensity you have worked on.

Keep duration of registration short and work on distance and duration separately. In other words, proof being able to get closer to the trigger without reactivity and with reinforcement for that neutral reaction before increasing the time observing the trigger.

Setting this dog up for success means reinforcing the pet's positive body language before they signal aggressive intent. If you do see that aggressive signal: look for "hard eye", squaring off, stiffening, freezing, hackles coming up, pursed lips, etc., you are going too fast. Go back to the last step that was working. A redirection followed by immediate praise for execution can get you back on track. And for when that doesn't work, you always have blocking or removing the dog from the situation.

It is paramount to keep in mind that, each dog is an individual and being able to read what is going on with that particular dog, along with their own distinct threshold and how to keep them under it, must be understood.

One step up at a time per modality is the next step to be considered. Increasing duration of registering the trigger at neutral response baseline may be a better option for the individual dog in this scenario. How long can the dog stay under threshold while aware of the distant trigger? Thirty seconds? Mark and reward the time. Can it be increased to 45 seconds? Mark and reward. Can it be increased to a minute? Mark and reward. Baby steps here, if you have a neutral response at 30 seconds, reward that over the course of days before increasing it to 45 seconds, etc. Mark and reinforce the instant of registration in place. Keep the voice neutral and repeat reinforcement or provide treats consistently. In that scenario the amount of time spent registering the trigger, without that step forward is being reinforced.

How many stranger/danger dogs can be registered at baseline, marked, and reinforced? There is your frequency. Work on only one aspect of modification at a time and do not push in repetitions, three is a good number for trials. Here is where the skill and fine tuning of behavior modification come in. It is paramount not to push past threshold at any point in the process, either with increasing frequency, duration or decreasing distance or to withhold reinforcement of any calm behaviors, no matter how fleeting.

When I work with biting dogs, I follow a standard protocol that includes knowing results from the dog's last veterinary visit and taking a bite history. I want to know the story around the dog's biting. Is there pain or illness here? How long have this been going on? Who gets bitten? What happens before? What is the response? How has that turned out? Etc., etc. The thing with these bite histories is, people may often not have all the answers or may not want to share them in entirety, lest one think less of their dog or decide that their dog is Cujo and no one will want to work with them. Still, who this dog is, comes out eventually.

Most often, given the dog is safe enough, as in not being easily triggered on the street, I will ask to start that session by meeting the dog and their person on a neutral corner from the home. Neutral, as in far enough away to allow the dog not to feel this is territory they are obliged to protect. There is attachment to space or group that goes along with time spent in it. Dogs, like most animals, are territorial, a trait we prize and have selected for. A dog may consider owned territory to be the home itself. They also may also consider the hallway in the apartment building and the street in front of your building, home territory as well. Incoming dogs or people are intruders to be evaluated for danger.

The neutral corner goes a long way towards establishing a new presence as not threatening, as does a side approach and limited eye contact. I also mostly ignore that dog first off, taking more pressure off them to interact. A quiet greeting by name is enough to acknowledge them and announce my presence as a newcomer to start and then a chat with their person. A short walk for observation and relief followed by return together to their place sets more of the foundation to work from. If the neutral corner is not an option, and or, if deemed necessary, barrier protection, as in the dog needing to be behind a barrier such as a gate and or wearing a basket muzzle (and muzzle training to accept these muzzles), and leashed during the session, may be needed for safety.

Mood can also change for a dog when entering the home. Throwing treats behind where the dog is stationed, so they can retreat safely past a flight distance to take them can relieve some pressure from the dog. Pushing an alarmed dog into stepping forward to take a treat does not benefit the dog. Making treats easier to associate with what is upsetting with less pressure, even for the barest of moments can shift focus. This can be most effective if it happens as close to entry as possible to work towards changing emotional state and tamping down arousal. A constant assessment of energy and body language, including that of the surrounding humans is needed.

Biting can also develop as a protective and or territorial measure or in response to overcorrection in dog parks and doggy day cares. I spent a lot of years in dog parks and day cares in the pet care technician course. One of the considerations with environment in such spaces, goes to dogs being kept in a fixed group and in a fixed location and the attachment that goes along with time spent it. Time spent in dog parks is often limited and extended in day cares. Limiting exits and protracted length of stay can impact aggressive behavior in compressed spaces. In a doggy day care, the territorial roles are set more depending on how long a set group of dogs has been kept together. Accordingly, custom introduction strategies need to follow in entering, handler ratio and mixing up the group. A case history:

What was happening at the doggy day care: A three-year-old female boxer, "Gladys" (not her real name) was reacting to new entrants to the play area. Whether dog or human, Gladys would menace the newcomer by nipping and biting. Certain staff would scold Gladys to end the behavior. If that was not effective, they would tie her to the wall. Other staff members would not wait for her to nip and grab a slip lead as soon as a new presence entered the play area and tie Gladys to the wall. Not surprisingly, Gladys was getting worse and was being blamed for her "aggression," as if she alone was responsible for it.

What I did: I had come to the location do a site and intern review and not to do behavioral work so I did think about turning around and leaving the situation but that meant leaving Gladys to progress and the treatment she was receiving to possibly progress in the wrong direction. Instead, I walked into the play area. Alone. Staff picked up on defensive signaling as Gladys approached me. She was warned off initially but soon as she felt eyes off her, she jumped and nipped, grabbing on to my shirt sleeve only, no body contact. This indicated some level of bite inhibition.

I ignored the interaction but retreated out of the play area and went to stand between the safety fences. This gave Gladys distance from the stranger trigger. I avoided eye contact with her but staying in her line of sight, interacted with friendly dogs, especially, boxers, who approached, hoping for friendly firing of play mirror neurons.

Next, I entered the play area briefly for intervals of two to three minutes and retreated for at least three times longer, further desensitizing her to my presence. If Gladys approached when I was in the play area, I dropped my eyes or turned away to take pressure off her to respond. She did rush the safety gate several times during this step in the process. When this happened, I offered a short, no more than two syllables, neutral verbal redirection and then dropped my eyes immediately.

After 15-20 minutes of this I entered the enclosure and directed staff to counter condition Gladys when she noticed me, by offering a ball to chase or enticing her with a high value toy. No scolding, no tying to the wall, just force free redirection. Gladys was now not stalking me but still stiff and interested. She continued to try and shorten the distance between us, to get close enough to investigate me, without warning signs. Seeing this, I asked for her familiar staff to approach, stroke, and praise her for any calm behaviors they would see in Gladys either when she was looking at me, approaching, or next to me. After the positive interactions Gladys would retreat and returned to investigate two more times with staff praising her calm behavior each time.

How it worked: **After less** than an hour Gladys was totally relaxed around me in the enclosure.

Why it worked: Gladys was being punished each time a new individual, human or dog, would enter the play area. She was reacting to both the punishment and the anticipation of the punishment. The animal being punished usually associates the pain with many signals not just the one behavior we are trying to extinguish. Look at Gladys staff's body language, stiff and squared off with a leash in the hand to tie her up with. That body language and leash would announce punishment and her being withdrawn from attention and the group environment. Being tied and restrained also was associated with new individuals making her reactions to them even worse. The stress of being punished can result in "displacement behaviors" in addition to increased aggression. Displacement behaviors are basically inappropriate responses connected to conflict and could be anything from excessive barking to breaks in house training.

Not punishing Gladys and rewarding her for her good behavior made all the difference. I took the pressure of Gladys initially by not invading her space when she reacted to my presence. No correction, one word redirection, followed by ignoring the threatening behavior. Staying in close proximity but far enough away not to push her past threshold, and interacting with friendly dogs, particularly same breed dogs, may have helped to stimulate Gladys' mirror neurons—those parts of the brain that would light up as if all that friendly interaction was happening to her. Going in and out of the play area at a distance, in short repeated increments of time also helped to desensitize Gladys to my presence. Redirecting Gladys away from me with a favorite ball kept her moving and redirected her initial energy in a positive manner. Of course, the most positive move, what did the trick, was counter conditioning by praising Gladys when she registered this stranger's presence while remaining calm and under threshold.

By soliciting Gladys' attention for praise and petting when she was calm and close to me, she was being mightily reinforced for good behavior and learning to associate my new and novel presence with good things happening instead of getting in trouble for them. If this protocol were to be repeated with new people and dogs consistently, over time, chances are good the biting and nipping might dissipate in favor of praise and petting to associate those new people with.

Remember, there is no distance at which it would ever be too far to praise the aggressive dog for not being aggressive. As soon as the dog visually registers the trigger is a good cue for humans to reinforce the sighting. Blocking a dog from seeing an oncoming dog is a management technique and a necessary one at times. Over rely on this and it becomes frustrating for the dog. The dog is aware of the trigger being blocked. There is little in being blocked that adds to desensitizing or counter conditioning. Better to work on behavior before you reach frustration or arousal.

An wholistic approach to target anxiety includes a sufficient period of decompression including enough rest, sleep, and downtime with necessary enrichment, puzzle feeders, routine and schedule followed in time by short frequent force free training sessions. Let them rest, the anxious and reactive dog is on high alert and often is operating with some level of sleep deprivation. Make sure there are places and occasions for restful sleep.

An often overlooked way of providing the dog the opportunity of practicing of alternate, calm, and intrinsically soothing behaviors is capturing them when they are spontaneous and reinforcing them. No matter how anxious, stressed, or reactive the dog is, there are times when they are not. Capitalizing on the moments where the behavior is softer, more relaxed, capitalizes on the congruence of softer, more relaxed emotions in these moments. For instance, a dog resting calmly, next to their person can be stroked and told how good they are to "rest," any word can do for the label, and any qualifying behavior can be labeled.

The key point is in identifying and acknowledging such behaviors as quickly as we might those we don't want, so the good behaviors can repeat more often and even on request.

If the dog is not being fed meals in puzzle feeders to chew meals out of, they can be introduced. Avoid "slow" food bowls and feeders that frustrate instead of satisfy, this is the opposite effect anyone is looking for. Chew toys and sniff it out or chew it out puzzle feeders, permit a dog to benefit from the opportunity to indulge necessary and natural behaviors, affording a needed change in emotional state.

Routine walks where opportunities to sniff are maximized are essential for all dogs to "see" the world and relax while doing it.

Working with the whole dog can have positive effects on welfare. It can be a longer road, but a safer one to traverse. Biting dogs have learned that biting can work for all the wrong reasons. Identifying the triggers/stimulti/motivations for biting behavior and managing the environment so these are not present along with interventions to modify behavior are often recommended. All this takes time, skill, care, consideration, and caution.

When then can we approach dogs that are not biting dogs? Look to wisdom in the old saying: "let them come to you." Never kiss or hug dogs, unless it is letting your dog safely do the kissing. Not your dog? Always, always ask first if you may pet someone else's dog. Only if the answer is yes and/or if the dog appears open to the interaction is going forward warranted. Follow by asking how the dog likes to be petted and remember your body position in relation to the dog. Turn your body to the side to be less threatening. With smaller dogs it is best to first lower ourselves sideways to the dog and then interact.

Look for signs of consent or stress signals on approach. Friendly dogs that approach us in a relaxed manner- wiggly and soft body tension, open, "smiling" mouth, ears and muzzle relaxed, tail neutral or wagging are looking for interaction. The best response is putting ourselves sideways to the dog so we are not towering over them, leaning into their space, or lowering our face into theirs. Patting the ground alongside our bodies or the thigh can visually invite the dog in. Do they respond by coming forward? Petting on the chest or behind the neck is best. Keep it brief, avoid long strokes along the length of the body. Remember, err on the side of caution, if the dog does not want to interact, do not force it. Respect signs given that petting or playing may not be welcome, just as with humans, dogs can be tired, anxious, or not feeling up to an interaction. That's OK.

At all times and with all dogs, be kind to them and always remember dogs have dog behavior and not people behavior. A dog that is pulling on a leash is pulling to get somewhere fast, with four feet that go faster than our two, and not because they are trying to be "in charge" of anyone.

Teaching our dogs what we want them to do with positive reinforcement lets our dogs know how we would like them to act. Never, ever, hit or be rough with a dog for any reason. They do not understand why they are being hurt no matter how much appeasement behavior they display. Physically punishing our dogs creates fearful dogs, increases stress and can lead to greater aggression.

References

- Chlopčíková, M. & Mojžíšová, A. (2010). Risk factors in the mutual relationship between children and dogs. *Journal of Nursing, Social Studies and Public Health*, 1(1), 102–109

- Davis, A. L., Schwebel, D. C., Morrongiello, B. A., Stewart, J., & Bell, M. (2012). Dog Bite Risk: An Assessment of Child Temperament and Child-Dog Interactions. *International Journal of Environmental Research and Public Health*, 9(12), 3002–3013.

- Dunbar, I. (2006). Dog Aggression: *Biting* (Video). United States: James & Kenneth Publishers.

- Center for Disease Control, Preventing Dog Bites. http://medbox.iiab.me/modules/en-cdc/www.cdc.gov/features/dog-bite-prevention/index.html retrieved January 15, 2024

- Coren, S. (2016). The Data Says "Don't Hug the Dog!" https://www.psychologytoday.com/us/blog/canine-corner/201604/the-data-says-dont-hug-the-dog retrieved January 15, 2024

- Elemental Media. In the Company of Dogs. https://youtu.be/tDiPzikyu04?si=tev7z89oowcQd0Cz retrieved January 15, 2024

- Humane Society, *Pets by the Numbers*. https://humanepro.org/page/pets-by-the-numbers, retrieved January 15, 2024

- Maddie's Fund Education. Canine Body Language in the Shelter. https://youtu.be/gFFto1XVtxI?si=OUW5MaDkF4_LlW5O retrieved January 15, 2024

- National Canine Research Council. Medically Attended Dog Bites. https://www.nationalcanineresearchcouncil.com/injurious-dog-bites/medically-attended-dog-bites/ retrieved January 15, 2024

- Rohee-Traore, A., Kahn, A., Khonsari, R. H., Pham-Dang, N., Majoufre-Lefebvre, C., Meyer, C., Ferri, J., Trost, O., Poisbleau, D., Kimakhe, J., Rougeot, A., Moret, A., Prevost, R., Toure, G., Hachani, M., De-Boutray, M., Laure, B., Joly, A., Kun-Darbois, J.D., (2023) Facial dog bites in children: a public health problem highlighted by COVID-19 lockdown, *Journal of Stomatology, Oral and Maxillofacial Surgery*, 101671, ISSN 2468-7855, https://doi.org/10.1016/j.jormas.2023.101671.

- Shelley-Grielen, F. Stress Signals in Dogs. https://youtu.be/o70LByEwe8o?si=A2aVHocQHb3DITX3 retrieved January 15, 2024

MOUNTING BEHAVIOR, WHY IT BOTHERS US MORE THAN THE DOG AND WORKING WITH IT

I learned a long time ago in this business not to offer unsolicited advice. People often have very good reasons of their own for what they are doing in the first place. When someone is ready to change something, they are the ones who will do the asking how to do that. And then there's not wanting to get anyone, especially a pet in trouble. You know what I mean here, that dog that jumps on you, mouths in play, humps your leg. We make all the excuses and just hope things get better when they get older, get training, keep going down the block, etc., etc. Then there is the exception to that rule. When someone is so upset at something a dog is doing, that the human getting the dog in trouble is scaring you with how upset they are with what the dog has done. What upsets some of us even more than reactivity or aggression, is something taboo, something that looks like a sexual overture even when it is anything but, humping or more correctly, mounting. Dogs can get in such trouble over mounting behavior.

One particularly upset owner at a dog park I was at with my dog and students, roughly hauled an offending mounting dog off the other dogs and launched into a severe scolding of "what is wrong with you" that every other dog owner looked away from. We were all sorry. Once he released his now very supplicating dog back into the play field, I went to sit beside him and started up a conversation about dogs in general waiting to work up to dogs and mounting behavior.

"I don't mind it so much, I just worry that other people will and it's so embarrassing," he told me. Totally relatable but getting over mounting as being sexual, when it is not, is the first step in recognizing what it is. Next, is looking at the motivations behind it and what interventions can help.

We can see a lot of mounting behavior in dog parks. These parks are relatively small spaces where excited dogs get to discharge energy and interact with other dogs off leash. Sounds like tremendous fun and can be, if you are the dog getting to play with another dog close to your size and with your same play style. But play doesn't always work that way.

Play in dog parks is fluid. Subject to the rules of play, dogs engage and disengage and dogs' behavior means play for dogs is counted in pairs or in "dyads." Three in dog play is a crowd and this is where you often see mounting behavior. The third dog that would like to be part of the chase or sword play can try and break into a pair and draw attention with a play bow, a bark, a chase and when not successful, repeat and repeat the request. Frustrated at no response, mounting is a perfect way to gain attention when you are being ignored. And that can work or not for the dogs but not for us.

Most of the mounting behavior we are seeing is not sexual and there are the times when it is. Reproduction between dogs is a consensual act and is female selective, in other words, intact female dogs who are in heat will select which intact males to mate with. Mounting in this context is seasonal and about reproduction. When mounting is sexual in nature, Karen Overall notes:

> "The form is very different from the affiliative form; sex and masturbation involve fast, repetitive motions, leaning with the head and neck on the object of desire, and facial signal changes…it's a normal behavior" (Overall, 2013)

Behaviors do not just occur by themselves; they happen with related behaviors in context and in response to changes in the external and internal environment. Mounting and jumping behaviors have been linked. A 2019 study looking at associated problem behaviors found that almost half of the sample size were reported to have fear and or anxiety problems, the sample also had male dogs who were more likely to mount than females, and of those that were mounters, they were more likely to be jumpers. Familiar dogs were shown to be the recipient of mounting over unfamiliar dogs and people. This finding is contrasted with an earlier study which highlighted people being the target of mounting behavior, perhaps indicating that many people may combine or mix mounting and jumping behaviors as being the same when they experience them. No doubt, both behaviors are attention getting for the dog and reinforced when they are responded to.

It follows that addressing mounting behaviors can be the most effective by addressing the overall dog along with defusing reactivity, stress, and anxiety. Punishing the behavior by scolding, grabbing, hitting, yelling or any other forceful responses can only increase reactivity, stress, and anxiety.

If mounting is looking for extended attention or a response, punishment gives it one, even if it is the wrong kind. You can always ask a dog to stop a behavior but what do you want them to do instead? Always remember that a replacement behavior, to be effective, must be as good as if not better than the one they were doing in the first place or in the mounting/frustration case, satisfy the frustration.

Lowering the excitement level by working on rewarding the occurrence of neutral and calm behaviors, as in praising, acknowledging, and treating the dog who has all feet on the floor or in a sit, whether you have asked for those behaviors or just notice them, can both reinforce calm behavior and work to prevent jumping and mounting. Having a dog who is responsive to the handler helps. Practice a solid recall and off request. Both will help to interrupt the mounting when it does happen and make sure when you are redirecting a dog to have a request for something else for them to do after the fact.

Looking at the effects of exercise, over the lack of it, on dogs, researchers, Zilocchi, Tagliavini, et al. compared 94 "active" dogs ("AD"), a group consisting of dogs performing agility, dance, search and rescue, obedience, etc. with 140 "sedentary" dogs. The study found:

> "The dogs of the AD group practicing agility show behavioral differences compared to the other subjects of the sample.

These animals have a greater tendency to exhibit "Staring an object" behavior but they show few behaviors such as "Turning on itself," "Mounting" and "Attempting to bite other dogs." (Zilocchi, Tagliavini, et al. 2016)

The number of problem behaviors was less in the active dog group, including mounting as related to the sedentary dog group. It is worth noting that 39 of the dogs in this active group were border collies, who have been bred selectively for "staring an object" with 14 border collies in the sedentary group. The study authors offer the conclusion:

"Physical activity, especially if carried out through sports, requires that the animal develops a remarkable ability to cope with frustration and to maintain the self-control. This could be the reason why these behaviors are less expressed by dogs belonging to the AD group. This effect is even greater in dogs that practice agility compared to other subjects of the sample, probably also because of the use of positive reinforcement during training for this sport. Moreover, dogs that practice agility show, with a statistically significant frequency, a lower tendency to be aggressive towards other dogs. A possible explanation may lie in the better intra-specific socialization to which these animals are subjected, having frequent contact with other dogs during sporting events."

Suggesting sporting dogs get the benefit of increased abilities of self-control and emotional self-regulation, the authors credit agility over other sports here due to the positive reinforcement used in agility training. These are worthy remarks, highlighting the benefits of reinforcement, to which can be added that these activities offset the frustration and boredom that can motivate mounting and are activities which are done in concert with the owner, one of the most fulfilling and enriching activities for all dogs.

References

- Dinwoodie, I. R., Dwyer, B., Zottola, V., Gleason, D., & Dodman, N. H. (2019). Demographics and Comorbidity of Behavior Problems in Dogs. *Journal of Veterinary Behavior.* doi:10.1016/j.jveb.2019.04.007 10.1016/j.jveb.2019.04.007

- Overall, K. (2013). *Manual of Clinical Behavioral Medicine for Dogs and Cats.,* p.166 Elsevier Health Sciences.

- Zilocchi, M., Tagliavini, Z., Cianni, E., & Gazzano, A. (2016). Effects of physical activity on dog behavior. *Dog Behavior.* 2. 10.4454/db.v2i2.34.

PLAYS WELL WITH OTHERS, UNDERSTANDING AND WORKING WITH DOG-ON-DOG AGGRESSIVE BEHAVIOR

We hear it all the time in the dog park, the remonstration to someone's dog that they are not playing nicely. And then there is the scolding and the threats. No one likes those. The dogs most of all.

Dog owners want their dogs to get along well with other dogs. But what about the times when we worry that our dog gets a little bonkers when they see another dog, is not playing so nicely or that the other dogs are being a little too rough? Or are they? Is it possible that a hard stare, growling, snarling, snapping or more at another dog in the dog park or on the street is OK? How do we tell the difference from dog warnings, threats, and impending danger? What should we be doing to make sure everyone gets along? And what do we do when they don't? Even knowing dog play is not like human play, how do we know when it crosses the line from play to aggression?

To understand and begin to work with aggression in dogs towards other dogs, we first need to recognize that what we are often seeing are effective communication strategies for dogs to increase distance or decrease another's behavior and not aggression. For example, a dog growling at another dog who is eyeing a favorite toy, or growling at another dog who has him pinned to the ground or biting another dog's flank and breaking skin, all have different motivations with the first, a warning, the second, a threat and the third, aggression.

The most widely accepted definition of "aggression" is action with intent to cause harm with "violence" being a form of aggression where the intended harm is severe or fatal. When it comes to human beings, we can further define aggressive behavior into "physical aggression" or "verbal aggression."

For all animals, threats and warnings are not aggression as they serve instead to prevent action intended to cause harm from happening if they are communicated effectively, that is "heard" and responded to. How we parse out our own warnings, threats and aggressive behaviors for humans comes from research and from our own direct understanding of human behavior as humans. We can also concur that aggression is defensive rather than offensive, as in we use bluster to perhaps intimidate and, in an attempt, to avoid rather than provoke violence. We too are more bark than bite.

All animals have in common the desire to avoid conflict and the ability to accomplish that through effective influence and communication. Physical fights are biologically costly for any species. Injury impacts the ability to gather food, seek and construct shelter, interact with others and when extreme can be fatal.

To avoid actual altercations, animals have developed highly ritualized bluffs and threat displays, which means they look the same and follow the same pattern no matter which individual in the species performs them. That squared off dog, whether Bassett Hound or Basenji assumes a warning posture that includes, the same low guttural growls, snarls, hard eyes, rigid body tension, weight forward on both front legs signaling intent to move forward if they must.

When considering animals, who experience the world in different ways than our own, we have no direct experience to draw from and instead rely more on careful observations of both behavior, context and studied conclusions of what those behaviors most probably mean. In a conflict scenario where dogs are involved, we can more effectively describe behaviors in these situations as "agonistic behaviors," which removes the motivation from either the actor or recipient in a struggle. Animal behavior experts, Camille Ward and Barbara Smuts do an excellent job in the following depiction of agonistic behaviors, possible motivations and contexts as follows:

"Examples of agonistic behaviors in dogs include threats like muzzle- puckering and growling submissive behaviors like crouching, lowering the head and tucking the tail offensive behaviors like lunging and snapping defensive behaviors like retracting the commissure (lips) while showing the teeth and attacking behaviors like biting. With the exception of biting that results in punctures or tears, none of these behaviors necessarily indicates intent to do harm. They simply reveal emotion (e.g., anger or fear), communicate intention (e.g., to maintain control of a resource or to avoid an interaction) or function as a normal part of play fighting (e.g., growling, snapping or inhibited biting). To determine if an interaction meets the criteria for "agonistic behavior," an observer must focus on an objective description of the communicative patterns displayed rather than automatically jumping to judgments associated with the use of the term "aggression."

If signals such as bared teeth and growling are not typically preludes to fighting, why do they exist? Paradoxically, such behaviors are usually about how to avoid fighting." (Smuts & Ward, 2011.)

Once we learn the content (or what they are "saying") in dog-on-dog adversarial communications, we can begin to determine whether we should step in or let the canines work it out on their own. To do this, we need to be able to tell the difference between warnings, threats, and aggression. We tend to overuse "aggression," especially when talking about animals, to the point where the word has become a catchall for each and every behavior, we may think is negative or are not comfortable with. This sort of thinking can lead owners to both think negatively and overreact as a result.

A more useful way to look at animal behaviors is in the context they occur in, as in what is happening, where it is happening, and categorize them as "distance increasing behaviors" or "distance reducing behaviors." In other words, what is the desired request or effect being communicated? The dog that is tensely rigid, ears back, feet firmly planted, lip raised and snarling is asking for increased distance from whatever it is in the environment that is threatening, while the dog that initiates a play bow towards another is asking for reduced distance and more interaction. Looking at the behaviors this way allows for the request, whether of more or less distance, or by a change in the environment to be addressed. So, if the squared off dog that is asking for space from the person or dog approaching and distance is granted, tension resolves and if the play bow is seen and welcomed, play can begin.

Much of how dogs exchange information with each other is lost on us as humans, who will never be able to hear or smell as well or fully process pheromones without a Jacobsen's organ or fully understand the significance of those abilities to dogs and other animals with these heightened abilities of sensation and perception. Dogs communicate with each other through vocalizations such as barking, growling, snarling, etc., visually through body language and body postures such as play bows, raised hackles, avoidance, etc. or through scents and pheromones actively exchanged as seen in butt sniffing or odors left behind through excreted urine or feces. All of these, more foreign interactions to humans. We can still become more sensitive to, and comfortable with, understanding how dogs correspond to disagree, agree to disagree, and resolve conflicts. To do this we have to first work on observing dogs.

We need to look carefully at all dogs, in different contexts: dogs at play, on walks, behind fences, tied outside the grocery store, asleep, sitting for attention, and so on and so on. Starting with our own dogs, we need to pay attention to what our dogs look like when they are happy, sad, excited, etc. What does their face and body look like? Being wary not to focus on only one aspect such as lips or eyes or ears but looking at the whole dog. How does the appearance of the dog change and in response to what?

Learning to read dog body language needs a basis in knowing the natural history of dogs along with paying attention to what is happening around the dog as much as what the dog is "saying" in response to it. Because dogs are a highly domesticated species through convergence in history with humans their story cannot be told without looking at their wild canine ancestors. Our modern dog's ancestors were those canines able to get close enough to scavenge from, hunt with or become companions to humans. This key point in learning about how dogs came to be dogs, is to understand that much of dog behavior is in response to human behavior and how that translates to what we influence when they are living in our homes, walking by our sides or at the dog park with us.

Dogs living with humans are dependent on humans for pretty much everything they get to do. This includes who they get to socialize with, dog or people wise, whether they have been taught how to act appropriately around dogs or people, if that training created good positive association or was painful, and whether or not they get to work out conflicts and potential conflicts on their own. As veterinary behaviorist, Petra A. Mertens writes:

> "Free ranging dogs can avoid conflict more successfully through avoidance of encounters with other dogs, making fights less common. Space restrictions and the influence of caregivers' attempts to control the behavior of the dogs may catalyze problems and prevent resolution using ritualized behavioral patterns." (Mertens, 2004)

Our discussion of dog behavior with other dogs then, is really about dogs behaving with other dogs around people. In fact, most studies in working with dog on dog "aggression" are primarily focused on humans redirecting the dog while the dog is attached to the human with a leash or an aversive device such as a head halter. These studies are really reflective of how well the human teaches the dog what to do in their presence. Interestingly enough, the same studies often find that without the human present or ongoing training, the dog reverts back to whatever the behavior was in the first place.

Studies on dogs freely interacting with other dogs' show that while dogs may "argue" with a full suite of distance increasing behaviors, threat displays or warning signs, dogs mostly do not attack each other:

- A 2011 study done at a dog park catalogued 127 agonistic interactions with none resulting in injury.

- In a paper on dominance and aggression in multi-dog homes, Petra Mertens (2004) noted that: "fights between co-habitating dogs represent between 5% and 18% of the overall caseload of veterinary behaviorists." Most of the fights being about proximity to food or toys 48%, proximity to caretaker 43% and defense of a preferred resting space 23%.

Mertens says if this is about resources, letting the dog with the most resource guarding potential have access to same when calm is most effective in defusing future fights. This is in effect giving them what they are the most worried about not having, which lessens stress which in turn creates a calmer dog.

- A 2016 study looking at multiple dogs and their interactions with other dogs in a doggy day care and in the home found that about half of the dog pairs observed at a doggy day care never interacted aside from mutual sniffing. New dogs were introduced into the environment on an ongoing basis and the study concluded:

"In these circumstances, establishing a relationship with every dog and negotiating dominance would require a lot of time, energy and social cognition. The ability to completely ignore a social partner who is in close quarters is probably another alternative strategy that dogs use to avoid conflict, and may have become more adaptive as the social environments of dogs changed from that of their wolf ancestors." (Trisko, Sandel, 2016.)

It is important to point out that those studies done in dog parks, doggy day cares and in homes with multiple dogs are most probably looking at dogs that are well socialized and already play well with others. These sites can be a filter and overly reactive and rambunctious dogs are often not welcomed or present at all. Owners, knowing their pets, mostly only bring dogs that behave appropriately to these environments. Rules are often posted or enforced by establishments restricting the presence of dogs that are prone to not interact well and often preventing the presence of intact males or females in heat.

So, having said all this, how and when do we know when too much is too much with dog-on-dog behavior? As much as this may be equal parts informed intuition along with a good grasp of dog body language and experience, it can still be a tricky determination. Here are some of my thoughts on the subject:

Never forget who the responsible party is on the other end of the leash- you. Think through possible scenarios and solutions ahead of time by reading and researching. Remaining calm to provide affiliation and social support for your dog is the most important thing to remember and to know how to do. Whether it is taking a deep breath, centering your body, reciting a soothing affirmation or whatever you need to do to keep your presence neutral is vital. Our dogs are experts at reading our body language and at sensing stress induced chemical changes (we smell different when we are nervous). It is highly likely that your dog senses your stress before you even realize you are experiencing it. A tense handler makes for a tense dog. As do punishment and force which increase aggression and fear and are counterproductive to containing aggressive encounters. They also severely hinder trust and damage relationships.

Survey the situation first, let them know what you think, ask, decide, and then act. Look ahead on your walk for approaching dogs or situations that you know rile the dogs. Do not wait until you are interacting with it. Decide on an approach or response as in, is this a situation or dog you want to/are comfortable interacting with?

Set you and your dog up for success ahead of time. Cue your dog in advance as soon as the approaching dog is in sight with asking for the kind of behavior you want and reinforce in a calm, positive and low-key tone as in - "Doodle, easy. Good Easy Doodle! Good Easy! Good Easy! (etc.)" Keep the dog engaged and keep reinforcing until you are past the trigger. That's your dog on the other end of the leash, no one but the dog, will care that you are talking to them.

Evaluate the potential interaction before it happens. We know that a loose, wiggly body on another dog is a good thing. We also know cross gender increases the chances for a positive interaction. Who is coming toward you on a walk? Does that look good on both counts? Ask the other handler if the dog is friendly, what sex and if they are good to interact. I used two sentences as I walked towards the dog and owner, I had determined was good to meet – "Boy or girl?" and "Can they say hi?" If the other dog checked those boxes, then I would let the dogs greet each other. If I decided not to, a polite nod and I kept walking.

Watch that tension on the leash. There is nothing more frustrating to a dog than being held just close enough to almost sniff another dog but not close enough. If your dog is straining at the leash with two feet off the ground you are creating and feeding frustration. If you are letting a dog interact with another, give enough slack in the lead so they can move freely and move around them yourself while monitoring the interaction.

It is sometime necessary and always good to know that you can walk in another direction, cross the street, and/or stand between two parked cars to avoid a situation. It is also helpful to learn how to calmly deal with one before it escalates.

Know that your body can always be used as a social support and a buffer against the environment and provide a safe zone by putting you between the hazard, dog, or person. Shorten, without tightening, the leash so your dog is closer to your side by running the hand that is not holding the leash down the leash and against your body, this will put your dog closer to you without pulling on the dog. If you need to move to put yourself between your dog and the oncoming dog, step around the back end of your dog to reach the other side of them and resist pulling on the leash to get your dog to move.

Working on desensitizing and counterconditioning dogs to triggers below threshold is the next step in working with reactivity around other dogs. Games like the "engage/disengage game" are good ones to start with. These exercises put desensitizing in place by exposing a dog, while the dog is not reactive, i.e., under threshold, to initial registration of the trigger of oncoming dog and counter conditioning in marking and reinforcing the calm behavior. Distance, duration, and intensity of the stimuli are targeted in small, gradual increments with much reinforcement. Over time, distance is minimally decreased to the trigger. How long the exposure is, needs also to be limited and increased in baby steps. And distance and duration need to be built up independently of each other, work on one aspect for several days before turning to the other. Timing in any training or modification is critical, also where our own human expectations need to be managed. Small, short, minor advances in exposure are how to move forward. The proverbial "go slow to go fast."

Learn the art of the neutral introduction: Having company over and that includes their dog? Meet your guests with your dog on the corner and walk together to your home. Dogs do not equate territory with just your house or apartment, the street and sidewalk in front of your home are also part of what they believe "belongs" to them. Meeting on neutral ground and going to your home together can help to remove the "intruder" tension.

Bringing treats with you to reinforce the neutral behavior for all dogs never hurts, just as long as there are enough treats to go around –make sure to have each owner treat their own dog. The dogs will be watching.

Allow for butt sniffing, particularly as walking a dog behind another dog, this sort of linear introduction is highly effective for familiarizing dogs without pressure.

Master entrances and exits that disperse energy: When entering the dog park with a do, make sure the dog is unleashed between the double fences before entering to guarantee an even playing field where all dogs can move around freely. Move directly to the middle of the space as soon as you enter to disperse the congregated dogs that have noticed your arrival. Your dog and the dogs in the park will follow.

Most squabbles in dog parks occur at the fenced entrances where entering dogs attract the attention of the dogs in the park. Multiple dogs lining up at a fence creates a confined space where excited bids for attention and exploration are easily intensified. Walking immediately to the center of the park allows for a full use of the surrounding space to both explore and retreat. Keep talking to a minimum to allow the dogs to interact with each other and not focus on you. Make sure to keep your hands at your sides, raised hands invites jumping and avoid direct eye contact with any dog.

For exits and entrances, be first, not to be "alpha," rather that you can see what is coming up ahead and cue your dog accordingly.

Remember separate by size and play style: wisdom for most, if not all, dog pairings. Dog parks often have two play areas, one for small dogs and one for large dogs. In a scenario with a small dog that plays too intensely for his same size companions, try the big dog park on for size and be sure to monitor for results. Safety first, if this looks like it is not working out for the small dog in the big dog park, go to the small dog park.

Become comfortable with asking your dog to return to you ("recall"). Get your dog also comfortable with this by training them to answer to your recall request, offering them a treat as a reward and immediately releasing them back to play. This allows them to learn a recall is never a punishment and you get to use the request when you need it and know it will work.

To train this, call the dog using their name and a request like "Come." This would sound like "Daisy! Come!" Use a happy and excited tone and try clapping your hands and backing away from your dog while facing them to get them to follow. (Be careful with this part, make sure you are able to safely back away first.) Use treats. Make sure when the dog reaches you to reward them with a treat and mark/acknowledge with a "Good Come!" so they know they have done what you want. Immediately release the dog with a "Go Play!" or "OK!" Let some time pass before you for the recall again.

Practice the recalls and releases at home and in the dog park. This is ideal to do without other dogs initially at home. Without outside distractions you can see more immediate responses. Shorten distances between you and the dog before asking if you are not getting a response. You can also do this with a partner where you can both take turns sending, calling, treating, and releasing the dog to each other. Once you have a few sessions of your practice recalls and releases down, try them out in the dog park with other dogs around.

In a dog park, you may want to use effusive praise as your reinforcement instead of treats, because again, other dogs are watching. Do not exceed three recalls and releases at any one session, either at home or in the park. You want to keep training fun. And always, always reinforce with treats, praise, and releases. Once you have this down, even if everyone is playing well, make sure and practice it at least once or twice during your park visits so it stays fresh, fun, and relevant.

Never, ever use this request to place a dog in a place of confinement such as a crate or a closed off room. Who wants to listen to that? Knowing where it ends up? If needed or necessary to send a dog to that space, always use a different cue or phrase.

Learn to offer a correction or redirection with a calm neutral tone, fluid timing and reward the response. In a scenario where you do not like what is happening in the dog park use your recall and redirect the energy with throwing a ball or a stick. Have an overly boisterous mounter ("humper") and the other dog looks ready to lunge in response? Refocus the energy on interacting towards something else beside the other dog. We can never just call a dog away from something and expect them to forget what it is that has gotten them so excited we have to provide a distraction that just is as exciting. An in the moment "uh habituating" followed by a "Good Dog!" when your dog responds followed by calling the mounter's name and then tossing a ball for redirection is a skill worth its weight in gold.

After some time at something else, let them go back to each other. Remember the rule of thumb with separating dogs at play: if you separate two dogs and the "victim" dog follows the "bully" dog for more play afterwards, you have just broken up a perfectly good play session. Learn from it.

In the dog park continue to reinforce the behavior you want. If the energy starts to shift allow for some time for the dogs to sort it out on their own, observing carefully. This is best followed with two dogs, with two dogs overly threatening one other dog, interrupting sooner rather than later is safer.

Get over "humping." First, the correct terminology is "mounting" mostly because this behavior which may look sexual is not (actual reproductive mounting has a different pattern of behavior). Mounting is an attention getting behavior that a frustrated or bored dog engages in. Dogs tend to play in pairs and mounting is usually seen when a third dog wants to join in and is ignored. Work on redirecting the energy of the mounter by engaging them in something else such as running after a ball, a tug of war, etc. For more on mounting, please see the preceding chapter in this book.

Know that most of what we do not intervene in will be resolved without us. It is a complicated determination that gets easier the more we know. Remember, as owner/guardians we never want to train out a warning. The old saying of not training out a growl is an important one to remember. We want our dogs to be able to offer that warning growl instead of learning that it is now forbidden and having to go right for the bite instead. *The Dogs Playing for Life* program advocates for group play for shelter dogs to increase welfare and in doing so help get more dogs out of shelter. The program offers excellent advice on the danger of "punishing the thought" as opposed to the actual behavior:

> "Don't focus on the minutia of body language. None of these signals tells us very much on their own. Try to take in the whole picture in order to best read the dogs in front of you. Allow dogs to communicate with one another. What is the other dog doing in response to those communication signals? Do not act right away if you spot signs of tension and stress. Do not "punish the thought", by correcting these communication signals. Always wait until there is an actual behavior that needs correcting."

> "the goal of play groups is for dogs to learn how to communicate with one another appropriately, which may sometimes include brief arguments, in order to establish themselves with one another. We do not dictate dogs' relationships with one another (the dogs decide who they like or do not like), but we do monitor their behavior, stepping in only when necessary."

Know when to interrupt an interaction: Use your words first, employing a calm and neutral tone. Make sure you are at safe close distance to the dogs you are addressing and not shouting or calling from too far away. Remain neutral. Do not add negative energy to the mix. Never yell at the dogs as this will increase arousal not lessen it. Avoid grabbing dogs by their necks or by the collar, this move can cause a dog to instinctively whip their head around and grab/bite what is in front of them protectively – your hands or arm. Call each dog by name, to get their attention and follow by a specific redirection. *Only, if it can be done safely*, and the dogs are not fighting, consider body blocking.

If you see a dog biting or breaking the skin of another dog or threatening behavior that is not provoked such as pinning or standing over another dog while vocalizing or if a dog displays serious warnings or threats towards a human such as snarling with bared lips, growling, aggressive bark, snapping or biting- the dog should safely be removed from the situation immediately.

They are fighting anyway: Breaking up a dogfight can be dangerous for the people and the dogs. Take the time to review possible strategies to respond. The simple act of thinking through response strategies ahead of an event can lay a foundation to put them into action.

The ASPCA's site formerly included a "Virtual Behaviorist" section. The American Pit Bull Terrier Assn Inc (NZ)'s Facebook page/notes excerpted the A's advice on breaking up a dog fight:

"Plan A: Startle the Dogs or Use a Barrier

Before you physically separate two fighting dogs, try these methods:

A sudden, loud sound will often interrupt a fight. Clap, yell and stomp your feet. If you have two metal bowls, bang them together near the dogs' heads. You can also purchase a small air horn and keep that handy. Put it in your back pocket before taking your dog somewhere to play with other dogs. If you have multiple dogs who get into scuffles, keep your air horn in an easily accessible place. If a startling noise works to stop a fight, the noise is effective almost immediately. If your noisemaking doesn't stop the fight within about three seconds, try another method. If there's a hose or water bowl handy, you can try spraying the dogs with water or dumping the bowl of water on their heads. Use a citronella spray, like SprayShield or Direct Stop®. Aim for the fighting dogs' noses. If you walk your dog in an area where you may encounter loose dogs, it's wise to carry citronella spray with you. If an aggressive dog approaches, spraying the deterrent in his direction may stop him in his tracks and prevent a fight. If he attacks, spraying the deterrent on or near his nose may break up the fight. Try putting something between the fighting dogs. A large, flat, opaque object, like a piece of plywood, is ideal because it both separates the dogs and blocks their view of each other. If such an object isn't available, you can make do with a baby gate, a trash can or folded lawn chair. Closing a door between the dogs can also break up a fight. Throwing a large blanket over both dogs is another option. The covered dogs may stop fighting if they can no longer see each other.

Plan B: Physically Separate the Dogs

If other methods don't work or aren't possible, it's time for Plan B. If you're wearing pants and boots or shoes, use your lower body instead of your hands to break up the fight. If they're covered, your legs and your feet are much more protected than your hands, and your legs are the strongest part of your body.

If you feel that it's necessary to grab the dogs, use this method:

1. You and a helper or the other dog's pet parent should approach the dogs together. Try to separate them at the same time.

2. Take hold of your dog's back legs at the very top, just under her hips, right where her legs connect to her body. (Avoid grabbing her lower legs. If grab a dog's legs at the knees, her ankles or her paws, you can cause serious injury.)

3. Like you'd lift a wheelbarrow, lift your dog's back end so that her back legs come off of the ground. Then move backwards, away from the other dog. As soon as you're a few steps away, do a 180-degree turn, spinning your dog around so that she's facing the opposite direction and can no longer see other dog." (ASPCA nd)

Another source on dog fights includes video clips to illustrate techniques: https://www.preventivevet.com/dogs/how-to-break-up-a-dog-fight, including a clip on the wheelbarrow move.

Know when time is up. Less is more. Dog park visits and play groups are always better on the short and sweet side and left off on a good note. There is too much of a good thing and prolonged interactions in constrained spaces can be too much pressure without allowing for down time.

Remember. There are many excellent sources to learn about canine natural history, communication, and behavior. The internet can be a great and freely accessible resource, if the site used for information, is well qualified and a credible one. Make sure to keep to experts who have well rounded, formal backgrounds –have studied animal behavior and learning theory at a university or graduate level and who are focused on welfare and science.

Videos are particularly useful as they can outline concepts and illustrate what they look like at the same time. A good example of what play looks like, is this one from our site and a nice example on canine body language, always good to review, is this other video from another great reference and source, Maddie's Institute.

The more you know about dog behavior, the more you eliminate force and punishment, and the more your dog can experience positive interactions and get through minor quarrels with other dogs is the best way for your dog to learn how to play well with others and for you to enjoy it too.

References

- Breaking up a Dog Fight – ASPCA. *American Pit Bull Terrier Assn Inc (NZ).*   https://www.facebook.com/notes/986272651839174/ retrieved November 8, 2023

- Dogs Playing for Life Manual (2014) . Retrieved from http://dogsplayingforlife.com/dpfl-manual/

- Capra, A., Barnard, S. and P. Valsecchi (2011) Flight, foe, fight! aggressive interactions between dogs. *Journal of Veterinary Behavior: Clinical Applications and Research*, 6-1, 62.

- Choose Positive. The Practice of Self-Interruption – The Engage Disengage Game. https://www.choosepositivedogtraining.com/single-post/2014/07/01/The-Practice-of-SelfInterruption-The-EngageDisengage-Game retrieved January 16, 2024

- Echterling-Savage, K., DiGennaro Reed, F.D., Miller, L.K. and S. Savage. (2015). Effects of Caregiver -Implemented Aggression Reduction Procedure on Problem Behavior of Dogs. *Journal of Applied Animal Welfare Science*, 18-2, 1-17.

- Orihel, J. S. and D. Fraser. (2008). A note on the effectiveness of behavioural rehabilitation for reducing inter-dog aggression in shelter dogs. *Applied Animal Behaviour Science*, 112-3, 400-405.

- Madson, C. (2023) How to Break Up a Dog Fight Safely. *Preventive Vet.*
https://www.preventivevet.com/dogs/how-to-break-up-a-dog-fight
Retrieved November 8, 2023

- Mertens, P. (2004) The Concept of Dominance and the Treatment of Aggression in Multidog Homes: A Comment on van Kerkhove's Commentary. *Journal Of Applied Animal Welfare Science*, 7-4, 287-291.

- Trisko, R.K., Sandel, A.A. and B. Smuts (2016) Affiliation, dominance and friendship among companion dogs. *Behaviour*, 153-6-7, 693-725-

- Ward, C. and B. Smuts (2011, April-May). Dogs Use Non-Aggressive Fighting to Resolve Conflicts. *Bark*, 65-68. Retrieved from http://thebark.com/content/dogs-use-non-aggressive-fighting-resolve-conflicts.

READING THE CANINE LADDER OF AGGRESSION

Ohhhh, that "guilty look." You know the one. You walk in the door and your dog sees you and greets you with what you just know is a worried expression. You stiffen, is something wrong? You raise your voice just a little bit. Did that dog do something wrong again? He sure is looking mighty embarrassed. Is that a true "guilty" look or is it highly effective appeasement behavior to stop the scolding, avoid punishment and get you to stop scolding?

When it comes to "reading" canine body language how versed are you in understanding just what your dog is saying? We can agree that dogs for sure know when a human is upset, but do dogs know that past behavior is what it is that is upsetting? Humans appease each other to keep the peace and dogs are no different. Dogs communicate with each other and with us all the time. While the messages may be clear dog to dog, are we able to understand what our canine friends are "saying"?

Both dogs and humans have developed sophisticated social communication systems. For each species, communicating feelings, intent and trying to influence each other (including reconciliation attempts) are highly effective ways to avoid conflict and maintain social order. Fighting costs big time and avoiding potential physical and emotional injuries is the best survival strategy.

The human words and gestures we use to avoid conflict differ from the canine signals used for the same purpose. To fully appreciate the conversation, we need to look at the whole dog, environment and/or situation and context at the time. Before proceeding to the behavioral, we need to first rule out the medical. Once all clear on that front, there is much to learn from closely observing what has just happened, is happening at the moment, and in the body tension, movement, posture, along with eyes, ears, mouth, and tail.

Kendal Shepherd, a veterinarian, and animal behaviorist, devised a "ladder of aggression" showing what this looks like for both species. A look at the ladder shows an escalation of aggression from the first rung to the last. What is of note is just how very different deference and reconciliation strategies look for a dog when compared with a human. For instance, those steps on the ladder of looking away or walking away on a dog's part is an attempt at peacemaking and should never be construed as an attempt to ignore the other party as it might be in a human context.

When we give human motivations to a dog, we lose the canine motivations in the translation. This also means that stress or discomfort being communicated with body language, such as the looking away the dog is doing, is being ignored and offense may be taken when none is intended.

Another example is the yawn, because as humans, we may yawn when bored or tired, we can think fatigue or disinterest is the motivation for why the dog is yawning too. Not so. For a dog, when we see a yawn out of the context of being sleepy or tired, what is being signaled by that dog and what is readily apparent to other dogs, if not us, is possible appeasement, self-soothing or preliminary stress.

It is important to be able to tell the difference between warnings, threats, and aggression. We tend to overstate and overuse the term "aggression," especially when talking about a dog's actions, to the point where the word has become synonymous for whatever we are seeing that we dislike, think is negative or are not comfortable with. This sort of thinking can lead owners to respond negatively as a result. Remember, the signals that the dog is offering on this ladder are simply communications that they are uncomfortable with what is happening in the environment, or that something is happening or someone is doing something they would like to stop or that they would like more space.

In Animal Behavior Science, we can classify behaviors into either "distance reducing" ("come closer") or "distance increasing" ("go away") behaviors. Such behaviors are also communications, requests or attempts to influence another. The next time you see a distance increasing behavior, try and figure out your dog is telling you and what you can do to help to both increase the distance and change what is causing the need for it.

For instance, the dog that growls at a stranger on the street wearing a hat and sunglasses is probably uncomfortable with something in an appearance they are not familiar with or someone he has not seen before. Acknowledge that concern (try saying "I know") first and then either turn away in another direction or move your body (not the dog's) so you are in between the dog and the stranger, creating a safe buffer and social support as you walk past.

Another instance: Are children playing with your dog in such a way that your dog is saying is not welcome with the behaviors they display such as lip licking, looking away, ears back, etc.? Then this is the time to stop the playing and remove the parties, taking the pressure off the dog, and taking advantage of the moment, explain to the children how to play nicely with dogs, etc.

Take a closer look at Dr. Shepherd's ladders of aggression – Human and Dog, compared to each other:

Canine Ladder of Aggression (Kendal Shepherd)

11) Biting (*most frightened and threatened*)

10) Snapping

9) Growling

8) Stiffening up, staring

7) Lying down, leg up

6) Standing crouched, tail tucked under

5) Creeping, ears back

4) Walking away

3) Turning body away, sitting, pawing

2) Turning head away

1) Yawning, blinking, nose licking (*least frightened and threatened*)

Human Ladder of Aggression

11) Slap, punch, kick

10) Push, throw something

9) Clench fist, threaten

8) Shout, scream, swear

7) "I've said I'm sorry! Stop it!"

6) "I'm sorry"

5) "Please calm down"

4) Walk away

3) Argue

2) Fold arms, frown, turn away

1) Smile, hand shaking

Oftentimes the rungs of the ladder are climbed directly whether we notice it or not. Other times, rungs may be skipped because in the past they have been proven to have little effect or a perceived threat happens too quickly or is too close for comfort. When appeasement signals are consistently ignored, they may cease and a dog may feel little choice but to defend more and more aggressively. This is one of the reasons a qualified trainer or behaviorist will tell you never to "train out" a growl -the warning and/or request for space the growl is expressing is before snapping and biting - pay attention to it and the next step on the ladder does not need to be climbed.

Understanding how these reactions are being used to communicate and handle stress can allow you to respond to them. So, the next time you see that yawn out of context or look away, etc., take the time to look at what else is going on in the environment. Remember to consider this in a hopefully canine relevant perspective. How does this matter to a dog and why? What is the behavior a reaction to?

A good grasp of what preceded the behavior and the context it occurs in are vital clues in determining how to provide distance, change the setting or otherwise mitigate the perceived threat. In our case of scolding the dog, we can stop the scolding so the dog no longer needs to appease us. If a raised voice brings on a yawn, we can lower the voice and hopefully alleviate the pressure from the dog and see the behavior change, etc.

References

- Shepherd, K. (2009) Behavioural medicine as an integral part of veterinary practice, in: *BSAVA Manual of Canine and Feline Behavioural Medicine*, (2), pp 10 – 23.

MINE!! UNDERSTANDING AND WORKING WITH RESOURCE GUARDING

Resource Guarding: "the use of avoidance, threatening, or aggressive behaviors by a dog to retain control of food or non-food items in the presence of a person or other animals." (Jacobs, Coe., et al.)

Gulping food at breakneck speed, playing keep away with the ball, growling at anyone near the food bowl, barking when somebody gets too close to their person on the couch, whining when the other dog gets the treat first the list goes on. Sounds familiar?

What do you call it? Resource Guarding? Possessive Aggression? Food related aggression? All of these terms have been used to define behaviors dogs display to gain or keep access to, or control of, something of supposed value in the presence or approach of another dog or person. Why does it matter? And can you fix it or at least manage it?

Depending on context or what is going in the environment when this is happening, these may be totally different things. A very hungry dog may gulp food to satisfy that hunger, a dog with a history of being deprived of food around others may gulp food to protect the opportunity to eat, and an anxious dog may gulp food as part of their general disposition.

Highly ritualized distance increasing behaviors serve dogs in avoiding aggressive interactions. Fighting is always a last-ditch effort from a biological standpoint, no one wants to get physically injured.

Most, if not all animals would rather the other party go away instead of engage in violence. We can be both misinformed and careless with the use of "alpha" and "dominance" when discussing dog society and forget that dog society is mainly a society of deference of "no, it's OK – you go first" in response to a warning or threat rather than "let's fight." And if that is how dogs "talk" to each other, if a growl is a warning to a back off, and if the other dog hears that and responds, what happens when we get it wrong?

Emotional states, motivations, and environments set the stage for much of what we are looking at with behaviors. We do know that what we call something is important, it colors how we think about it and from there how we respond to it. Putting the aggression label on to Resource Guarding, or any behavior, can muddy the waters unless we are in fact talking about true aggression here – the intent to cause harm as opposed to bluffs, threats, stare downs and the like.

A 2018 study reviewed expert opinions on what to call those avoidance and guarding behaviors of protecting perceived assets. The expert consensus was to drop the aggression label (unless that is actually what is being seen) and to use:

> "resource guarding over possessive aggression due to the potential for motivation to be interpreted more accurately by owners." (Jacobs, Coe, et al., 2018)

Here, study participants mentioned concerns not just with the use of the word "aggression" but also with using "possessive" for fear of additional misinterpretation that resource guarding behavior might be viewed as challenging an owner coupled with the then potential of owners resorting to force in retaliation – a strategy that makes things worse not better.

While words matter in how we view behaviors and respond to them, for those actions that guard or protect resources, valued objects, places, or people, how do we assess and manage them? And what might the possible concerns be for the dogs for guardians to mitigate?

Researchers De Kuster and Jung discuss possible motivations and emotional states:

"competitive disputes over resources (including puppies) may occur with family members of all ages, but tend to occur with children more than adolescents, with adults being least frequently involved. The disputes may also occur with other familiar animals in the home, including other dogs and certain pets, such as less fearful cats…the dog feels threatened and potentially frustrated and so the under- lying emotion is negative. The choice of strategy (whether or not escalating into a bite) will depend on many factors, including actual mood state, perception of the situation, and previous learning experiences (aggressive episodes) and their outcomes…Dogs displaying aggression over resources should be screened for signs of generalized anxiety" (De Kuster & Jung, 2009)

Motivations for this behavior may vary from territoriality, uncertainty over place in the group dynamic, fear, health issues, etc. Canine behavior experts agree that aggressor dogs are also found to have high anxiety levels and benefit from routine, schedule, and owner predictability. And some of that predictability if leading to negative consequences, can also make things worse. In our homes we tend to admonish the aggressor dog and support the victim which may increase tensions between the dogs.

Dogs not living under human control, such as free ranging dogs, or perhaps even free roaming dogs, are better able to circumvent conflict, most probably due to independence and agency, along with spaces to retreat to after deference.

Jacobs, Coe, et al in another study, also looked at what factors are frequently associated with resource guarding around both other dogs and around people. In this study, they found that when it came to the presence of other dogs, this was a relatively fixed pattern and high impulsivity and fear increased the chance the behavior would occur. When it came to resource guarding around people, the finding was resource guarding to be a more flexible behavior and that an often recommended intervention, like removing the food dish during meals was associated with increased resource guarding. Adding more palatable (yummy) food during meals when people were around was shown to reduce the behavior as was teaching dogs to drop items to reduce resource guarding both around dogs and around people. While asking a dog to release a valued item may have an immediate effect, a better intervention would be to introduce a "trade" instead (more on that below).

Looking at circumstances surrounding resource guarding can help us in both better understanding the occurrence and addressing it. Dr. Petra Mertens writes that research shows that:

- Most fights between dogs in the home occur under states of high arousal/excitement, next to food or toys or next to the caregiver, all of which are valued resources.

- Volatile contexts include; defending favored resting space and responding to threatening postures such as hard stare, tensed, and leaning forward, growling, etc.

- For who is the responsible party or the instigator in the same home, younger dogs or new additions to the home, are found to be aggressing first.

- Female pairs represent the majority of most physical conflicts.

- Fights were found to occur more frequently when owners are present or close by, as opposed to when the dogs are alone. Such a finding, suggests this may follow either as it relates to proximity to owner, whether as such presence is a resource, or to owner as perceived social support emboldening a response.

Working with resource guarding behavior and/or possessive aggression whether with a single dog or in the multi-dog home is possible. **It must be stressed that there are never guarantees when working with any behavior or intervention.** Individuals, history, dynamics, adherence to, and or successful applications, to behavioral interventions are complex, varied and cannot be certain. Best possible outcomes require a strong commitment and follow through to, appropriately best managing the environment to prevent the possibility of conflict, devotion to a consistent, every day, force free, welfare and science based behavioral program of counter conditioning, desensitization and retraining and the time to do it.

Even with the best of scenarios there are latency periods – when things stay the same, such as conflicts between dogs and opportunities for human injury, before they change. Well educated and appropriately qualified, behaviorists and trainers can help. Chances for success are compromised by: the history of the problem persisting for an extended period of time, if fights lead to injuries, or if fights are not predictable for the owner.

How owners deal with resource guarders can add to the problem not lessen it, notes Mertens. Not punishing unwanted behaviors can be too much of a challenge for many owners, and punishing aggressive behavior exacerbates it. Resulting stress from punishment or resentment over not punishing from an owner can impact an owner's connection with the aggressor dog which in turn may increase pressures between dogs.

Mertens suggests providing the dog with the greatest resource guarding behavior precedence to resources when they are not in conflict. This is the best way to satisfy the dog's need to feel secure with those resources in a more positive, less stressed, emotional state. For instance, letting the dog have the tennis ball when they are not frantic over desperately needing to have one. Following this advice is another struggle for owners. Humans can miss that for the dog and their congruent emotional states, having access to a valued resource when not reactive is not a reward for overreacting but might be a reward for not overreacting. Possession of that valued tennis ball when calm about it can reward the calm behavior.

Dr. Karen Overall, writing on treating, what she calls possessive aggression with dogs, offers the following:

- "This condition is most easily treated by managing the environment so that the dog cannot gain access to possessions he or she might want to control.

- No one should reach for anything the dog is guarding.

- Walking away from the dog and trading, when needed, are the preferred management strategies.

- There are very few dogs who are affected with possessive aggression alone. Accordingly, teaching dogs to sit calmly and rewarding them for being non- reactive when people reach toward them, if this can be done safely, will benefit a treatment plan for reactive dogs.

- If the clients have just noted that this condition is developing, teaching the dog to sit, take a deep breath, and then "trade" could prevent the condition from fully developing."

Having good information and cautions in place are necessary. When considering working with resource guarders or food aggressive dogs individually, in a group, or even with appropriate professional advice, make sure your own knowledge and understanding include a familiarity with the right approaches to be used and why. Protocols should consider and include:

1) Removing all punishment. No scolding or anger no matter what. **None**. Dogs never, ever understand a lecture, they just act what looks like contrite so we stop scolding them. All punishment creates fearful associations, interferes with learning, damages relationships and needs to be escalated to be effective.

2) Managing the environment. Avoid access to those locations and situations where the reactive explosions are taking place. Redirect dogs from confined spaces. Feed and play with toys and dogs separately at a far enough distance possible. This can be in separate rooms, floors, or opposite sides of a room.

It is key to remember, that we too are part of the environment, and the most important part that our pets react to. Keeping our mood, tone of voice, body language positive or neutral are important when around or when interacting with our animals.

3) Behavior Modification. Become a keen observer of dog behavior and be able to identify signs of stress and how they escalate along with suitable mitigations. Whether working alone or with a professional make sure you can follow along with the "why's" and "how's" of

behavior modification protocols for resource guarding. Jean Donaldson's "Food Aggression and Resource Guarding" and Patricia McConnell's "Resource Guarding, Dog to Dog" are good ones to learn from.

- Work with the whole dog. Adding in small bouts of positive, force free training can help anxious dogs through structure and enrichment, build the bond and trust between dog and owners and encourage better behavior to meet human expectations.

- Make sure the dog gets sufficient opportunity to walk and sniff. Three to four walks a day are optimal.

- Keep walks and mealtimes on schedule to offer the sense of control routine can provide.

- Puzzle feeders that channel chewing and sniffing are good to provide for satisfying necessary natural behavior and reducing stress, as long as the dogs can be provisioned separately and at a distance from each other.

- Teach alternate behaviors like "trade" when asking dogs to surrender a valued space or item – as in have a just as good, if not better, replacement object in hand, obvious and ready to offer in exchange when asking for that trade.

- Never forget that changing a behavior requires replacing it with one of equal or greater value. Be generous and consistent with rewards when training and reinforcing behaviors. One of the best doggy day cares I worked with, had a floor literally covered with tennis balls.

- More is more. Remember, always keep dogs at a distance they feel most comfortable when working with them.

- If it can be done safely, hand feeding is an excellent exercise for impulse control, reactivity, affords structure, control, and safety. Incorporating this into my exercise to teach "Off" and

"Take it" (see notes on how below) and using those cues for releasing objects or spaces or engaging with them can be a valued tool to have in your behavior toolbox.

This is an exercise that takes practice to get the timing down. And timing is essential to the exercise. Remember, the dogs cannot get it wrong since they will only be following your behavior. Work on your timing and keep it loving the entire way through, this is very important.

For dogs who can be SAFELY hand fed: Off and Take it:

1) Hold the food up to the dog's nose while saying "(Dog's name here) Off." Do not move your hand.

2) When you see the dog pause for even a second or pull their head back, right away say "Good off!"- you are marking/acknowledging the "Off" and that the dog has done what you asked and rewarding them with praise.

If they move towards the food instead, use a short "uh uh" as a redirection and ask again "Off." Say "Good Off!" when they do what you ask.

Remember to keep your hand with the food in front of the nose and not to pull it away -if you do that you are "asking" them to follow it.

3) Next immediately holding the food, in the same place, in front of the nose, say "take it."

4) Let them take the food. Release the food quickly.

5) As they take the food: Say "Good take it!"

And repeat the whole portion this way: Hold food to front of nose/muzzle. Say "(Name of dog) off." Mark or redirect with either "Good off!" or "uh-uh." **(Do not move your hand).** If "uh-uh" ask for "Off!" again. Say "Good Off!" when you get it and offer food. Say "Take it"-now allow for them to take food and say"Good take it!"

Keep a diary of what is going on, what is happening each day, make sure to note any small conflict, behavior changes and around what resource they are happening in. It is far easier to change an emotional state or redirect an agitated one before it escalates too far up the ladder.

References:

- De Kuster, T and Jung, H. (2009). Aggression toward familiar people and animals. In Horwitz, D.F. and Mills, D.S. (Eds). *BSVA Manual of Canine and Feline Behavioural Medicine* , (pp. 182-210). Gloucester, England: British Small Veterinary Association

- Jacobs, J.A., Coe, J.B., Pearl, D.L., Widowski, T.M., Niel, L. (2018). Factors associated with canine resource guarding behaviour in the presence of people: A cross-sectional survey of dog owners. *Preventive Veterinary Medicine*, 161(1) 143-153

- Jacobs, J.A.. Coe, J.B., Pearl, D.L., Widowski, T.M., Niel, L. (2018). Factors associated with canine resource guarding behaviour in the presence of dogs: A cross-sectional survey of dog owners. *Preventive Veterinary Medicine*, 161(1) 134-142

- Mertens, P. (2004). The Concept of Dominance and the Treatment of Aggression in Multidog Homes: A Comment on van Kerkhove's Commentary. *Journal of Applied Animal Welfare Science.* 7. 287-91

- Overall, K. L. (2013) . Abnormal Canine Behavior and Behavioral Pathologies Involving Aggression, In Overall, K.L. . *Manual of Clinical Behavioral Medicine for Dogs and Cats* , (pp. 172-230). St. Louis, MO: Elsevier

SEPARATION ANXIETY

When Daisy was still a puppy, we moved from urban New Jersey to suburban Florida. She was my first dog as an adult and at the time, I was still very much learning about living with and training dogs. When we left Daisy to run an errand, she could not be part of, go to dinner, or visit friends without her, we would just leave. That was it. We did not say "good bye" or tell her we would see her later, we just walked out the door. That had worked just fine in our house in New Jersey but it was not working in Florida.

Moving house is one of the top three stressors in life. For human beings who contemplate, plan, and execute a move, the stress is a major one. For pets who get no say in the matter, there is also major stress. Maybe it was the move, maybe it was the new and unfamiliar space, maybe it was her age. Daisy would not make any noise, she didn't bark or whine. What she did do was scratch at the door frame. So vigorously did she scratch alongside that point of exit, that the paint came off. While I had no name for this behavior at this point, I did know that she was clearly upset by us leaving her behind.

I told this story to an uncle-in-law who had dogs. "Did you tell her goodbye?" he asked me. No, I did not, I was so busy with everything else going on, it just had not occurred to me. After all, dogs cannot go everywhere with their family in a new environment, no matter how much they would like to. Or how much we would like to bring them with us. I just made sure to lock the door and left. No goodbye.

After that conversation, from then on, I did just that. When we were leaving, I told Daisy goodbye. I also told her that I would see her later and to be a good girl. Sometimes, I would add that she should watch the house, as that was her only job. Why it worked I cannot say but it did.

This is not to say she was thrilled with our leaving, even with saying goodbye, it was plain to see that Daisy was not happy when we left her behind. It showed, she would retreat to her bed or the sofa, exhale with a woosh, and lower her head and watch us leave. But she did stop scratching at the door and when we came home, she was thrilled to see us. I would ask her to get a favorite toy to channel all that exuberance.

Targeting a door, or a point of exit, to get out like Daisy was doing, is one of the signs of separation anxiety or separation related problems. "Separation Anxiety" in dogs, is characterized by behaviors indicating severe distress when apart from an owner or other a dog is attached to. Second to aggression, Separation Anxiety is one of the most common issues owners list when seeking professional help or rehoming dogs. While we know, separation anxiety affected a good number of dogs pre-pandemic, it will no doubt affect an even greater number, post pandemic, as more dogs who have never been home alone or have grown accustomed to continual owner presence, end up there.

This anxiety disorder is additionally known as "Separation Related Problems" depending on what behaviors are presenting. This suite or set of behaviors is typified by reactions that include excessive vocalizations, destructiveness (especially at points of egress such as doors, window frames, crate doors and walls), rearranging objects, inappropriate elimination, depression, restlessness, self-mutilation and more. As such behaviors are mostly happening when a care giver is not home, not seeing evidence that they have occurred, does not rule out the dog that suffers in shut down silence. We also do not know the extent to which separation anxiety may affect cats who may be experiencing the same anxiety but not displaying signs we are aware of.

Research notes apartment dwellers are more likely to report this problem, possibly due to neighbor proximity alerting them to an event a more distant homeowner may not be made aware of. My applied work in animal behavior modification is mostly in New York City, while I do also work in the surrounding metropolitan area, most of my separation anxiety clients are in the city.

Researchers, Lenkei, Gomez, et al. (2018) note that frustration versus fear and or anxiety, may be the most overriding emotional state for the dog experiencing separation anxiety. What is so important to remember here, is that separation anxiety points to extreme anguish in the emotional state of the animal. Dogs are highly social animals, and being deprived of attachment figures can evoke an extreme response. Frustrated, fearful or anxious, these dogs are absolutely beside themselves and much of what we see in their behavior is how they are coping with their overwhelming panic and anxiety over being left alone.

Classic approaches to treating separation anxiety in the literature, are limiting owner absence, removing punishment and behavior modification with counter conditioning (creating good associations with the bad thing) and desensitization (increasing exposure through minimal increases in duration or decreasing distance to the bad thing) while under threshold (the amount of time with or next to the bad thing which is not stressful or not so bad).

The challenge with working with separation anxiety, aside from not leaving the dog alone, which in itself can be a monumental one, considering just how very many "no dogs" places there are, is timing the formation of positive associations close enough to threshold (in this case, the point where being left alone or knowing they will be left alone evokes a phobic reaction) and not exceeding threshold in duration.

How close to the scary event of leaving can you go, to make the prospect of what is coming up not so scary? At what point when the dog has picked up on your imminent departure, is it too late for them to be comfortable enough, to still take a treat, or chase a ball, or be past threshold?

When anxiety does takes over, the brain is flooded with stress chemicals and learning or relaxation is not possible. Periods of time left alone in the desensitizing process must be measured in each scenario for the individual and counter conditioning must occur with the dog under threshold so that anxiety cannot take over in either phase. This means when to offer reinforcers and desensitizing periods can be counted in seconds and minutes depending on the dog. It also means that there is no small a period kept under threshold that is not a big enough accomplishment to build on.

In scenarios where these routines and behaviors are believed to be motivated more by frustration, some researchers place responsibility for this behavior more squarely on the dog's owner rather than a dog's nature. Lenkei, Gomez, et al., writing about the role of frustration in separation related behaviors with the "lenient owner", the owner's "permissive and inconsistent behavior" and the "dog's demanding behavior", found barking to be more indicative than whining, adding additional emphasis needs to be placed on the role of routine, schedule and training to modify behavior. Leaving off the pejorative characterizations, it is agreed that such a fix applies to all companion dogs, especially the anxious ones. There is a tremendous burden in anxiety, and no doubt a frustrating one, to have to constantly demand attention, meals, walks, playtime, toys, snacks, etc., etc., etc., in world where you cannot control any of it.

For maximizing possible successful outcomes, knowing how to tackle Separation Anxiety, and determining threshold is the first step. Looking at the research on applying behavior modification approaches can yield a wealth of information and can be inspiring in keeping owners on track and motivated with protocols of baby steps and consistency.

Generic plans offer guidelines to start from and can be helpful. Tailoring for the individual dog in treatments with specifics of duration, rewards and more can increase desired results. A study by Blackwell, Casey, et al., compared a generic treatment plan with a customized plan. The study found owners of 56% of the dogs in the generic plan reported significant improvement, while an additional 25% said the dogs showed slight improvement. This is compared with all of the dogs in the customized plan who were reported to have improved.

As any owner or separation anxiety dog would tell you, we should not diminish improvements for those dogs no matter which group they are in. Still, we all want to do better for our dogs and perhaps we can. The more we look at the research, comb the findings, the more inspiration we can take that efforts, even the few, simple, erratic and haphazard ones, can be effective and have beneficial outcomes.

Scientists, Takeuchi, Houpt, et al., looked at treatment outcomes and owner compliance with 52 dogs seen at the Cornell University Animal Behavior Clinic for separation anxiety. Owners given less than five instructions were found to be the most likely to follow them. Removing punishment, increasing exercise and provisioning a chew toy when leaving were the most implemented directives. According to the owners, 62% of the dogs improved.

Another study on interventions and acquiescence to treatment plans, done in 2011 by Butler, Sargisson, et al., looked at a small number (eight) of separation anxiety dogs where owners were instructed to leave dogs in isolation with food treats three-four times per day. Starting with five-minute segments, increments were increased by five minutes until 30-90 minutes was achieved without phobic reactions. Food was to be provided immediately before leaving and on return. Leaving dogs alone otherwise was discouraged, as was punishment, even as the plan called for ignoring a dog and withholding food for 30 minutes on arriving when evidence, such as defecation, rearranging, etc. was found showing the dog had been distressed. Exercise was instructed for at least 15 minutes daily.

Results showed six of the eight study dogs improving even as compliance was uneven: Food treats were provisioned to only three of the dogs, praise and toys to only two of the dogs, exercise provided to only five of the dogs, no exercise given to three of the dogs, and punishment ceased for all dogs, save for two who had never been punished. The researchers concluded:

> "systematic desensitization was a consistent factor in the improvement of separation related problem behavior" and "The consistency with which systematic desensitization was applied did not predict the speed of progress or final success." (Butler, R., Sargisson, et al., 2011)

Again, removing punishment, limiting owner absence, and increasing exercise are fairly universal in recommended treatments. It is not uncommon to also see plans which introduce unwitting or negative punishment such as ignoring a dog on leaving and arriving, as in the Butler, Sargisson study above. With a dog who is anxious or frustrated to the point of panic over losing owner presence, this is often counterproductive and can add to and not lessen anxiety.

Scientists Amat, Camps, et al. write on offering just the opposite in their paper on predictability and contextual fear in separation anxiety dogs:

"Predictability is one of the main psychological factors that modulate the stress response" and "we recommend increasing the predictability of the owner's departure." (Amat, Camps, et al., 2014)

A review on predictability of events and control as they relate to welfare in captive animals by Bassett and Buchanan-Smith further supports these benefits throughout the research and concludes in touching on the additional benefits of the assurance of structure and routine for animal welfare:

"Studies investigating the effects of predictability of aversive events on behavioural and physiological responses are complex, confusing and often questionable in terms of experimental validity and in their generalisation to common practice. However, they consistently show that animals actively choose (both signalled and temporally) predictable over unpredictable aversive events. Whilst the physiological stress response data are less consistent, and hence further research is required especially with common-place negative events, they also suggest that aversive events should be made predictable, at least if the events are to be present over a short duration only. Further, the evidence suggests that signalled predictability is more critical than temporal predictability." (Bassett & Buchanan-Smith, 2007)

Adding comfort and affiliate touch to departures can also be valuable for separation related behaviors or anxiety. Mariti, Carlone, et al., compared the effects of gentle touching for one minute before dogs were separated from their owners for three minutes. The dogs tested were not separation anxiety dogs and the petting prior to departure was a new routine for them. Compared to dogs who were not petted prior to owner departure, the dogs:

"displayed behaviors indicative of calmness for a longer period while waiting for an owner's return and their heart rate showed a marked decrease after the test." (Mariti, Carlone, et al., 2018)

Despite the limitations of applying the study to SA dogs, the finding is a noteworthy one and adds support to the value of not just saying goodbye but saying it along with offering the reassurance of social support, especially, when such support is offered through contact with an attachment figure.

Comfort can sometimes also be found in food. Environmental enrichment may also be helpful with separation anxiety depending on timing, value and method of delivery. Not accepting food or treats is an indicator that a dog is too stressed to focus on anything beside a heightened emotional state seeking relief. Research shows and I have seen in my own practice, establishing a prior pattern of providing food and treats, in particular meals, with intrinsically satisfying behaviors like prolonged chewing can lessen boredom, frustration, and overall anxiety. A benefit for all dogs, especially SA dogs. Switching food bowls for food puzzles that dogs sniff or chew meals out of over a prolonged duration with their benefit of intrinsic satisfaction for natural behaviors, as opposed to those frustration inducing puzzles or bowls that tip, slow, wait for, or poke, is an intervention that goes a long way to reducing reactivity.

Two studies published in 2022 support the use of food as environmental enrichment in reducing separation anxiety. In the first study, Ok-Deuk Kang found "problematic behaviors" to be the greatest in the first 15 minutes of owner departure with the provision of a food puzzle significantly lessening those behaviors. The Kang study used a variety of food puzzles. In another study, Flint, Atkinson, et al., compared a toy that dispensed dry food, a smart dry food dispensing device, a person speaking to the dog, and a chew toy that lasted for a long period of time. The study found that the chew toy was the most effective at reducing arousal and engaging the dog for the total time observed.

Other approaches to treating separation anxiety include pharmaceuticals which are rarely, if ever, 100% effective, difficult to calibrate dosing and can have severe reported side-effects. It is worthy to note the inherent difficulties in deficiencies of observation and monitoring complications of side effects in species that cannot adequately communicate certain internal states directly to us when considering treatment with psychotropics.

A study on Fluoxetine (sold under the name of Reconcile for dogs or Prozac for humans) showed 42% improvement in medicated dogs undergoing a behavior modification plan at the same time. Side effects in 45% of the medicated dogs was "calm/lethargy/depression", "anorexia/decreased appetite" was even more common, with reduction in dosing required in 20 of the dogs to reduce vomiting and anorexia. Three serious adverse effects of seizure were seen in the medicated dogs. With this or with similar studies, while there are definite side effects from medication studied, it is difficult to tease out the effects of the medication separate from the behavior plan applied at the same time, or to know if that behavioral plan decreased or contributed to stress.

The Fluoxetine behavior study plan consisted of: ignoring the dog when home should the dog initiate contact, keeping the dog in a sit, down or stay at a distance, giving departure cues at times other than departure, ignoring the dog for 20 minutes before and when leaving, ignoring the dog on return and when the dog had initiated contact. Punishment was otherwise to be avoided and a food toy was recommended on leaving, which was to be taken away on owner return.

None of this means that it is not possible to get a benefit from drugs, just that research, extreme caution, and monitoring should be used when drugs are introduced. We need to additionally know that even when drugs are recommended by veterinarians, those recommendations include pairing them with behavior modification. Just as important, we also need to carefully evaluate the behavior plans we are putting into place. Good behavior plans make a difference, similarly, so will those behavior plans that are not so good.

Another key aspect to behavior modification, which can be overlooked in conventional treatments, is targeting the frustration and anxiety on multiple levels to treat the whole dog. We already know that research shows increasing exercise and removing punishment helps separation anxiety dogs but there is more.

Here is where we go back to the "magic" of life rewards on dependable, routine schedules, good force free training and structure building, impulse control, and reactivity exercises, like hand feeding one meal a day with "off" and "take it" and games like engage/disengage.

When we talk training, we need to remember this needs to include reinforcing the good behaviors we see when they occur, whether we ask for them or not. We are very much on our game in redirecting and often oblivious for those naturally occurring behaviors we want to see more of. Even lying quietly by your side deserves praise and reward. Key note for an anxious dog.

Because reassurance and predictability is so vital to managing the stress response, I advise saying "good bye" or "see you later", followed by gentle petting goodbye to a dog when leaving. And because we do not want to punish or ignore the positive emotional response of either dog or owner being happy to see each other again, acknowledge and channel excitement on arrival with saying hello and asking them to get their favorite toy.

Add schedules as another way to offer comfort and control to the separation anxiety dog. Feeding, short frequent training sessions, walking and interactive play routinely on schedule can help relieve stress and anxiety. Such consistency provides the dog with a sense of control over some of the most meaningful life rewards and events to anticipate– food, walks, social bonding with owners and more.

To start working on other interventions for reducing separation anxiety, departure cues are an excellent place to begin. After a walk that allows for maximum sniffing opportunities and a highly valuable chew puzzle feeder set aside for departure times, portion the diet to allocate some to this feeder and an extra savory of whatever the dog really likes and give five minutes before leaving. Begin determining how comfortable the dog is when you are getting ready to leave home by noting the dog's responses to leaving preparations. This step will also include desensitizing and counter conditioning but will be broken down with pausing between each stage of the process you go through when you get ready to leave. We may not give much thought to the process when we get ready to leave or even be aware of our own set and standard routines, or know how much of it is automatic for us, but very little gets past certain watchful eyes where every move is monitored by our hypervigilant separation anxiety dogs.

The key to relieving separation anxiety is desensitizing and counter conditioning to owner absence while an owner is gone and before they leave. Start paying close attention to what is happening with your dog when you begin to think about leaving and the motions you make as you prepare to do so. Can you stand up to get ready to leave, walk to the coat closet, pick up your briefcase, look for your keys without your dog alerting and stressing? The dog's reaction to each part of the getting ready to leave process counts. Being able to recognize stress signals in your dog can help to establish the difference between under threshold, approaching threshold, and over threshold.

Pay close attention to body language, it is always telling you something. Especially when escalating signs of distress present such as:

- Lip licking (not around food)
- yawning and blinking (not when tired)
- panting
- furrowed brows
- eyes hardening or whites of the eyes
- tensing of facial muscles
- turning head away
- turning body away
- ears back
- standing crouched, tail tucked under
- stiffening up, rigid, body tension increasing

Your dog is reacting emotionally to stress. How long did it take before you saw that first sign of stress? Was there a second sign? What were you doing when you saw these stress signals? This is where keeping time and noting movements comes in. Once you can determine the time period the dog remains calm and where this sits with what stage of the leaving process this is paired with, you have a baseline to begin to work from. Additionally, the very process of desensitization and counterconditioning becomes a routine the dog can depend on. You can then begin to get the dog more used to or desensitize them to the steps of your leaving along with introducing and pair them with good things or counter condition.

It is important to know what span of time exists for the individual dog to stay under threshold/not react/stress to being left or the idea of being left. The trick is to keep the duration time of leaving cues or owner absence very short in the beginning while adding a positive event (food puzzles, long lasting chews, stuffed toys, etc.) at the same time and build minute by minute from there.

It is helpful to clock the amount of time and keep a log of this so you know how to progress in duration. And for those who can walk out the door before the separation anxiety begins, remote video options offer a wealth of information for establishing just how long a dog can be alone without being stressed. Remember, not to leave out your farewell-for-now petting and to say your good bye.

Whatever the amount of time you have determined the dog can be comfortable, whether it is 30 seconds or five minutes, this is the line to hold, from which you can approximate threshold and begin timing the duration of the exercise or trial just under, or in this case, 25 seconds or four minutes to work from.

To note:

- Do not increase progression, steps, or durations in one day

- Keep drills/steps/durations to less than five times a day

- Allow for pauses or time between steps or durations. Wait at least 15 minutes or longer between trials.

- Vary durations - as in alternate with lesser time periods interspersed with longer under threshold duration periods

- Behavior modification is dependent on creating positive associations through food or play. Always remember acknowledge/thank/praise as with any requests or training, and before and after these interactions, steps, or durations.

A lot of separation anxiety plans include stationing or place exercises. Good **IF** used correctly. Do **NOT** use this exercise before leaving or when coming home. You want to create the option the dog can choose on their own, not command it. This is never to be forced or a punishment.

What I call the "go to bed" exercise needs to be practiced with loving encouragement to go to a wonderful and comfortable place to relax with good memories and things in it. To do this: Be close by to your dog's bed (you may need one for each room), working or watching TV or reading are all good. Offer high value items to the dog to savor in the bed, including soft voiced encouragement, gentle touching, reinforcement and chew toys like bully sticks or pizzles in their bed. Consider feeding meals in puzzle feeders there. You are working towards creating positive associations of a safe and cozy place to go to with the comfort of the people the dogs love in sight of them. This helps to super power the association of this spot.

Doing this exercise more often when you are without company makes it easier for the dog to recall the comfort of the spot when company does arrive or when they are eventually alone. With this or any training, the more you do in a low distraction environment, the greater its power in a high distraction environment. Work on increasing distance from the dog in his/her bed gradually. And alternate that with increasing duration. Be aware of increments and modalities. Never increase distance and duration at the same time or in the same day.

While more research is needed, natural stress remedies that have also been clinically studied without deleterious side effects, as in the use of supplements such as fish oil, probiotics, melatonin, and valerian, can be considered.

Studies also support the effect of being stress being lessened when classical music is played. I suggest leaving a classical music station being played softly in one room. Leaving a radio on amplifies the benefit of the soothing qualities of the music with the added comfort of the dulcet tones of the human hosts and guests, hopefully such voices along with the music, are additional comfort.

Remember, with much of this work, we can push too fast in our hope for results. When things are not working, it is often by going backwards we get to go forward. Go back to the last step in the plan that was working and go from there. Attention to the right plan for your individual dog that includes consistent routines and schedules, limiting owner absence, removing punishment, desensitizing, counter conditioning, targeting anxiety on multiple levels by allowing for natural behaviors for the whole dog, increasing force free training with praise for the good stuff too - consistency and commitment to all this can be tall orders, but doing any plan starts with beginning them.

Baby steps.

References

- Amat, M., Camps, T., Brech, S. L., & Manteca, X. (2014). Separation anxiety in dogs: The implications of predictability and contextual fear for behavioural treatment. *Animal Welfare*, 23(3), 263–266. https://doi.org/10.7120/09627286.23.3.263

- Blackwell, E., Casey, R. A., & Bradshaw, J. W. (2006). Controlled trial of behavioural therapy for separation-related disorders in dogs. *The Veterinary Record*, 158(16), 551–554. https://doi.org/10.1136/vr.158.16.551

- Bassett, L., Buchanan-Smith, H.M., (2007). Effects of predictability on the welfare of captive animals, *Applied Animal Behaviour Science*, 102, (3–4), 223-245, ISSN 0168-1591.

- Butler, R., Sargisson, R.J., Elliffe, D. (2011). The efficacy of systematic desensitization for treating the separation-related problem behaviour of domestic dogs. *Applied Animal Behaviour Science*, 129, (2–4), 136-145. https://doi.org/10.1016/j.applanim.2010.11.001.

- Flint, H. E., Atkinson, M., Lush, J., Hunt, A. B. G., & King, T. (2023). Long-Lasting Chews Elicit Positive Emotional States in Dogs during Short Periods of Social Isolation. *Animals*, *13*(4), 552. https://doi.org/10.3390/ani13040552

- Kang, O. D. (2022). Effects of Environment Enrichment on Behavioral Problems in Dogs with Separation Anxiety. *Journal of Environmental Science International*, *31*(2), 131-139.

- Mariti, C., Carlone, B., Protti, M., Diverio, S., & Gazzano, A. (2018). Effects of petting before a brief separation from the owner on dog behavior and physiology: A pilot study. *Journal of Veterinary Behavior*, *27*, 41-46.

- Lenkei, R., Gomez, S. A., & Pongrácz, P. (2018). Fear vs. frustration–possible factors behind canine separation related behaviour. *Behavioural Processes*, *157*, 115-124.

- Simpson, B. S., Landsberg, G. M., Reisner, I. R., Ciribassi, J. J., Horwitz, D., Houpt, K. A., Kroll, T. L., Luescher, A., Moffat, K. S., Douglass, G., Robertson-Plouch, C., Veenhuizen, M. F., Zimmerman, A., & Clark, T. P. (2007). Effects of reconcile (fluoxetine) chewable tablets plus behavior management for canine separation anxiety. *Veterinary therapeutics: research in applied veterinary medicine*, 8(1), 18–31.

- Takeuchi, Y., Houpt, K. A., & Scarlett, J. M. (2000). Evaluation of treatments for separation anxiety in dogs. *Journal of the American Veterinary Medical Association*, 217(3), 342–345. https://doi.org/10.2460/javma.2000.217.342 , 158(16), 551–554. https://doi.org/10.1136/vr.158.16.551

THE USE OF SHOCK COLLARS WITH DOGS

I am forever grateful when people stop using shock collars, that they have realized that pain and punishment was not helping their dog, that there was a better way. It is a lot to process. Recognizing the effects of shock on dogs, removing the collars, speaks to a person's capacity for empathy and is to be lauded. Without empathy there can be no significant relationship with dogs or with people.

Too often we shame owners for putting shock collars on their dogs. What we leave off here is that they were masterfully persuaded that these devices were the best things for their dogs. Well intentioned dog owners search online, turn to trainers and pet store personnel on advice to control a dog that pulls or barks or gets up on furniture or any other unwanted behavior. Someone, something they read, saw online has told them a shock collar is the answer. It is not.

Smooth presentations and nods to some people's minds that perhaps "spare the rod and spoil the dog" is the most fitting way to train dogs. Perhaps these persons were brought up to believe or convinced that certain breeds require a "strong hand." Perhaps they think the electronic fence that dispenses shock is not so bad. Maybe they are rural or suburban dog owners with dogs that chase or worry livestock, chase cars and joggers and attack dogs off lead or otherwise have a whole set of behaviors that users contend justify the use of head halters, choke chains, shock collars and electronic restraint. Such equipment, figures prominently in use in certain board and trains guaranteeing results. Punishment after all, has an immediate effect along with its fallouts. Celebrity trainers use them, like the monks of New Skete, highly lauded in the media, with a recent bestseller and putting shock collars on German Shepherd puppy dogs at three months old. If we are to shame anyone here for their use of shock, it would be the owners, trainers and manufacturers that continue to market, champion, and never reconsider the use of these devices.

People use electronic collars, whether calling them e, training, or shock collars (all such collars work by delivering an electric shock), there are dogs out there right now wearing them. And someone with a remote in their hand to power them up.

I have worked with enough owners coming to me for help with fearful, aggressive behavior in traumatized dogs that had been forced to wear shock collars that I truly abhor those devices for what they do to dogs. The extreme and heightened vigilance and wariness these dogs exhibit, a defensiveness in constant expectation that pain is imminent, that the world/a person can hurt you without provocation, without any meaningful warning. The superstitious cues reacted to, each different, each lurking around any corner unseen. Working with these dogs requires rehabilitation before we start asking them to see the world differently. Rest, recharging, play, balance, and confidence building exercises help. And then the path to modifying behavior and short bouts of force free, fun training.

Electronic shock collars or "training" collars are readily available in much of the world but that's changing. In 2018, Scotland banned the use of shock collars. England committed to ban the collars by 2024 but proponents of shock collars are lobbying for the ban not to go in effect. France was expected to ban shock collars in 2023. The devices are already banned in Wales, Denmark, Norway, Sweden, Austria, Switzerland, Slovenia and in parts of Australia. In the United States, Petco stopped selling remote electronic collars in 2020. Legislation was introduced in 2022 in New York State to make that state the first in the United States to ban the collars should the bill become an actual law. As of this writing, it hasn't. Whatever side of the electronic fence you are on the use of shock collars on dogs, they are still in use. Collars that activate shock with electronic fencing are not affected by this legislation and are increasingly used with not only dogs but with equally sentient, farmed animals such as cattle and goats, nationally and internationally.

Where does this leave us with what we know about shock collars? How negative is their impact, their effectiveness and how much do they hurt the dogs? And are they better at controlling problem behaviors than positive training methods? Advocates of shock collars insist that they have certain benefits. Do they?

Newer versions of shock collars now feature vibration and noise settings (in addition to electric shock settings) that manufacturers advertise as aids in "training." Theoretically, the "training collar" would never be dialed past those noise and vibration settings and such settings would be set at the lowest possible levels. Vibration settings on these collars can be increased sevenfold in intensity, which can harm the neck area, (an added concern for cats (also now being targeted in online ads) and small breeds with fragile and sensitive trachea's) in addition to causing great discomfort. But again, just how effective are these collars? Do they work at stopping problem behaviors? And does hurting your dog to change a behavior make sense scientifically?

When I originally wrote about shock collars, reader response was high and controversial. England's current pro shock contingent fighting the ban, points to strong or well-organized support for their use. With the growing popularity of shock collars, it makes sense to look at them again:

Electronic collars work by delivering positive punishment or negative reinforcement. In the case of negative reinforcement an aversive or unpleasant stimuli such as pain or discomfort is caused, for example, with the application of shock, or the tightening of a choke chain or the spikes of a prong collar due to, for example, a dog pulling on a leash. When the dog stops pulling, the leash is relaxed, the delivery of shock terminates and the pain ends. The goal is for the dog to learn to associate the pulling with the pain and not pulling with the lack of pain. Positive punishment introduces the unpleasant stimulus when an unwanted behavior occurs, such as kneeing a dog in the chest when it attempts to jump on a person or applying an electric shock on approaching a sheep or running after a car.

In a review of shock collar literature commissioned by the RSPCA, Emily Blackwell and Rachel Casey write that shock collars are used as both agents of positive punishment and negative reinforcement, a feature that is of benefit to proponents of their use. Misuse of the collars has continued welfare concerns.

Blackwell and Casey note inherent difficulties in the use of shock collars along with unintended associations. Delivering shock simultaneous with targeted behavior is rarely achieved and even when it is, the shock may not necessarily be linked with an unwanted behavior, rather a trainer, location, or situation. Therefore, an attempt to control excessive barking at the dog run using a shock collar may turn into a situation where the dog run is now associated with the shock and not the barking. Even when a shock collar is no longer used, trauma persists, dogs have been known to continue to associate the shock with the trainer or situation and exhibit signs of stress.

A separate study by Schilder & van der Borg observed that when trainers were working with dogs wearing shock collars:

> "the command was followed by a shock so quickly that the dog was unable to prevent a shock. This leads to unwanted conditioning: the dog has learned that getting a command predicts a shock." (Schilder & Van der Borg, 2004)

In reference to this study, Blackwell and Casey, write that if this occurs with professional dog trainers, who have hopefully trained to have degrees of finesse in the applications of learning theory, it is even way more likely to occur with the general public.

Timing is the essential component in training, whether it is positive or negative reinforcement being delivered. The marker that the desired request has been fulfilled must occur at the exact moment of the behavior to create an association. Blackwell and Casey say the danger of unwanted associations exists in both scenarios. The difference with unintended associations caused by unpleasant or painful stimuli is that the avoidance learning (demonstrated by our fictional dog that now refuses to enter the dog run), lasts a very long time and is very difficult to correct. Simply put, behavior that is rewarded tends to repeat and one trial learning with aversives can be equally effective at stopping behavior.

The use of a shock collar as a successful training device is problematic at best. Mounting studies continue to show that training with positive reinforcement is more successful than the use of negative reinforcement (only positive reinforcement is used training assistance dogs in the UK). The study also found that the use of punishment correlated with a rise in the number of unwanted behaviors.

Another study by Cooper, Cracknell, et al compared training methods using shock collars and those using positive reinforcement with dogs exhibiting problem behaviors with recall and livestock (sheep and chickens) worrying and chasing. These are the sort of behaviors that are notorious for justifying the use of shock collars. However, the Cooper study found that the rewards-based training was equally effective in changing those unwanted behaviors. The study further found that the difference between the methods resulted in shocked dogs exhibiting increased known responses to pain with vocalizations such as whines and yelps. No such vocalizations were observed in the group trained with positive methods. Stress related changes in behavior observed in the shocked dogs and not in the positively trained dogs, included an increase in body tension, yawns, paw lifts, panting and other:

> "distinct changes in behavior including sudden changes in posture, tail position and vocalizations that are consistent with pain and/or a version in dogs."(Cooper, Cracknell, et al, 2014)

When the researchers compared citronella collars with shock collars the citronella collars were found to be more effective (and preferred more by owners) and when dogs are trained with other severe training methods such as physical punishment, they were found to exhibit fewer stress related behaviors than when trained with shock collars. Owners instructed in continuing the training methods used in the Cooper study expressed greater confidence in being able to implement the rewards-based methods as opposed to using the shock collars.

Shock collar advocates contend that there is an art to the use of the collar with some researchers calling for a licensing system for qualified users and that properly used, the dog responds in a manner that would indicate no apparent discomfort. Blackwell and Casey raise two issues in response: working dogs with high levels of excitement may tolerate shock more readily in order to perform the task at hand such as herding sheep or tackling criminals. Additionally, calibrating the appropriate level of shock for an individual dog, requires knowledge of breed, nature, present temperament, past experiences with shock, thickness of the coat and how moist the skin is, at the moment of delivery.

The Cooper study observed five experienced shock collar trainers at work using the devices. Results showed only one trainer out of the five attempting to determine an initial setting for a dog's response to the device. Two trainers fitted the collars onto the dogs without any attempts to test a response to any settings or determine if the devices were working. One trainer started with a low-level setting while the remaining trainer started with the highest level of shock available on the collar. It bears emphasis that these results are with qualified users in terms of their experience.

The potential for negligence and intentional abuse exists with shock collars as well. The devices are able to deliver extremely high levels of shock which have been found to leave neck lacerations particularly in inclement weather. Deliberate abuse, due to anger or for entertainment, has also been observed such as repeatedly shocking a dog for running away when the animal returns or is captive.

The authors point out other issues including the use of shock collars to control aggressive behavior countered with numerous studies showing:

> "that pain caused by an electric shock is a well-documented stimulus for aggression in a wide variety of species." (Cooper, Cracknell, et al, 2014)

Fear can fuel aggression as well as anxiety over separation from an owner. Shock collars have also been shown to exacerbate fear and fear-based behaviors. Extensive research supports prevention of and treatment of aggression in applied animal behavior practice by first removing all force and punishment.

The number of countries and municipalities limiting the use of electric shock collars is now increasing along with their availability and use, knowing more about them needs to be the first step to consider in taking either step further.

<u>References</u>

- Blackwell, E. and R. Casey. (2006). The use of shock collars and their impact on the welfare of dogs: A review of the current literature. *Royal Society for the Prevention of Cruelty to Animals.*

- Cooper, J. J., Cracknell, N., Hardiman, J., Wright, H., Mills, D. (2014). The welfare consequences and efficacy of training pet dogs with remote electronic training collars in comparison to reward based training. *PLOS ONE.* 9, e102722. doi: 10.1371/journal. pone.0102722

- Electric Shock Collars. (2018, February 10). Retrieved from https://www.thekennelclub.org.uk/ our-resources /kennel-club-campaigns/electric-shock-collars/

-Schilder, M.B.H. and J.A.M. van der Borg. (2004). Training dogs with the help of a shock collar: short and long term behavioral effects. *Applied Animal Behaviour Science*, 85, 319-334

WORKING WITH THE FEARFUL DOG

I work with fearful dogs, their behaviors; anxiety, aggression, reactivity and more, are ones that get them in trouble with us. What breed this dog might be or even what sex, is immaterial. This dog could be any number of fearful dogs on the island of Manhattan. You may have walked by one dog this morning on your way to the subway at West 4th Street or passed by on your way home getting off the M116 cross-town. Fearful dogs are in no short supply, this city, your town, or suburb has its share. Such behaviors often stem from impoverished or arrested development, socialization is complex and what is missed in limited and lack of exposure to complex environments, can leave a lasting impact, so can trauma and abuse. Certain settings, past associations and situations can be overwhelming and reassurance can be hard to find.

Fear is a universal, necessary, and useful emotion for all species. Nature has its reasons. Fear serves to protect animals as a warning system that caution needs to be exercised in the face of possible danger or the unknown. Fear can abate once an animal feels secure in a setting even as such security can be elusive or fleeting. In addition to socialization deficits and trauma, the imposed isolation of the pandemic limiting development and socialization has contributed to an increase in the number of fearful dogs now dealing with all the strangeness and newness of environments and individuals they have never met before. For relocated and rehomed animals fear surfaces as they face worlds turned inside out and upside down.

Most aggression proceeds from fear. Much has been written on addressing behaviors working with the individual aggressive dog, aggression between dogs, and bite prevention in dogs, all of which include basic guidelines in approaching all dogs, including the overly fearful dog. Following are some specific "hands on" strategies and approaches for the fearful dog. This is no way a guarantee to achieve results, exhaustive, nor can it deal with every specific situation. Every animal is an individual, every situation, owner and dynamic unique and every approach need be individualized based on circumstance, history, and presentation:

- Go slowly, more slowly than you want to or than you think you should. The dog is afraid, that lip licking, yawning, looking away, freezing, trembling, urinating, or barking is all that the dog has to communicate this to you. Respect it. Take infinitesimal baby steps forward in training, behavior modification and relationships.

- We animal people can want every animal to connect with us in instant intimacy. We can be convinced they get us, can easily read how very much we love animals, they just must know. It does not work that way. This is not about you it is about the dog. Relationships and trust take time to develop. With the uncertain, tentative, and fearful, time can take even longer. And rest, time off, sleep, can be seriously underrated in this process.

- First and foremost, give them time. Time to rest. Soft and many beds to sleep on in calm and quiet environments. Time to decompress. Time not to perform. Fearful dogs may need even more time to trust in and feel safe in new environments and situations. Give it to them.

- Give them the safety of routine and schedule. In a world where the humans are in control of pretty much everything that dogs get to do or not do, the assurance of fixed times for breakfast, dinner, walks and play dependably delivered on a routine basis provide structure and a sense of control over meaningful life events.

- This is not the time to ask for compliance, "Calm, submissive" is the antithesis of what you want from this dog. Forcing any dog ("flooding") to do anything they are seriously afraid of or reluctant to do, will exacerbate the situation, making the dog's issue larger. Such practice is additionally cruel and inhumane.

- Be aware of how your own body language and presence affects the dog. Keep your form loose and relaxed. Breathe.

- Do not rush forward in approach initially and avoid eye-contact. Give them space.

- To help in creating a feeling of security for the dog, toss treats from where you are standing to just past where the dog is. This allows the dog to retreat to an even safer space by offering that treat in the dog's safety zone and not asking it to advance towards you.

  Even if the dog might take the treat thrown in front of them, the fearful dog is probably taking it under pressure evidenced by creeping slowly or darting toward it. The toss behind the dog removes the pressure. When the dog takes the treats, throw two more. Wait.

  Advance one step and pause. Observe the dog's reaction to your forward movement – do you see signs of stress, as in lip licking, yawning, looking away, stiffening, retreating? If yes, take a step back and try that treat throw again. Look to establish the comfort level of the dog first before moving forward.

- Never tower over or lean into the dog. There was a time when an outstretched hand or downward pat towards the head was thought to be welcome to the fearful dog. It is not. That hand advancing towards a scared animal can be seen as threatening instead. Do not. Wait first for interest from the dog. When the dog indicates consent with softer body language, ears not flattened, eyes relaxed, no tightness around the muzzle, a nostril flare or two, sniffs even better. When and only when the dog indicates they are feeling safe, can contact be explored.

Watch to see for when the dog takes steps forward, sniffs the air. If you pat gently the side of your leg or the ground next to you as an invitation to come closer, do they respond? If yes, pause, pat and repeat.

- Sitting at a comfortable distance, to be determined by how relaxed the dog is in relation to the person, reading out loud can add to reassuring the dog of your presence as a harmless one. Positioning the body sideways on the floor alongside the dog, choose a book you are interested in and go through a few pages. Focusing on the reading instead of the dog, takes the pressure off the dog to keep vigilant or to interact and allows them to be more accustomed to your company in a safe and non-threatening way. Keep story time going for several days, added benefit for keeping the time consistent.

- Play the sort of classical music that soothes for the dog in the background or when you need to step out without them. Respect superhuman hearing and keep the volume low. Classical music radio stations have an added benefit of dulcet toned radio announcers, more soothing for doggy breasts.

- When the dog has calmed and indicated consent, try an approach. Make sure to move laterally (sideways). Direct eye contact and frontal approaches are aggressive behavior in a dog's world. Speak softly, turn your body to the side, keep eyes downcast, avoid direct gaze, and when ready, lower yourself to the dog's level.

- Ask where this dog prefers touch. Pay close attention to how your proximity and movements are welcome or not. Does the dog retreat, ears go back and start to flatten, yawn, lip lick or more? Slow your movements or stop. Watch how your response to what they are "saying" matters.

- Stay away from full body stroking, this movement can excite and arouse, while downward strokes in the front of the chest can be more soothing with a dog that has established trust with a handler. The chest and behind the neck are often good spots for touch but confirm this with owners who know their dogs best.

Allow for the time the process will take. Making space emotionally and with our own expectations for the time it takes to happen, is the greatest benefit in any intervention and the one we have in the most supply and ration out the least. Every animal we work with is an individual with their own history, personality, and timing. It may take several weeks or months for your dog to accept novel situations or people and only after consistent, repeated positive and reinforcing attempts.

Reassurance is not a dirty word. You can reassure your dog. Please, keep it to one or at the most, two short sentences in a calm, even tone. No little red hens needed please, the sky is not falling, it's OK to point that out. Just remember not to act as if it is.

Your dog knows the difference between what you are saying and how you are acting.

- Set your dog up for success. If a street-cleaner, larger dog, group of toddlers, etc., is coming towards you, start an even and reassuring conversation ahead of time.

  Tell the dog what a good dog they are at their very first glance towards the oncoming trigger. Once they have registered the scary thing and you have marked and rewarded their neutral response, give them space. Put your body between them and the scary thing. Your physical presence as a buffer will alleviate stress in these situations.

- Take micro steps forward. Very small baby steps very slowly. This may be the most difficult for humans to apply. We need to pay attention and take our cues from what the dog is saying with body language. Going too fast or too far in this process can have the opposite effect and make things far worse. Think of throwing the individual with fear of water into the pool to learn to swim. A move like this, can get someone to somehow get out of the water while flooding them with such fear and anxiety that water is poisoned permanently in their mind. The concept of "flooding" in behavior is aptly termed and not recommended here or in humane and ethical methods.

Gauge the distance and time the dog can be comfortable with the scary thing coming forward. Do not push distance, duration or contact at first. Let the notice the dog takes of those scary things last a bit longer over time. Time measured at the dog's pace. How much longer is going to depend on how scary and how long the dog can stay calm when they register whatever it is that is disturbing.

Staying calm can best be seen in a loose and relaxed dog, a dog with no lip-licking, no yawning, no shake-off's, stiffening, no looking away, no whining, etc., no body language that says stressed, anxious, or reactive. Much of this will mostly depend on distance. We all have a flight distance we feel safe in and when that distance is breached, we can feel vulnerable. Dogs are no exception.

Find the closest distance and the shortest span of time the dog feels safest at registering things that can cause fear to set in and reassure, praise, tell them what a rock star they are. If you have treats, and they are relaxed enough to take them that is even better. Not taking treats is an indication of anxiety. Increase distance and decrease duration. Treats or not, always remember you have your words to praise and reassure with.

Over the course of days, can you go forward in distance two inches, six inches? A foot? Signs of stress, as in lip licking or yawning out of context of eating or sleeping, shake-off's, stiffening, freezing, etc., show you have gone or are going too fast. Go backwards from what was last working and go forward from there.

Do not increase distance, intensity, and duration at the same time. Work on one before proceeding to another. I have seen with dogs that working on distance before duration can be more effective.

When looking for and working with professionals to assist, make the necessary effort to find the right behaviorist and/or trainer, one that is formally educated in the field of animal behavior and learning theory, experienced, well qualified and utilizes science based, force free positive methods.

Except for veterinarians and veterinary technicians, no education or license requirements exist for pet service professionals. Make sure to ask which handling philosophies a prospective veterinarian uses, you want low-stress or fear free practitioners or where a trainer has learned their methodology and how they put that into practice. Ask open ended questions and wait for the answers. Avoid filling in the silences. What equipment is being used? How is direction and redirection accomplished? What rewards are offered? What happens when things go wrong? Be part of the process to both learn and advocate for your dog before and if force or misunderstanding present.

Beware of red flags; avoid sites and trainers advertising "balanced" as in, they will balance force and punishment, or "results" as in, force and punishment can achieve a cowed, submissive result, or "alpha" or "pack leader," as in outdated thinking that research has shown is not the way dogs learn or interact socially.

Knowledge is power. Make sure you know about how your dog learns too. Read a good book on dog training from the right science based, force free sources. Steve Mann, Ian Dunbar, Patricia McConnell, Sophia Yin, Pat Miller, Paul Owens, Victoria Stillwell, and Suzanne Clothier, are all wonderful to choose from for starters. Much material is now free for the reading online.

Choose wisely, ask yourself who stands to benefit most from what you are looking at. Beware of quick fixes or guarantees, both of which are impossible with all the confounding and myriad variables of individual, history, application, etc.

Let "ignore the bad behavior, reward the good behavior" become your mantra and apply it always. It's not bad, it's what they are doing out of fear, protective or coping, the behavior serves a purpose. Look at the context, remove the pressure, give distance. Give time. Direct and redirect. Keep in mind that when asking to change a behavior, offering a good or better behavior as an alternate is the most successful way to go.

Work on actively noting the good behavior. This much underutilized aspect of training can be the most effective. A dog simply lying quietly can be praised for it ("Good Dog to lie Quiet!").

Your dog loves praise, especially coming from you. Capitalize on the positive moments, this way you get to build much needed confidence in your dog on a more frequent basis.

Figure out what amplifies the positive for your dog. High value treats may be welcome at other times but are hard to process when on the defensive. The dog I am working with responded from day one to tons of praise in a sing-song happy voice and this dog loves to chase the tennis ball.

We play a lot of ball.

CANINE VOCALIZATIONS-WHAT IS YOUR DOG TRYING TO "SAY"?

When I brought Daisy to the pet care technician classroom, she spent the day with me there. In the five years I worked in the program, she barked once in the classroom. That day, neither one of us noticed an unannounced visitor, until Daisy told all of us, they were at the door. Daisy was well trained and a registered therapy dog, but she was still a dog with dog behaviors to express. Daisy was not a reactive or alarm barker otherwise. There were times when I would want to elicit vocalizations to illustrate playfulness, significance, or context. Especially those vocalizations we can get confused or concerned over. Like growling, which can vary in meaning and cover a range of emotions depending on the situation. There are different growls for different settings. Play growls and warning growls can both be categorized as growls but they do not sound the same and they do not mean the same thing.

Dogs will growl as a distance increasing behavior in response to a threat perceived in the environment. This growl is a defensive signal and can be preceded or followed by barking. Play growling is softer in tone, lighter, breathier and accompanies play. When I pulled out Daisy's tug toy and she grabbed the other end, she would play growl delightedly. Such vocalizations were made in play along with loose body carriage, lowered fore legs and often a wagging tail.

These play growls Daisy was demonstrating, appeared, and sounded clearly different from any other kind. If you looked and listened. But using the same word to describe a warning signal asking for distance or heightening aggression and for expressing joy can be problematic. Or set us to thinking the wrong things about what they mean. One of my students told me when her dogs played at home, they called the sounds the dogs were making "roo rooing" instead. That made more sense to them. I can see the usefulness of coining a phrase there. How else are we making sense of what our dogs would like to tell us? And how often are we more interested in what they have to say as opposed to how well they understand what we have to say?

Human beings are fascinated with the possibility of dogs speaking, especially if the language they are using is ours. Bunny, Stella, Bastian are just some of the top canine social media stars now using language boards to "communicate" with humans.

Language boards or Augmentative and Alternative Communication Devices ("AAC") were developed to augment communication for humans experiencing language barriers. These boards have powered our pursuit of understanding how well other animals can learn and communicate in our own very specific human language, such research that has been going on for years with multiple species, including great apes and dolphins in addition to cats and dogs. How valid this work is for determining what a dog wants to convey versus what they can learn or what we can teach them about certain buttons is an important and perhaps often overlooked point; who gets to choose the buttons and who's vocabulary are we giving them?

In a *Salon* article on AAC devices and their use by cats, an observation about the difference in cats interacting with the boards and the limitations of the boards being meaningful for cats in the scarcity of buttons as we have provided them, is telling:

> "What's interesting is that they [cats] tend to not do that much in the way of multi-button presses, but there's like a lot of single-button presses," Trottier tells Salon. "With cats, you kind of have to find things they really want, and there are just fewer of those than with dogs." (Karlis, 2021)

Adding a few more cat relevant buttons as in "bird," "mouse," "play," "scratch," etc. might make a difference. Still, both cats and dogs already speak volumes to each other and to humans perceptive enough to understand their body language and chemical signals in scent and pheromones without using buttons or human terms.

Animal behaviorists usually group canine vocalizations into barks, growls, whines, whimpers, or howls. Observations, on these signals, show that dogs employ them to communicate a request (or solicitation), an alert, a demand, an alarm, or some sort of declaration. What signal is used depends on the context of the situation. For example, a whimper or whine is often used to express an intense desire for closeness or for something deeply significant while a bark, growl or howl is often used for attention, alert, alarm, or emphasis.

If "communication" means an attempt to influence or exchange or relay of information between sender and receiver or dog and human, in this case, how well are we able to decipher what our dogs are telling us? A study looking at how successfully humans were able to interpret dog barks found that we do a pretty good job. The study also found that there is no one sort of bark:

> "barks can communicate both types of basic emotions (aggressiveness versus submissiveness/ friendliness) because humans were able to associate bark sequences with certain acoustic parameters as being either aggressive or submissive." (Pongracz, Molnar, et al., 2006)

The researchers discuss that mammals will respond to qualities of vocalizations in all animals such as pitch, tone, and intensity as signifiers of emotional content along with the timing between barks, for dogs (or "inter-bark intervals"). Human listeners score low pitched bark sequences as highly aggressive and "high pitched and tonal barks were scored as non-aggressive, but as fearful and desperate." How pressured or closely, barks are spaced from each other, also adds to what they are conveying. The shorter the space in between a bark, the higher the listeners scored them on aggression, while "long inter-bark intervals were considered non-aggressive, and identified as either fearful/desperate, or playful/happy."

Thanks to humans, dogs also have a lot more to say when it comes to barking than their canine relatives. We know that adult wolves typically confine vocalizations such as barks to specific aggressive situations. Our domestication of dogs has selected for barking in a wide range of scenarios from positive to playful to dangerous. We have selected, asked them to do it a way, valuing dogs for their guarding and alerting abilities and interacting vocally with them. Partly, no doubt, because we are a visual and verbal species. Which is one of the reasons that acknowledging a bark with a "thank you" or "I heard it" in response, before asking for quiet, is only fair and gets better results.

On the other end of the scale, play behavior has its own vocalizations. How dogs breathe in play is different, "panting in short bursts" defines "huffing." And there's even laughter. Dr. Patricia Simonet defined a "dog laugh" as a "breathy forced exhalation" that a dog offers before and during play. Dr. Simonet's work included using recordings of dog laughs to soothe dogs in shelter environments, which worked no doubt because the dogs understood the significance of the sound.

In broad terms, we can speak of behaviors having two purposes, to either increase distance or to decrease it. There can be little doubt that dogs are well aware exactly what the intent is of whatever vocalization they use with each other. To get a better understanding of this, scientists tested the responses of dogs approaching an attractive bone set next to a covered dog cage containing speakers which would play one of three separate growls; a play growl, a food guarding growl, and a growl expressed in the presence of a stranger. They found the quickest reaction from the dogs was in response to the food guarding growl as opposed to the stranger growl. Perhaps the time to react to keeping hands off a bone is more significant than stranger danger.

For how well humans understand similar dog growls in threatening and playful contexts, a separate study tested human reactions. The study also looked at whether or not being a dog owner and a woman improved the chances of being correct. Results found humans were able to identify the play growls 81% of the time. The threatening contexts were more difficult for us. Food guarding was classified 60% of the time and threatening stranger 50% of the time. Dog owners and women did better at the test than non-dog owners and men.

Play vocalizations or affiliative vocalizations that dogs use with us and each other are easily misinterpreted by humans with generic labels as "growls" or "barks" for instance, with all the aggressive and or negative connotations of those words. Remembering that barking and growling are also used in play or to alert us to notice something is helpful to respond and notice what our dogs are asking of us.

Qualifying the vocalization in context is also helpful—"play growl" used to identify that growl in play or "talk growl" used to specify that friendly growling your dog may do in a social context around you can better put each of these vocalizations in perspective.

References:

- Faragó, T., Pongrácz, P., Range, F., Virány, Z., Miklósi, A. (2010). "The bone is mine": affective and referential aspects of dog growls. *Animal Behaviour*, 79(4) 917-925

- Faragó, T., Takács, N., Miklósi, Á., & Pongrácz, P. (2017). Dog growls express various contextual and affective content for human listeners. *Royal Society Open Science*, 4(5), 170134. http://doi.org/10.1098/rsos.170134

- Karlis, N., (2021). A "talking" cat is giving scientists insight into how felines think. *Salon.* retrieved 1/29/23. https://www.salon.com/2021/12/12/a-talking-cat-is-giving-scientists-insight-into-how-felines-think/

- Pongracz, P. Molnar, C., Miklosi, A., (2006). Acoustic parameters of dog barks carry emotional information for human. *Applied Animal Behaviour Science* 100, 228–240

- Simonet, P., Versteeg, D., & Storie, D. (2005). Dog-laughter: Recorded playback reduces stress related behavior in shelter dogs. In *Proceedings of the 7th International Conference on Environmental Enrichment* (Vol. 2005).

WHEN YOUR DOG BECOMES AFRAID OF THE WALKER

Just the other day, I saw another one, a dog running solo at top speed through the streets of Manhattan. This time it was on the Upper East Side, at the intersection of 79th Street and Third Avenue. Following closely but not nearly close enough, on two legs, no match for those four at full speed, ran a man after him, no doubt, his walker. Possibly, another dog slipping his lead, not wanting to be in the situation he found himself in, home his only goal and another dog walker desperate to have a different ending to this story.

The dog came running straight towards me to cross the avenue. I stood still, arms out in hopes of stopping him. No such luck. He was smarter than that, he ran right around me, no break in his stride, and west toward the Park. The man followed. A car making a left turn almost hit him.

Runaway dogs are not the sort of stories that get a lot of press. Stakeholders and responsible parties are keen on keeping them off the radar. But if you take notice around town, you will see parts of some of these tales unfold. Months before seeing the Upper East Side dog, I wrote about another runaway dog, Oliver.

It was when walking home through the North Woods in Central Park, I came upon a dog frozen in fear and hurt in the middle of a stream running from The Pool under a stone arch. The dog had leapt the bridge over that stream in his escape from his walker, landing in the water below. A small group of people were gathered and watching him. Not responding to anyone crowded around him, scared and trembling, he was immobile. We all stood there until a tall and well-meaning man got into the stream and started towards him. For the dog that was motivation, the kind that meant move away from the tall well-meaning man and towards one of the waterfalls in the stream. Not a good idea.

That was when I spoke up, asked the man to leave the water and please let me try to help. I asked the crowd if anyone had anything to eat, a sandwich, a cookie, anything. I knew this dog would be too terrified to take food but also that my offering would bring some positive association, no matter how remote. Treat in hand, lying prone on the side of the bank, keeping up a steady and cajoling outpouring of soft, kind words, and telling him what a good, good dog he was, I was able to convince him to move towards me, able to persuade him in those moments that things were OK. And then to touch him, always keeping a hand on him, and finally, pull him by the side and then hoist him up to the bank and then on path aside me. Dog out of the water. And that trembling mass of so much fur and beseeching eyes and top knot hairdo, just shook and shook against me.

His walker and the day care manager were in the crowd of onlookers. I found out his name was Olivier. What came after the day care manager collected him, I do not know and that left me in tears. Sometimes, the more we know about things, the harder it makes everything else, hard to think on the unfamiliar or untrained walkers and scared dogs wanting to just get back home.

Working in the field, I have mostly seen information given in scenarios like these that are majorly skewed in favor of the business establishment doing no wrong. Most doggy day cares wanting to stay in business, will be reluctant to tell a client their dog is perhaps better off going home after an hour or two or does not want to be there or that they do not have the staff, in number or training, or enrichments to engage even the dogs that tolerate it. More likely the business will tell the owners the handler/walker is "suspended," "terminated," "being investigated," and worst scenario -some spin along the lines of the dog liking his run so much he wanted to swim…

No one sends their dog to a sitter or day care or walker without wanting the best for their dog. These owner/guardians are not to be shamed, they are the ones who are spending the money to provide what they believe and are told are the best services for the pets they love.

But it doesn't always turn out that way. The dog that runs away from the walker, is tied to a bench for a timed walk while the walker goes through their phone, is shushed, and scolded at a crowded daycare with nothing to do or worse, so much worse too, does not belong there. Bored, or surrounded by chew toys, waiting at home with a relief walk or two — now that is where so many dogs are running to get back to.

These are worthy cautionary tales and considerations. And with all of this, there is ever present real life, with its demands of trips and traveling again, work, dinners, events, commitments, of juggling, of so many, many things we cannot do. Our dogs needing to get out for regular walks, for relief, for exercise, to sniff and all that with someone they like, who will be good to them. Just another of the many things we are still trying to figure out, and not an easy one.

One of the most frequent requests I get from clients is about finding good dog walkers. A version of this chapter, about what happens when things go wrong, and not the post on how to find the right walker, on my website, consistently gets high traffic.

But you are not reading this for sad stories, you are looking for answers. How do you balance your own busy schedules and regular dog walking for your dog? Sourcing out walks to the right professional dog walkers to help is one answer. How to choose? To know if you have found the right walker when, no matter what, at the end of the selection process, the only individuals on that walk will be your dog and the walker and only one of them gets to tell you how it went?

One of the first jobs I did out of behavior grad school was to walk dogs and sit cats. Cat sitting then, was a definite niche and untapped market. Cats benefit from routine and schedule like dogs but can be fed as early as daylight with most owners taking the morning shift on their own for their dogs. That left midday relief walks for dogs with the biggest demand. Dog walking was not the most lucrative, everyone wants their dog walked at lunchtime and everyone wants to pay the same rate. This works best if you can develop clients in one neighborhood or better yet in one building. It is difficult to cover more than two adjoining areas for the same time spans, much less the city. This is also one of the reasons you will see groups of dogs walked together, from a financial standpoint this makes sense for the walker or the service but not necessarily for the dogs.

Pack walking is not often the best idea for dogs no matter what you hear. Think about it. How does pacing work with dogs of different sized bodies and leg lengths? How many dogs can one person realistically keep an eye on at one time? How does one dog get to take the time to sniff while another dog needs to target a different tree pit to leave pee mail? The list goes on.

There are a lot of dogs, clients, and walkers in a city like New York. And a lot of those services telling clients what they would like to hear about how great their services are for client's dogs. You know all that, but you need a dog walker. And this is one of the reasons you are still reading, and you would like for me to get to the point. This must be navigable. How do you make sure a dog walker is the right one for your dog? What do you look for and what questions should you ask? And what happens if your dog becomes afraid of the walker? How can you help to fix it?

First, read closely the chapter in this book on "Home Alone, What to know about Pet Sitters and Dog Walkers," then read it again. Take the time to find the right pet sitter.

Take the time to find the right fit here, this is important. Not going for a walk is way less harmful than going for a walk with the wrong walker. Do interview more than one walker to compare and find the best fit. There is always newspaper and wee pads.

Do not overlook the importance of installing and monitoring dog walking visits on remote cameras. You need to know what is happening with your dog and their walker or sitter when you are not home. The dog cannot show you until it is too late.

Fast forward, you believe you have found the right walker and with all your best efforts, there comes a time when the walking or sitting did not go as you or your dog would have liked. What then? Perhaps time for proper introductions were not made or sufficient time taken for the walker to develop a relationship with the dog or the walk was rushed too often or the handling was too rough behind the scenes.

A sufficiently negative experience or an accumulation of not-so-great experiences can prompt a strong fear response and elicit a behavior that is self-defensive -your dog may not want to go out again with walkers and communicate this by hiding when the walker comes and growling and snapping if coerced.

When your dog starts telling you something is wrong, try and figure out exactly what it is. Take an inventory of any recent changes in your dog's environment that might be responsible to evaluate their impact. Make sure to consider the things that might be happening that you cannot see, including medical and what may be happening when you are not around. Again, and again, and again; investing in a hidden remote camera and observing your walker when you are not around will allow you to make sure that your dog is not being scolded, handled in the wrong way, or otherwise harmed. It also can tell you if your instructions are being followed as to what walking equipment is being used, when walks are happening and if other dogs are part of the party when you have requested a solo walk.

What changes in behavior are you seeing? Does your dog appear more anxious before or as you are leaving your home? Now reluctant to leave your home or go on walks? Are they more reactive than usual around strangers? Are they more anxious/shy/aggressive than usual? Are they hiding in corners of rooms? Close to or under your bed? Following you around more than usual?

In cases where the dog has been seriously traumatized by the events, pause the arrangement immediately and do consider working with a good, force free, science-based behaviorist or do the appropriate research to learn how to help the dog get over the trauma. Time off in between to rest, decompress and rebuild is needed to build back up from trauma. Make sure the dog has multiple, comfortable resting spaces in the home. Allow for plenty of sleep, relaxed walks with plenty of time to sniff, chew toys, classical music when you are not home or even when you are. All the enrichments count before, during and after this period. This is time not to be demanded of, as in time not to rush to another new relationship and time to make sure the next walker is better suited for the situation and the dog.

Once you have determined that everything again is as it should be when you are gone, revisiting walking protocol from step one is in order. And then there is remediation. The introductory process with a new walker needs to be extended and revised, as in five joint walks instead of three and walks of shorter duration, paying close attention to the body language and comfort level of the dog on each walk.

One of my clients went through this and no one was happy about it. After one walk with a new walker, the dog took to hiding under the bed to stay away from the walker on their return. Here's my advice (I have changed names to protect the innocent):

Oh Valerie, I so wish we could fix this right away, really, for you and mostly for Lawrence who is petrified to be doing all this. If the first walk with the new dog walker went well, whatever approach the walker used for the second walk should have been the same. What did the walker do the second time that did not happen the first time? Something was different that was scary perhaps speed, restraint, caution, noise, attention, different handling. But if the walk was not a good one - the dog is adamant that the experience not be repeated, we need to step in to make sure that happens.

The walker may not have been aware of it at all, but Lawrence reacted to the difference and once he was scared and defensive again another strange human (the manager then went over to try and walk Lawrence himself) entering would not help unless accompanied by a trusted person and that human did remedial work -softly announcing their presence upon entering the apartment by using Lawrence's name, lying parallel on the floor to the bed, at a good and safe distance to begin, only gradually, slowly, inching closer, no talking at first, looking away, then speaking soothingly, then offering treats, squeaky balls, etc., that's the protocol and it might take repeated attempts over a period of minutes, built up incrementally over days or even weeks to bring back trust at that point. It may take time and investment in relationship building before Lawrence can walk with a new walker. Until then, new walker on pause.

Lawrence does need to trust the stranger entering his home, it is natural and necessary behavior for him to be defensive if he is fearful of being hurt. Meeting the walker before the first walk with you there for one visit, where lots of treats and no pressure to interact is a must. No walk yet.

A second visit, after a day or two about play and relationship building, treats and squeaky tennis balls or whatever your dog loves the most. Still no walk.

Next, in a day or two, a third visit where you and the walker both do a short walk together, followed by a fourth visit in several days, or even a week where you both walk together. Fifth visit, you do want to be home, and gauge the comfort level of your dog around the walker, what is the body language of the dog telling you? If all is going well, the fifth visit can be one where you are at home and the walker walks the dog independently, again for a short walk, this is about relief and sniffing and not about timed walks.

On introductory and remedial walks, dogs need to be able to go home when they ask to, never when the clock say. Walks need to take as much time as the dog wants for sniffing but otherwise, no forced time walks. Monitor subsequent visits remotely via the remote camera you have installed for this solo walk, a walk alone (leave his collar on for this visit, less handling that way -some folks leave the collar off a dog when they are home, leave it on if you are expecting a walker so that new person in your dog's life does not have to be putting the collar on when they arrive, less stressful for your dog).

Evaluate the footage and the dog's behavior when you get home to determine where your dog is with both. You may need to more joint walks if the dog is still not relaxing with the new walker. And obviously, if there are behaviors of concern on the human end. Start your search again.

FOR MY DAISY

There is a certain particular and added aspect of the grief we experience when losing companion animals. Such loss is an unsupported grief, a grief that is dismissed as not being justified by much of society as it is "only" for a companion animal. As if these animals in our lives are somehow less than, and the grief felt, the holes in the fabric of days and lives are a failing, a loss of perspective, a thing to be remedied with getting over it and getting another replacement animal. Such sentiments are easily shared. If you have lost like this, there is no easy road back and going it alone is even colder, harder, lonelier.

I never wanted another dog after Daisy. I wanted her back. Just her. I still do. The years soften the pain but they don't erase the loss. Writing about her helped. Telling her I love her helps. Write their stories and tell them you love them, even when they are not there. They still hear it.

As I write this it has only been a few days since losing Daisy. Saying it out loud seems impossible but writing about her seems only fitting. I don't know if this paying tribute to the dogs who leave our lives would help each of us who lose them but I know they all deserve one.

14 years ago when Olivier brought Daisy home as a "surprise" I told him he changed our lives forever – now our lives would forever be punctuated and scheduled by house training, accidents –ours in not figuring out the house training fast enough, dog training, the eternal search for the right vets, pet sitters, dog walkers, what time to be home

for the dog, whether she could come or not, -"what about the dog?" – with everything we would do.

I complained he hadn't told me before bringing her home. People told me to give her back (except for one of my sisters who believed as I did, that dogs are not disposable). I told them, no – now we had a dog. And what a dog she was. Sweet, playful, loving and a spirit that was wise, soulful, and kind. She was family, teacher, and more than best friend. She gave more with love and open heart than she ever took.

When we took our first puppy training class with Daisy, the instructor teased Olivier and I about babying her and called her "Baby Daisy." We loved that and it was one of our favorite nick names for her along with "Doodle."

Daisy taught me so much about dogs. I learned from her about how to listen to them to know what they wanted, to give them what they asked for or were saying. I learned how dogs cannot stand heat. When we lived in Florida, an outdoor trip in the summer meant only a matter of minutes before it was time to bring my hot dog home. We turned around so many times with Olivier saying: "we're not taking that dog the next time," we always did and we always turned around when it got too hot for her.

Daisy taught me about she felt about certain groomers, about paying attention to what she was saying when she was so frightened, she stood still in a groomer's doorway and peed on a return visit. She never eliminated like that until then, until so abjectly terrified, she lost control of her bladder. I turned around, whatever happened had obviously hurt her in some way and was not to be repeated. We discovered "express grooms " -55 minutes in full view – I missed these so when we moved from Florida. I always wished they had these in NYC.

When I took the pet care technician teacher position at a vocational school, one of the first changes I made to the program was to bring a cat and a dog into the class room so the students could learn how to observe and work with companion animals in real life. Daisy and I took therapy dog classes, so when we went to school, we could add in animal assisted therapy and quell the rigors of working in an office building banning dogs.

For six years in a row Daisy was a registered therapy dog. She passed with flying colors two Pet Partners evaluations and her third with New York Therapy Animals. I passed too but it was all Daisy – the last test we got our highest evaluation, even as her hearing was almost gone, she aced that test being all over checking in and responding to visual cues. She was sharper than ever.

We were a team. Daisy and I taught students how to walk a dog, how to observe behavior and read what body language meant, how dogs play, how to put on a leash and halter, how to press the gum to check capillary refill time, how to tent skin to check for dehydration, the parts of a dog's anatomy, what a dog's vocalization mean, how dogs learn and so very much more. She was groomed countless times for grooming demonstrations in front of the class room and at the sites my students interned at.

Her mid-day walks were field exercises in how to walk a dog. And what behaviors were of note for the dog walkers too. Breed demographics counted too for being tested on. Students worried about remembering a pet's name but I assured them it was the breeds we need to recognize. We love our breeds. There was much to observe, remember and discuss out of the classroom with the downtown crowds and lunch time dog walkers, owners, and local dog park dogs.

Daisy loved coming to class with me. She was the most perfect dog when we worked together. I would bring her downtown on the subway in a pet carrier, the only time I ever used one with her. Every time she would see the bag come out of the closet her tail would wag and she would do her "happy dog dance" – a series of play moves and jubilation. I used to say when she was no longer happy to see the carrier, when her behavior changed on sight of it. I would know she did not want to come to school anymore and that would be it for her school days. It never happened. She was always over the moon happy to see that bag.

Daisy loved ice cream, food, especially cat and people food, chasing a cat that thought it could slink by another cat or her without being noticed, raiding the litter box for cat poo, her toys, especially her tennis balls, car trips, being with her people 24/7, trapping lizards in Florida (she was very confused about where the lizards were when we moved to NY but she never gave up searching for them).

Daisy loved the kayak, the beach and swimming, she was a cocker spaniel after all, a true water dog. One time in Florida, she was knocked over by a wave and some water got in her lungs. She was wary of water after that. We worked with her to soften and overcome that fear, holding her in our arms chest high together for short periods in the water, she liked that. Then, we used a boogie board she could lie on to hold between us. She liked that too, soon she was swimming on her own just fine again. She loved being on the kayak so much she would shiver with anticipation when she saw it.

I think the highlight of kayak trips for her was one very early morning in Clearwater, when a pod of dolphins found us and swam alongside us jumping over our kayak as they went along. We were their toy. Daisy wanted to join them, whimpering for it. I didn't let her go, I was worried she would be their football –maybe I was wrong and I deprived them all of a perfectly good time and swim together.

Daisy had dog friends in every neighborhood we lived in and a decided preference for fluffy white dogs wherever she would meet them. More than once when I mentioned this to an owner, they would tell me that they had heard this before (I believe there is a study in that).

When Daisy really liked a dog, she would give them little patented cocker spaniel hip bumps. Once on West End Avenue, the receiving dog's owner asked me if Daisy was spayed. I assured him she was; he shook his head -"I'd get that checked if were you" he said. Even towards the end her dog friends would give her so much joy with her hip bumps and doggy smiles you would never know there was anything wrong -especially in those moments - because there wasn't.

And Daisy had people friends too. When we lived on West End Avenue the little children on our floor loved Daisy. Whenever they would see us together, they would plead to pet her. I would let them take a tennis ball and throw it down the hallway for her to fetch. One little girl was distraught when her parents decided to move to Brooklyn. She told them she was worried she would never meet another dog that liked her the way Daisy did.

We had play dates with that little girl for a while and then years later saw each other again. When Zoe and her parents pulled up to the house, we were outside waiting. Zoe told her parents she wondered if

Daisy would remember her. As soon as the car door opened, Daisy lifted her head and ran across the open field in between, right to Zoe.

Once, when we were walking down the hall in the last building we lived in, on the way to a walk, I realized I had forgotten something and told her we needed to go back to get it. Another little girl was walking by with her mother and overheard me. "Mom! Dogs can talk!!" she exclaimed with joy. I agreed with her.

Daisy also had pet sitters and we always wondered if they took as good care of her as we wanted them to. After a snow storm in Manhattan, the snow is plowed into high banks parallel to sidewalks. Crossing the street can be challenge, those banks can be several feet high and spots to access the sidewalk are sparse. On one of those days, Olivier saw Daisy and Paulette, her pet sitter, from a distance. When he told me about this the first thing I asked was, how were they together? Then, did he say hi? He told me was too far away but he never worried because when they got to the corner, Paulette picked Daisy up gently and carried her over the drifts and to the other side of the street. That was the best report I could have had. I was very happy that Daisy had so much love and care in her life. Once she was your friend, you would never forget her and she would never forget you.

The happiest parts of Daisy's day were when we came home to her and saddest when we left her. She was so thrilled to see us home. I trained her to get a toy to channel all that jubilance. When we would leave, she would retreat to a resting spot with eyes downcast and head down, not happy.

When we found out Daisy's kidneys were failing, we knew we had less time than she or we wanted. It is always this way – God does not give them to us for long but he charges us to care for them with the utmost love while he does.

I can never know what she bore inside her but I knew her well enough to know there was discomfort in how she would hold her body at times, a stiffness that came and went. Jumping on things became harder and she would place her front legs where she wanted to be and turn and look at me for a boost up. And in those same days, along with the moments of difficulty, there would be contentment, that full sigh as she settled down next to me, that way of lying with hind legs rotated out -her frog posture, that has to, must feel good, if they do it.

Even so, there was so much that I went through with giving her all that she wanted in the end. When she would not eat, I tried everything I could imagine to tempt her palate and when she went from eating two of those impossible to come up with meals a day to one, the bar got higher. At the end I was up to just Haagen Dazs ice cream- she still loved it enough to eat some.

Mostly Daisy wanted to be with me every minute. So, I spent as much time with her as I could. Every minute and told her how much I loved her. She got very weak and had a hard time walking down the long hallway in our building. I would carry her to elevator and she would take it from the lobby. Every walk at the end was to Petland around the corner, even when she couldn't finish the biscuits they would always give her. We went anyway and she still liked going – she walked faster in that direction and she still made sure I asked for those biscuits for her.

Bladder control happened and we slept with wee wee pads and she gave no argument to wearing baby diapers in the end - my little noble girl.

This saying "you'll know when it is time" is never so. And the times you pick are the ones you have decided on, not them. It's not you that knows, it's them and as close attention as you may pay to what they are saying, you can never know what they are thinking. And you know they are aware of this, of what is happening inside of them. In ways you cannot know.

I can only try and do some justice to her with this. And I can only pray that I can somehow be close to the person she believed me to be. I hope we did not take too much of the good moments she had left or of the life she still wanted to live. Because no matter what we think of knowing whether or not they are ready, we know we can never know this. The body keeps going. I wish that it is true that we spared her pain and gave her a peaceful passing. I know how quickly she slipped over; I think her spirit was so much stronger than her body in the end.

I don't have the right words for how much my heart aches for her or for how very much I cannot keep but still loving her. Part of me does not believe this is real, I catch myself thinking I need to hurry up to get home to walk her, expect to see her bouncing ahead of me at the beach, on a walk, at the park, turning around to check and make sure we are there – with her doggy smile and a little bounce, making sure no one is left behind and asking us to come along with her.

I believe our dogs love us so very much, maybe even more than we can ever love them, that's how they're made. And that our hearts and homes cannot but miss that very love and them when they leave us. In Spanish when you say you miss someone you use the phrase "te echo de menos" which means "I am less without you."

Without Daisy, I am infinitely less. I do not know if there is a rainbow bridge but I do know if there is a heaven Daisy is in it.

CATS AND DOGS

In a natural environment, cats and dogs are typically not sharing living quarters, but when they are, as in our homes, just how well can dogs and cats get along? How similar are these two species in how they relate to us? And when does the research looking at those differences do so in meaningful ways to help us in our understanding of the companion animals in our homes?

Researchers, N. Feuerstein and Joseph Terkel studied the relationships of cats and dogs living with humans in an article published in a 2008 edition of *Applied Animal Behaviour Science.* The authors' analyzed questionnaires distributed to 202 households in Israel and observed cat and dog interactions in 25 of households surveyed. The scientists found, that, yes, cats and dogs can live "amicably," read happily, together.

According to the study, cats and dogs are more than adept at reading each other signals and dogs will even adapt their own greeting behavior to accommodate the cats in the house. When encountering each other, cats tend to sniff nose to nose while dogs prefer a nose to tail sniff, followed by a muzzle sniff. When cats and dogs live together, cat greeting behavior or the nose sniff becomes the universal greeting. Other key points made, were that dogs and cats seem to be the most successful in adapting to each other when the cat is adopted first. Cats that are brought into the family before a dog, appear to become more readily accustomed to the dog, whereas when a dog is an established family member prior to a cat's introduction, the dog may exhibit greater aggression or indifference to the cat. The researchers posit this is owing to the dog's greater dependence on humans and what might look like, and perhaps be, jealousy on the part of the dog.

Should we be blaming the dog here? Human domestication of dogs has created the possibility of such a trait. Not only do we ask our dogs to work very closely for us as farm dogs, hunting dogs and service dogs (all heavily dependent on human direction and interaction). We also breed our dogs for what we consider acceptable temperament-namely sociability with humans. In addition to what the researchers suppose, basic territoriality is bound to figure into all this as well. Cats, perhaps even more so than dogs, are very attached to territory. No matter who, the original resident, is far more entrenched in their history of living in the space as their home territory is being infiltrated, while the newcomer has more work to do in acclimating to the new space and assimilating with the others in it. There is pressure on both sides.

Not surprisingly, the age of the dog or cat can be a factor as well. Babies get along well with other babies. The researchers found that cats under six months of age and dogs less than one year tended to have the lowest levels of aggression and indifference to each other. This suggests integration with puppies and kittens to be more fluid.

In a separate 2015 study, seeking to study attachment in cats and dogs with their people, researchers, Potter and Mills, compared attachment styles of dogs and cats to humans. Twenty cats and their owners were recruited to take part. The experiment called for an attached pair to be placed in a new and strange environment, separated the two, introduced a stranger, then reunited the pair and evaluated what behaviors happened at each turn. Attachment in the test used, is shown when an individual strives to remain close to the other, is upset when unintentionally separated, pleased when reunited, returns to the other when frightened and feels safe enough to explore in the others presence.

The use of this test, commonly known as the "Strange Situation Test" or "SST" was first developed in the 1950s to measure how attached infants were to their mothers. The test was subsequently used with chimpanzees and has been quite popular in studies with dogs.

Researchers drawn to the similarity of the bond/dependence existing between the domestic dog and the owner and a human infant and the mother have found parallel results in attachment levels. This no doubt contributes to how well the Strange Situation Test works with both sets of species. But how cats encounter their environment along with the differences in how they establish relationships impacts how useful the Strange Situation Test is to measure how attached they are to their owners.

In the Potter and Mills study, the cats were not found to respond with the same behaviors as infants and dogs to the SST. This does not necessarily prove that cats are not similarly attached to their humans, both cats and dogs are social animals and both have developed successful domestic relationships with humans but how they live in a natural environment and with people is markedly different. The results shown in the test, as administered in this study, is not a useful one for feline attachment to people. This is a key point and ties into how we can confuse comparisons when looking at how differently cats and dogs value territory and relate to others. The home environment and the security of its familiarity is more salient for the cat while for the dog the security is more familiarity with the human attachment figure.

A new environment for cats is akin to a four-alarm fire. Cats are seriously territorial and owned cats are famously under socialized to places they are not housed in or choose to frequent on their own terms. Even cats given access to the outdoors roam in a specific and familiar territory in a radius close to home, typically, a radius of one mile for females and four miles for intact males. New environments are serious stuff and need much time and vigilance to acclimate to. Francis Galton, scientist, and cousin of Charles Darwin, writing on cats and domestication, noted that staying home is part and parcel of what made the domestic cat, domestic:

> "The cat is the only non-gregarious domestic animal. It is retained by its extra-ordinary adhesion to the comforts of the house in which it is reared." (Galton, 1865)

The SST starts with placing an attached pair in a new and strange environment, for the test to be effective, there must be a level of stress and discomfort in an unknown environment. To successfully compare attachment in a new environment between individuals there needs to be a somewhat similar level of stress and discomfort as baseline in the new environment and to gauge attachment to familiar persons from there. Ostensibly what is being measured is response to familiar individuals not horror at change in environment. However, there is a serious difference between stressed and terrified, extreme vigilance precludes the display of natural behaviors in lieu of seeking safety. Stressed allows for comforting and terrified calls for survival first and foremost (think of Maslow's hierarchy of needs where safety comes before social).

Terrified and survival strategies are what a new environment means for most cats, that is why in shelters and re-homing we provide cats with beds, boxes, and places to hide.

The cat relies on scent and hearing as primary senses. New environments are processed through these pathways with great consideration and vigilance. The Potter and Mills study finds the greatest behaviors that the cats display in the test to be ones where the cats are scanning the new environment or sniffing the air while still or actively exploring it and moving around in it. These behaviors are not described as vigilance, because according to the authors, they are apparently not accompanied by fully open eyes or ear flicking. Such a conclusion does not fully take the significance of the cat's full suite of behaviors, and of even greater concern, dismisses the cat's great reliance on scent, for instance, to process information and changes in environment and how this translates to vigilance. Worthy of additional consideration, is then, how vigilance is defined (and this disconnect is not confined to this study). A strong argument can be made that the study cats were in fact vigilant while they scanned the environment or moved around in it. Additionally, it can then be said that vigilance occupied much of their time to ascertain the safety of this new space and which precluded the display of other social behaviors that would occur in a familiar setting.

It is difficult, if not impossible, to measure a degree of natural behavior, including natural social behavior, when extreme vigilance is needed to encounter a new environment. In a 2013 study by Drs.' Saito and Shinozuka, showing cats' abilities to recognize their owner's voices, the author position their choice of conducting the study in the home environment because:

"Visiting owners' homes allowed us to observe the cats' natural behaviors which might be disrupted in a laboratory because of vigilance against a novel place."
(Saito & Shinozuka, 2013).

Looking at how cats and dogs relate to human beings as distinctive species is a worthwhile and valuable endeavor. To do well with it, we must measure the natural behavior of the cat as a unique species and not as a feline version of a dog. That never works.

<u>References</u>

- Feuerstein, N., Terkel, J. (2008) Interrelationships of dogs (*Canis familiaris*) and cats (*Felis catus L.*) living under the same roof, *Applied Animal Behaviour Science*, Volume 113, Issues 1–3, Pages 150-165, ISSN 0168-1591, https://doi.org/10.1016/j.applanim.2007.10.010.

- Galton F. (1865) The first steps towards the domestication of animals. *Transactions ethnological Society London* 3:122–138.

- Potter, A., Mills, D.S. (2015). Domestic Cats (*Felis silvestris catus*) Do Not Show Signs of Secure Attachment to Their Owners. *PLos ONE*.

- Saito, A., Shinozuka, K. (2013). Vocal recognition of owners by domestic cats (*Felis catus*). *Animal Cognition*.

MANAGING SUCCESSFUL CAT AND DOG INTEGRATIONS

Daisy was the first dog I got to live with as an adult. When she came to us, we already had cats, two adults and one kitten. Daisy was a puppy and as such had little definite opinions, was very responsive to any off-putting cat body language and decidedly more interested in her people than the cats. The cats felt differently about this. None had ever lived with a dog. One of the cats was upset enough by her presence that he hid behind a sofa, hid behind that sofa for days. Until, I managed to coax him out of hiding with lots of attention, treats and play.

Daisy coming home as a "surprise" (a whole other story), meant there was no time for careful consideration on introductions and strategies before she arrived. True, that even had there been preparations for her arrival, most questions on managing cat and dog interactions come when an issue arises and not before.

Either way, knowing how to best introduce a new cat to your dog or a new dog to your cat is good information to have no matter where you are in the process. Questions like, how long does it take? How difficult is it? Does it matter if the dog has cat experience or the cat, has dog experience? What about what kind of breed the dog is? Or how old they are? Or is it easier with a puppy or is it a kitten? Answers to managing successful cat and dog introductions are often found in paying careful attention to several key elements: individual differences, the right environment for each species, resource distribution, behavior monitoring, behavior modification and time.

Individual differences: While it is certainly helpful to know if a cat or dog has lived successfully with a dog or cat in the past, it is also necessary to realize that this was a unique relationship where that certain cat, lived with that certain dog. Not all cats and dogs are the same and not all past relationships equal new relationships. Each pet, whether cat or dog, has a different personality, level of socialization and history. Also worth noting, that cats and dogs that have had positive experiences with each other in the past, are more likely to anticipate positive experiences. Similarly, being aware of a history of negative experiences or no experience means a new situation can be stressful.

Whatever the individual story, you still want to help broker the best possible way to integrate a cat and dog that are new to each other. Factor in personality and history always, and keep in mind the canine and feline perspective are not always the same where communications are concerned, e.g., butt sniffs work for greeting in the dog world and discrete nose sniffs are greeting behavior in the cat realm.

Owner attention is the first resource. Bringing a new pet home will mean less attention for the original pet or the wrong kind of attention in being told what not to do around the new addition. Who we pay attention to and how we pay attention matters. Regular and positive contact for both animals needs to happen. This is even more important during an integration process.

No matter how adorably cute or needy the newcomer is, greet the original resident first when entering the home and or a room they are in. Offer first pets to the original resident pet or pair interactions closely in time with the newcomer and then the original resident to assure the original resident of their place in allocation of owner affections.

Separation. During the initial stages of integration, cats and dogs should be physically separated. This allows for each species to get used to the new space and inhabitants in the safest and most comfortable manner. You are dealing with two new things to get used to here: new environment and new animals.

Dogs are more comforted by access to the people they have secure relationships with, as opposed to cats, where familiar territory offers security. Keeping the dog close to humans is the least stressful for them. For cats, in new places, less space can be more, especially with all the new smells, sounds and sights to conquer in multiple rooms. Smaller spaces, as in a bedroom or office, where the resident humans will be spending a good amount of time, are easier to adjust to and be comfortable in initially because cats benefit from a more discrete space to become familiar with, again, one with us in it too.

Avoid basements, garages and even bathrooms, as an initial space, if possible. Cats, too are social, and the bathroom is a cold and lonely place when you are not in it. Installing the cat in a bedroom along with all necessary cat furniture, toys, litter box and food allows the cat to get acquainted with a new space on a more reasonable scale. The more time the cat can spend with humans or the room we spend most of our time in, the more they acclimate to the environment, the more social support afforded by your presence and your scent when you are not around.

This means keeping dogs, for that crucial first week or so, out of cat spaces during the day. Living rooms, offices, and watching TV with us after work, are ideal places and times when dogs can be close by as long as cats are safely elsewhere. The dog should not have day time access to the bedroom or office the cat is in, for at least, the first week or two. If there is limited space and the bathroom, the only safe other option, make this space as comfortable as possible. Along with litter station, food, and water, add bedding, boxes to hide in, scratch pads and classical music.

With only one bedroom which the dog customarily shares for nighttime sleeping, do not disrupt this practice, rather address comfort and safety when the cat gains access through raised, cat reachable resting spaces (such as cat shelves and towers) and sleeping with the dog attached to a leash attached to you. This allows cats to feel secure and you to monitor and redirect dog movement.

The right environment: Physical space must be set up so that it is enriched appropriately for both cats and dogs, allows for comfortable, safe interactions including exit routes and refuges. For dogs, this means enough dog beds or places on the people furniture in every room to be used as resting places. It also means enough toys. Think at least ten, varied and consistently available, including puzzle toys and chew toys to interact with independently and with people. With the dog that does not interact with toys or respond to play, ask first what other toys or play might the dog like, using natural behavior to inform your choices and try them out, giving enough time (more than you think) to see results. Demonstrate play and keep your own reactions more subdued for the truly reluctant. Think also puzzle feeders to get them targeted on objects along with other benefits. For instance, the oral expression of chewing with dogs leads to prolonged interactions with objects to chew on and is an intrinsically rewarding activity.

Dispense with food bowls for most meals. For dogs, substitute meals fed in bowls for stuffed puzzle feeders to gnaw dry and wet food meals out of. Avoid slow feeders that impede rather than satisfy natural behaviors. Make sure to consider the subject, licky mats are best suited for geriatric dogs while teething puppies and active dogs need to chew. Snuffle mats are another good choice to indulge sniff time and well suited all around. Giving a dog a much needed something to do with their time aside from us, engages the dog in a natural behavior while having a calming effect working on a pleasurable food puzzle or chew and in doing so, lessens vigilance and frustration.

Puzzle feeders are a must for cats too, allowing for innate feline behavioral needs that help to lessen stress and boredom. The portion of the cat's dry food diet can be provisioned in puzzle feeders. Puzzle feeders for wet cat food are not currently available. In the initial stages of integration, when cats and dogs are separated, a rolling puzzle feeder is a good choice. The batting to initiate the rolling action that releases kibble, simulates batting and playing with prey cats naturally employ. When cats and dogs are sharing the same spaces, use a flat based puzzle feeder that only the cat can access by placing it on a raised surface the dog cannot access, like a counter top. Use the same process in determining how the cat puzzle feeder engages natural behavior versus obstructing it. Select the feeder carefully, flat feeders with a larger visual field that cats can survey and access food to fish out are more appealing than mazes or tipping vials or any other slow feeder that irritates in preventing access.

The more a dog or cat must do in line with their natural behaviors, the less bored, the less behavior issues and the less entertaining or worrying the cat or dog, as a distraction or menace will be.

In their natural environment, cats are arboreal and use trees and other raised areas to survey their environment, detect prey and to escape from predators which translates to a necessary refuge from unwanted attention, canine or otherwise. A well-placed cat tower, against a wall and ideally, next to a window, is a great way to add raised vertical spaces cats can get to and dogs can't. A cat bed on top of a dresser or desk is another nice solution when working with limited space. Cat shelves are another good option, as long the shelf has adequate space to allow the cat to turn around and stretch out on.

Think cat and dog, in thinking how easily a cat can access a raised resting space and how difficult that might be for the dog. Exit paths and access routes should be effortless for the cat to figure out and obtain when it comes to vertical space. Cats also need cat beds with at least three raised sides to curl up in. Add cat beds to a cat tower's enclosed plane or cat cubes as welcome added appeal to the space. Think against walls and under chairs for beds placed on the floor, always keeping in mind the cat wants the hide as a safe refuge and not a place to be trapped in. Rub powdered valerian root or catnip on beds and resting spaces to raise their attraction value and reduce stress. The best cat beds can come in the mail, cardboard boxes turned on their side can work too. The raised sides often give kitty a feeling of security, sometimes, even away from a wall. Seeing what your cat uses, taking in preferred cat locations, and giving a few days to acclimate to any new object, will tell you what they like.

The right kind of music can assist in an integration process and adds to an enriched environment for both cats and dogs. Several studies have shown positive effects of classical music on various species and are especially effective in shelter environments and group housing situations. Classical radio stations are a good choice to leave on at a lowered volume when you are not home (and even when you are). In addition to the relaxing tones of the music, the soft announcer's voices are calming which adds to the comfort. Look for smoother classical sets vs operas aimed towards background music for studying, etc.

Interactive Play and Schedules. There is a definite and certain benefit to introducing and keeping routine and schedules for companion animals. Cats and dogs are subject to when their humans decide they will eat, what they will eat, where they will sleep, on what, what they play with, when a dog will walk, who they will interact with and on and on. Putting most of these events on a routine schedule has been shown to lessen stress for cats and dogs and is one way to afford control over events these animals have little control over.

Both cats and dogs need attention and daily interactive play time with their people. Play reduces stress, releases beneficial hormones associated with pleasure, boosts trust and relationships across the human animal bond, creates positive associations with the change in the environment, and is just plain fun.

With cats, begin by working with a fishing wand toy that can be dragged across from or away from the cat's line of vision and allowing the cat to follow and pounce on the toy.

For dogs, choose tug-of-war with a designated toy –nothing else, no hands or clothing please. "Find –it" or other games that are fun for the dog, like fetch, should be indulged in. Hand feeding "off and take it" – more on this below - can qualify when done correctly.

Avoid laser pointers for both cats and dogs, which can be harmful vision wise and are frustration inducing with nothing to "catch."

Behavior monitoring. Closely monitor interactions, including your own. Body tension, lectures, scolding or any forms of force or any other types of punishment need to be avoided. We are trying to create positive associations with the changes in the home. Punishment creates fear, distrust, interferes with learning and prevents forming affiliative relationships with all species. Always remember that training and interactions need to be force free, positive, consistent, and consider individual relationships with residents and newcomers. Your affiliation and interactions are a valued resource to distribute along with food and resting areas. This is why, recommending greeting the senior resident animal when you enter a room or your home first, before the newcomer or chaining exchanges with the new pet, followed immediately to acknowledging the first pet, goes to keeping the priority of the original pet foremost in your attention. They have not forgotten who was there first.

When the animals are sharing space, be aware of the quality of the attention you are paying to both animals. And work on improving it. We are very quick to remark on the things we do not like to see and often let positive events go unnoticed. Those positive events are some of the best times to train and reinforce just by marking their occurrence and praising them for happening. Treats are fabulous for rewards but do not let not having one cause you to lose the moment.

You always have on hand, your words -   Saying a pet's name and telling them what a wonderful creature the universe has sent you for whatever it is they have just done you would like to remark on, is always available to you (tone matters – happy and uplifted for dogs, soft and quieter for cats.)

There will be some low-level stress initially, just as long as this does not escalate, allow for it. Pay attention to what the cat and dog are "saying." Relaxed bodies and lack of intense interest are good signs. Intense vigilance for either cat or dog by body tension, extended staring or fixed gaze alerts us to discomfort. Lip licking or yawning out of context of eating or fatigue, for either cat or dog are examples of stress signals for both species. These lower-level stress or distance increasing behaviors can alert another to a need for distance or to cease a behavior. Watch to see that they are acknowledged by the receiving party.

Look at body language, stiff and rigid are more about defending territory or preparing to mount an attack. Vocalizing for dogs or straining at the leash are both signs of frustration and attention getting. Hissing, growling, flattened ears, whiskers drawn back with cats, are some of their signs that attention is not welcome or of displeasure with the environment. Stiffness and yowling are feline escalations. Keep your own body relaxed and reaction neutral. Do not fuel this with excess energy. Keeping dogs on leashes even when you are at home will both allow for safety and afford opportunities for alerting us to intention movement through leash tension.

Think on how to manage the environment or increase distance to lessen triggers. Seeing this means time for redirecting a dog with a simple request and/or relocation to a neutral, safe location. Fishing wand toys can be great distractors and guide paths to alternate places for cats, as long as the dog can be relied on not to give chase. Why every room requires a raised resting space that is easily accessible for a cat to safely escape to.

Multiple, short-lived, positive interactions with no conflict are the goals you are shooting for. Those interactions can be brief. Less can be more. Ending on a good note is better than waiting for a bad one to terminate contacts. A series of momentary, frequent positive interactions are what builds the basis for longer interactions.

Always keep in mind it is way more effective to defuse an eminent situation from happening rather than dealing with one underway. Redirecting sooner rather than later is easier and less stressful for everyone involved. Having a strategy in place beforehand can help with dealing with a volatile encounter that might occur. Force free training, for both cats and dogs, affords dialogue, relationship building, and can lessen reactivity. It also has the benefit of increased acceptance in direction and redirections, both of which can only work if someone is listening or wants to, in the first place, being able to ask for someone to "leave it" or "off" takes practice.

To work on this, especially for dogs that are not particularly responsive to "off" requests, begin training this now with "off and take it":

Make sure you can safely use the morning or evening meal to hand feed. Offer a piece of dry food placed directly in front of the dog's nose and say in a soft, calm voice- "Off." Your voice will cause the dog to momentarily pause in going for the food. Make sure not to repeat the request when the dog has stilled. In the very same second, you see that pause say "Good Off!" in a happy voice. You will see the same pause. You are fine to repeat the "Good Off!" more than once. You have just marked the requested behavior and reinforced it at the same time by saying "Off" -the name for the behavior and "good" – reinforcing with praise.

Next, offer the kibble a bit closer while saying "Take it" in a happy voice and let them have the kibble.

If the dog is not listening at all, and going straight for the food, try starting with giving a piece of kibble to begin without any requirements for a while and then begin with the "Off" request. Still, no response? Try an "uh uh" in a bit of a higher tone but NOT more than one or two syllables and when you see that pause to the "uh uh" reward with "Good Off"! and "take it (give kibble)!"

Do not scold or reprimand in any way if you think the dog is not getting it right. Timing is important and there is a learning curve in mastering it. Missteps mean the human part of the exercise most probably has a timing issue, so work on that by you trying to get the timing of requests, marking, and reinforcing closer to behaviors.

Practice, practice, practice always remembering it is about YOUR timing so the dog can get it right. This exercise can also be done with cats.   Pay particular attention to changing your tone of voice depending on the species you are interacting with. For cats, eliminate higher pitched and louder voices; keep speech softer, briefer and more melodic. For dogs, unless they are shy dogs, a jolly happy, not too loud happy camper voice usually works.

Avoid punishment, it is not as effective as positive methods, makes things worse not better and damages relationships. Think instead of how you can work on getting the desired behavior through environment and meaningful direction. If you have made sure that the cat and dog have safe spaces to retreat to and can get to them easily in an antagonistic situation, ask next how can you facilitate that redirection safely happening? After getting attention using the name of your pet, can you: Offer a feather toy to chase after? Throw a ball to run after?

More behavior modification: Start now with encouraging and praising all the behaviors you do want, even if they are simple ones. Notice and remark on the "good" behaviors and not just the "bad" ones. Everyone needs validation. A sleeping, calm dog can be praised for "good quiet" or "good sleep." Getting in the habit of praise and reinforcement for what we want our pets to do offers an opportunity to develop more of the good behaviors happening more frequently because we are rewarding them. Not to mention, this builds a better human animal bond.

Include encouraging and praising those behaviors that include registering, glancing, or looking at the new addition, cat, or dog, especially at the very onset of the behavior. This is when the animal is mostly likely to be under threshold and when you want to reinforce that calm behavior that has not escalated. Desensitizing animals to each other can only happen when arousal or excitement are below threshold. Counter conditioning with positive reinforcement needs to happen below threshold as well.

Watch for when distance increasing behaviors start to escalate and go unheeded. Body tension, posture, and vocalizations are some of the red flags. Vocalizing, that has escalated to guttural yowling with airplane ears and tensed body, typically signals time for intervention with cats, as does squared off, weight forward, ears back, snarling, and low and sustained growling, for dogs. When you do pass threshold as in, a dog that is getting ready to chase a cat, the redirection needs to happen before the pursuit begins. This requires monitoring and attention so you can see those anticipation movements, whether they be registered in body tension with heightened vigilance, fixed gaze, flattening of ears, etc. and respond. Always keep it positive.

Think of redirecting as a sequence of a direction such as "off " or "leave it" (avoid "no" which can carry a tremendous negative emotional charge by people when delivered). Follow direction by redirection, what you would like the dog to do instead: "Off Rover," "Good Off Rover!" "Go to your bed/get your ball," etc. Throwing that ball for your dog to chase is even better in refocusing the energy. Similarly, a cat that is getting overly riled by a dog needs to be redirected.

For cats, that feathered fishing wand toy is a great lure to change the energy in tense moments and can aid in moving them away from the dog. Not handy? You can softly say "off" and **only** if the cat can be handled safely, is tractable, pick them up and place them in the closest of their raised resting spaces. If the cat offers consent, reassure with one or two strokes alongside the muzzle or behind the ears. If the cat is not tractable, placing an object (not your hands), like a newspaper, pillow, or other object to block the visual stimuli and forward motion between cat and dog are usually the most effective. If you must use your body as a visual barrier, and **only if it can be done safely**, the lower half of your clothed body either from the front or the back is preferable to hands which should never be put in between contentious animals.

Separate fighting animals with loud noises, if that does not work in three seconds, water, if that does not work in three seconds, a bath towel or large denim shirt thrown over one animal can be effective if done safely.

Time and consistency: As much as we would like to sit everyone down and explain how we all are going to be one happy family; we don't have the words to do that with our cats and dogs who cannot and do not benefit from them.

What we do have is what we show them by the spaces and experiences we create for them. Based on the quality and safety of these interactions and given time for them to allow for the belief that their environment is stable they can come to believe in and trust that one big happy family exists.

NEW BABIES, PETS AND HOW IT CAN WORK

I have had friends tell me their plans on getting a puppy as practice for their futures as mothers. The idea being, the puppy would bolster/test maternal instincts and caretaking skills. Dogs might be harder than babies I have said, they never learn to talk, at least in English, never get to be old enough to go to school, spend the day with a friend, walk home alone from play dates or school, bring home classmates, develop much in independent interests and activities, go off to college or come home for the holidays. And such a plan, as ill-considered and inconceivable as it might be, could only work as a true test of how family works, if the dog as family member, got to stay forever - no matter what happened and was never, ever disposable or subject to surrender.

When babies do come into our lives, they bring a sea change in the lives of the parents. When that family has pets, the change affects everyone in the home, humans, cats, and dogs.

The arrival of a newborn baby to a household with pets can bring concerns, some of which are not so easily resolved. A study of 12 animal shelters in six states, including New York and New Jersey (Scarlett, et. al. 1999), looked at the personal and health reasons listed for the surrender of 520 dog owning households giving up 554 dogs and 384 households relinquishing 488 cats. "New baby" is among the top three reasons given for giving up cats and in the top four reasons for dogs. The study also showed allergies as the number one reason for cat surrender, even as 15% of these households still had dogs at home and 11% still had other cats.

Such a study is a valuable starting point as a source of information. It also may suggest more questions than it answers. For one, with data drawn from interviews given to survey questions on occasions of surrender, answers may be less than forthcoming. Whether or not some of the reasons for relinquishment such as "allergies" (especially where other pets are retained), are offered as socially acceptable reasons for surrender, is indicated but cannot be determined. People dropping off family pets at animal shelters are understandably reluctant to be negatively judged or in certain cases, have their pets negatively judged. The study also does not compare people who have new babies and retain their pets, so we are not able to determine if similar issues exist for the pet retaining group and why it is not an issue, or if it is, how it is overcome.

There is no other information offered when "new baby" is listed as the sole reason when the addition of the baby prompts the surrender of a family pet. The study also shows, the shorter the time a pet is in the home with a baby, the more likely the pet will be surrendered. This finding, by itself, may suggest any number of things; from a relationship with a pet that is tenuous for any number of reasons, including a lack of history, apprehension over how to successfully integrate a new baby into an existing pet owning family group, fear, and anxiety on the part of the pet or owner, and, or a response to general and or specific input from surrounding community and family members that pets and babies don't mix.

When we look at the response of families surveyed citing a new baby as the reason for relinquishing a pet, we see that a good number of the pets given up, had not been with the families for very long at all. More than 33% of all cats and dogs had been acquired during the previous nine months and 40% during the previous year.

A more developed relationship may influence the decision to retain a pet and possibly their resulting relationship with a child. "Conflict with a child" was reported for more than 3% of all dogs given up and less than 2% of all cats given up, with close to a third of these surrenders occurring within one month of ownership. Again, the timing is notable.

When "conflict with a child "is reported, we cannot know how "conflict" itself is being determined. Is the conflict one of feeling, sentiment, or actions? In the case of action, is the behavior appropriate, misunderstood, or provoked? Is there appropriate supervision to determine what is going on?

What then are some of the things that can help pet owning families keep pets and babies safely together at home? Understanding, preparation, and behavior monitoring of both species all play a part in the answers.

Understanding how our pets fit into our families: One of the reasons our companion animals fit so well into our families is their similarities with humans in a shared social system. Molly Love and Karen Overall compared the convergence between human and canine social systems and advise on how heightened awareness of the differences prevent disasters:

> "Canine behavior may seem unpredictable and complex to untrained observers. Dogs communicate with each other, in part, by use of body language. They also communicate with people this way, but because of similarities in canine and human social behavior, people may interpret canine signals incorrectly. For example, when a dog presses its paws on a human's shoulders, we anthropocentrically call this a hug; however, in communication between dogs, such behaviors are often challenges. The same behavior can be a challenge in canine-human interactions, or it can be a learned behavior that humans have encouraged and rewarded. Similarly, dogs, even those trained in obedience, may be confused by all but a specific set of verbal and hand signals they've been taught; other verbal or hand signals may interpreted as provocative. This confusion may create only frustration in the relationship between an adult human and a dog, but can result in a bite to a child. Children can be equally confused by a dog that uses a signal for which the child has already formed an association. This is particularly dangerous when a dog that is fearful, confused, or anxious gives mixed signals. For example, children may view a rigid, highly held, wagging tail as a signal that the dog is happy, or a lifted lip as a smile, without realizing that these behaviors can be potential signs of danger. Even children who have lived with dogs cannot always be expected to perceive subtle differences in canine signals or to act

appropriately in potentially dangerous situations." (Love &
Overall, 2001)

Our increasing knowledge of cat social behavior can extend this
convergence to feline social systems as well. In other words, cats,
dogs, and humans share a social system with extended family groups
caring for young, highly ritualized visual signals, including a
communication system that relies mainly on non-verbal
communication or body language and social deference to avoid
conflict. How we enact these, differs for each species, and this is
where confusion, stress and conflict can arise if we expect human
behaviors from non-humans.

A look at greeting behavior for dogs shows us a specific approach
which averts eye contact, uses approaches from the side, not frontally
and gains important information from butt sniffing. For dog affiliates,
a joyous jump on each other celebrates the occasion of meeting and
greeting. For cats, friendly intent is signaled by a raised tail, soft eyes
with a slow blink and a nose touch. Good cat friends may bunt heads
or do a quick side to side body rub.

Human greeting behavior is culturally dependent but often
includes forward facing body stance, direct eye contact and close
physical contact in the form of hugging, kissing, or grasping hands -
none of which cats and dogs do. And while we may share deference
with companion animals in withdrawal, looking away, and walking
away, much of their stress signaling differs markedly from ours. For
dogs, early signs of stress are yawning or lip licking out of the context
of hunger or fatigue. For cats, displacement grooming, flattened ears,
whiskers, and tail flicking for cats, are some stress related behaviors.
All of which, for cat or dog, may be misinterpreted by humans, who
may find such behaviors offensive as in not being listened to by that
look away, yawn or displacement grooming.

Clear communication is one of the most important things we can
do and understand as it relates to our pets in general and most certainly
when a new baby is on the way, or has arrived. The next most
important thing is to understand how we are communicating with
them and to make sure that our interactions are appropriate and benefit
the humans and the pets.

Getting your pet owning family ready for a new baby: A new baby is a major change in household routine, environment and impacts how everyone in the family will be acting and reacting going forward. Instituting optimizing physical spaces, training protocols, and keeping schedules and routine can benefit each species in the home, especially pets, affording a sense of control over most events they have little control over – food, walks, playtime, etc.

Environment counts, this includes for both cat and dog, being able to easily access, use and locate a safe retreat from a stressful situation or unwelcome attention. This is especially effective for anxious or fearful dogs or cats. A review of the home surrounds in addition to baby proofing should make sure the home is comfortable for the pets as well with an enriched environment (see more on this in the chapters for cats and dogs), providing for varied raised refuges and ground level hiding and resting places with raised sides for cats as well as multiple refuges and resting places for dogs to retreat to at their choice and when asked.

If the dog in the family has not been trained, this is the most opportune and necessary time for evidence based, force-free training as well as working with any cat behaviors that might be problematic. Putting such structure into place allows the dog to learn alternate behaviors and be rewarded for new behaviors and appropriate responses to new situations. Behaviors such as jumping, barking or those early morning feline food requests that may have been ignored or tolerated prior can be addressed or managed as well (training and free access to food puzzles offer benefits). It is key to look at the individual breed characteristics and personality of each pet and tailor training and modification programs accordingly. While there are no one-size-fits-all solutions, there are a wealth of good strategies, guidelines, and tips that families can utilize with their own pet in mind.

Adding to individual knowledge is a necessity. A good number of valuable resources for pet owners detail the how-to's of safely getting your home ready for a new baby. Make sure to look at credible sites such as the ASPCA (for dogs), Blue Cross (for cats), and the American Humane for both, all of which are linked here. Working with a well-qualified, force free, educated behaviorist is also highly suggested. The only trade organization for no force professionals working with companion animals is the *Pet Professional Guild.*

Even before and after the baby arrives: It is important to be aware that the relationship with pet and owner predates that of the relationship with owner and baby. This means for the pet that they will respond primarily to the owner and form whatever associations the owner shows are meaningful in relation to a child in their presence. If an owner is anxious, angry, or fearful around a new baby when the pet is around, this is the association the pet is exposed to and what influences them most. Love and Overall discuss the concept of "appropriate guidance" to anticipate, manage and supervise the right interactions between child and dog.

It must be stressed that we cannot teach what we have not learned, adults also need appropriate guidance to make sure we understand dog behavior, cat behavior, humane handling, attitudes, and interactions so we can apply them and make sure our children learn them as well (also see the chapters on bite prevention and safe cat handling).

A 2021 study on at risk interactions between children and dogs in an Inuit Village in Canada made the following observations on findings as to children learning what they live:

> "Regarding family factors, insufficient adult supervision during dog–child interactions was an important contributor to risk-taking behavior. In addition, the idea that children learned at-risk behaviors toward dogs from family members was brought up: "children do it if their parents do it." Some interviewees believed that children were afraid of dogs because their parents were and thought "that some of them learn that dogs aren't useful."

> "Among socio-situational factors, a perception that dogs have little worth was identified as a contributing factor to risky behaviors. The high number of dogs in the village was thought to reduce their importance in the community. Children learned that "it's normal that the dog is just a piece of furniture outside." (Gouin, Aenishaenslin, et al., 2021)

Development milestones of humans contrasted with canines are also highlighted in the Love and Overall paper. When we look at the human infant less than six months, we see reflexive behaviors along with sitting up and creeping. The behaviors of babies typically affecting dogs (and cats) are new noises such as crying, screaming, and babbling. All of which is novel stimuli for the pet to react and respond to, including any owner anxiety at baby vocalizations or concern over pet responses. The presence of the new baby will also generate a host of new smells from baby and mother that a pet is aware of.

How infants react to pets is age dependent. When less than six months old, they are prone to grabbing body parts or fur of a pet. The typical response on the part of the dog is sniffing, licking, and initial avoidance. A cat will typically remain, tense, tail flick and retreat. As children develop and begin crawling, walking, and running a family pet becomes a natural target of curiosity and investigation including using hands, mouth, and teeth to do so. Typical behaviors expected in response from a pet are freezing and avoidance with the same cautions to allow for retreat. For both a cat and dog, ensuring the opportunity and allowing for avoidance and retreat is key. Distance increasing behaviors when thwarted can result in defensive behaviors on the part of the animal over time.

While toddlers and younger children may be fascinated by animals, they are developmentally still egocentric and empathy skills are not developed. This is significant for caregivers to be mindful of. We can and should explain how a dog or a cat might be feeling and what constitutes humane handling. Because of where children are developmentally, we need to remember that this will bear continued repetition and supervision to be absorbed and applied.

Cats and dogs need baby safe areas they can easily access to decamp to. Again, raised resting spaces for cats in every room and baby gates with dog doors are essential for pet welfare and a must with babies. Forcing an interaction or asking an animal to submit to one they would rather avoid can only create stress and reactivity. For dogs with a diagnosis of predatory or fear aggression or a cat with a history of fear aggression, unsupervised interactions put both baby and animal at risk.

Supervise interactions and be mindful of extra consideration for fearful or anxious pets. Now is the time to begin to educate children how to interact with consideration and safely with animals. Very young children may not have sufficient motor skills to stroke pets, tending to pat repeatedly or lay hands on an animal instead. And while this may often be endured by the pet it does not mean it is welcome. Teaching a child, the correct way to approach and pet a dog or cat means having the knowledge to both be able to model the behavior for the child and to shape it with appropriate intervention if necessary.

A look at two to three-year olds from a French study showing how young children and their pet dogs communicated (Millot, Filiatre, et al. 1988), finds that children approached their dogs twice as frequently as the dog approached the child. The researchers found that agonistic –social behavior relating to fighting, were associated more frequently with two- and three-year-old children. These behaviors relating to fighting were usually met (61%) with the dog usually retreating or showing appeasing behavior. Similarly, children showed retreat more frequently than returning a threat when a dog displayed threat or aggressive behavior toward them.

Research has shown that children are the recipients of most dog bites, with male children being the majority. Personality also figures into this.   Yet another study done in 2012, observed children interacting with unknown dogs to determine relationships between temperament and risky behavior around dogs. Finding from the study show:

> "Those children who were rated by their parents as more shy— those rated as more cautious and fearful in novel or uncertain social situations—tended to be somewhat more cautious and fearful in interacting with the dog. Those parents who rated their children to be less shy—whose children were rated as bolder and less fearful in social situations—were somewhat more likely to take risks in interacting with the dog." (Davis, Schwebel, et al., 2012)

Teaching, cautioning children, and monitoring their behavior around dogs and cats, can prevent unwelcome advances towards pets and remove pressure from the pet to respond with distance increasing behaviors, such as hissing or stiffening. Ritualized displays such as these, may not be welcomed in general but are useful to read and protective for the animal as ways to avoid conflict and injury.

Warning signs and what to do about them: Signs that an animal is not handling the addition well may be: sudden changes in behavior, including withdrawal – absenting themselves from persons or places in the home, vigilance, patrolling behavior, watching and being on alert rather than relaxing, and increased or different vocalizations, changes in sleeping patterns including duration – less or more sleep, and locations – especially away from persons in the home, increased reactivity in specific or general circumstances, anxiety when around the child or signs of fearful or defensive aggression around the child. These are clear and significant signals of stress, discomfort, and unease.

It is vitally important that the owner first and foremost, manage the situation by removing the stressor, which is most probably inappropriate interactions between child and pet. Avoid punishment where either pet or child is concerned to reduce stress and lessen aggression and begin managing the environment providing safe retreats, informed supervision and guidance, and a careful program of behavioral modification.

Research and working with an appropriately qualified, well-trained professional is the most ideal solution. Taking the time to learn, appropriately supervise and manage the situation needs to be immediate to ensure everyone's safety.

References:

- Davis, A. L., Schwebel, D. C., Morrongiello, B. A., Stewart, J., & Bell, M. (2012). Dog Bite Risk: An Assessment of Child Temperament and Child-Dog Interactions. *International Journal of Environmental Research and Public Health*, 9(8), 3002–3013. https://doi.org/10.3390/ijerph9083002

- Gouin, G.G., Aenishaenslin, C., Lévesque, F., Simon, A. & Ravel, A. (2021) Description and Determinants of At-Risk Interactions for Human Health Between Children and Dogs in an Inuit

Village, *Anthrozoös*, 34:5, 723-738, DOI: 10.1080/08927936.2021.1926713

- Love, M., Overall, K.L.(2001) How anticipating relationships between dogs and children can help prevent disasters. *Journal of the American Veterinary Medical Association*, 219 (4), 446-453.

- Millot, J.L., Filiatre, J.C., Gagnon, A.C., Eckerlin, A, Montagner. (1988). Children and their pet dogs: how they communicate. *Behavioural Processes*, (17) 1-15.

- Scarlett, J.M., Salman, M.D., New, J.G., Kass, P.H. (1999) Reasons for Relinquishment of Companion Animals in U.S. Animal Shelters: Selected Health and Personal Issues. *Journal of Applied Animal Welfare Science*, 2(1), 41-57.

HOME ALONE, WHAT TO KNOW ABOUT PET SITTERS AND DOG WALKERS

As the strict restrictions and cautions of the pandemic have waned, a lot of us are back to traveling or going into the office every day or alternate days. Back too, are travelling for work, vacations, day trips, and weekends away, all sometimes with, and perhaps, mostly without, our pets. Pets who have gotten used to, having us around or who have only known us to always be around.

Getting there is just not be as fun as it used to be these days. Staff shortages, longer lines, security concerns, extensive delays and the current "less is more" philosophy in airline service. Still, there is relief in being able to do it again. For us that is, maybe not so much for our pets. Leaving the rest of the family behind can be less than ideal for some of us, except for the cat who likely prefers to stay at home. For those dogs who will endure the flight stuck in a travel bag under the seat just to be with us, weight limits vary airline to airline.

Then there are those recent FTC guidelines, banning emotional support animals from airplanes and allowing only service dogs. Specific service dogs only that is, namely service dogs with training and elimination control certifications. Not to mention, airports are rarely pet friendly enough to accommodate sufficient relief stations. And flying in the baggage compartment is a risky option, for any pet, please investigate carefully other pet transport options for welfare and safety issues.

Add to this, not every destination or trip, works with or welcomes pets. Unfamiliar vacation rentals and hotel rooms rarely equal where your pet wants to be while you spend the day or dinner away from them. All of which, further limits the possibility of pets traveling with owners.

And if they can't come along with you, what then are the next options, what are the most important points to consider? Where does that leave their care?

The "how to find the right pet sitter" question, is one I get asked a lot and one I think about for my own pets. A question I wish I had better answers to. Pet care is an unregulated industry. Certain aspects of brick-and-mortar establishments may have to fulfill local municipal requirements as to size of kennels or even staffing ratios, depending on location. But those specifications rarely, if ever, go to welfare standards. Local, independent, come to your house, pet sitting is mainly provided by private services or independent contractors with mostly no prerequisites for formal education, training, or licensing as to qualifications. Some pet sitting and walking services advertise their employees pass background checks but these checks are limited in what they are looking at. Bad or good credit rarely affect pet care.

Pet sitters for pet sitting and dog walking were staples during those work weeks when everyone left home for the day. I relied on them for midday walks, had more than one walker and sitter, back-up walkers and sitters too. For extended time away, there were cat visits by sitters for the cats and boarding for Daisy initially. Pet care when away from home seemed sorted until I started working in the industry. There is a definite shift in understanding services when the view changes from "provided to" vs "provider."

I started freelance dog walking and cat sitting jobs after grad school. I liked the companionable dog walks but not being a pack walker, could never realize any true earnings, especially when most dog walking all needs to be happening in the same two-to-three-hour lunch time window. With so very many dogs in NYC, dog walking is a competitive business price wise. Many walkers and services make the cost affordable for guardians. Such competition also drives more dogs per walk for a walker to make ends meet. More dogs might be better for the walker but not necessarily the best arrangement for the dog when it comes time to sniff around. Cat sits were easier to schedule, with visits that can be in the off hours and then, less cat sitters to compete against.

My entry into the corporate end of pet services began when I took the pet care technician teacher position, where weekly visits to the internship sites at doggy day cares, groomers, rehabs, and shelter sites were required. There is a lot going on behind the scenes, a world most guardians rarely get to see.

There are good pet services establishments and there are ones that promise care that does not get delivered. Boarding and day care facilities provide a service in keeping pets safe and fed. This is significant and not to be overlooked. Where other, more preferable, options do not exist, the value of a place that will mind a pet and keep them free from harm and provide for basic needs cannot be undervalued. Where welfare concerns come in, there are additional considerations that may be as equally valid.

Visits, to evaluate sites and measure intern/student performance took me from the front of the shops with all the puppy and kitten posters and toys on display, to back rooms and staff doing what they do when no one is watching.

Because I cared deeply about the quality of services, the pets themselves and the value of students learning the applied components of pet care in the places people are paying for it, I worked on placing students in the better internship sites. The challenge was that these good sites were too many times the exception and not the norm. Most states, including New York, do not regulate a handler to dog ratio. Multiple dogs in one room can test the observation powers of even an expert observer.

There are entire books written on how to accurately code behavior of multiple subjects at the same time. It's not easy. Now add five, ten, fifteen, twenty or more dogs to one room with one person tasked with other additional responsibilities, like picking up after the dogs, etc. and see how that works. And what about knowledge of animal behavior, welfare, or safety practices? New York City does mandate a two-day training course in animal care and handling for businesses that sell, board or groom pets and requires a single person who has completed that course to be on site but not necessarily present with the animals, when the business is open to the public.

I found in field observations to multiple sites, management of groups of dogs done by keeping them doing as little as possible in barren, empty rooms which prevent contact with objects or furniture and lessen the interactions with other dogs. The thinking perhaps being, the less there is for the dogs to do, the easier the job is for the handler. The lack of toys and furniture was explained to me at several sites by staff as preempting the dogs from fighting over them (more toys and furniture or redirecting the dogs before they squabble would be a possible fix).

Without furniture to lounge or play on or toys to play with and activity monitored, little if any play is seen going on in a day care environment. Play was often prevented simply because most staff members did not recognize what play looks like, mistaking it for fighting. I saw even dogs looking for attention, jumping gleefully and barking, being scolded. There is often not much for the dogs to do for hours in most dog day cares. Boredom and overcrowding can result in any number of stress-related behaviors ranging from internalized stress, to displacement behaviors, including, but not limited to, coprophagia (feces eating), fighting, or repetitive mounting. Barking was discouraged to the extent that most places had a "no barking" rule, typically enforced with spray bottles or scolding.

Dogs in the day cares the program worked with, and as is routine in most day cares, are typically kept there all day. Some places kennel the dogs for a portion of the day to give them a break from being on the communal floor, allow for napping, or to feed them. The dogs are often not walked unless the owner has requested, paid for the additional service, and the staff can provide it. For the dog that is house trained, not being walked is highly stressful. These dogs have been trained not to relieve themselves indoors and dogs really want to do what they been trained to do. When a dog does break down and relieve themselves, urine and feces can only be cleaned when a worker recognizes it and with a lot of dogs in a room, admonishments usually accompany that process.

On one visit to a day care, I talked up benefits to the dogs in another site, unique in having scores of tennis balls all around for staff to throw for the dogs. "But you would have to throw those balls all day!" was the response. Yes, you would and the owners paying over $50 (the then current going rate) a day for daycare for someone to play with their dog, probably thought that was happening. But it wasn't. And it wasn't even happening at the day care with all the tennis balls on the floor. Some of the saddest dogs I would see during my visits, were the dogs who sat on the perimeter of the floor, not interacting, and fixedly looking towards the exit door. If I could get in those dog's heads, I bet there would be one thought only, "I want to go home."

For my own dog, we had yet to find the doggy day care with enough of the elusive appeal of the comforts of home and family. Enough toys for each dog to play with, enough spaces for dogs to find playmates and disengage from them, enough floor area for beds and furniture for all dogs to lie on and around, enough skilled handler to dog ratio and truly knowledgeable force free staff to interact with dogs (including relationships, walks, and knowing when a dog wants to go home and bringing them there), etc. There is so much more to consider than we think. Until then, each neighborhood we move to begins a new search for the right pet sitter.

In the search for the right persons to provide pet care, prepare to dig deep in the looking, interviewing and monitoring yourself. Word of mouth, always a good source, can begin the search for the right sitters by asking neighbors, local vets, pet stores for referrals. Who do they use? Who do they like? Observations count too. If you see a dog walker on the street or better yet, in your building, who looks like they know what they are doing in how they handle their charge, ask them if they pet sit. And search online.

A definite must is to do a live, in person interview. Take the time to make sure the sitter/walker is the right one by meeting them first. No matter how busy you are, make space for this. No matter what the reviews say, despite what a potential service may advertise online, no matter what they tell you about how wonderful their sitters and walkers are, nothing, repeat, nothing can substitute for seeing and making that determination yourself.

We met sitters who have never walked a dog but would like to, sitters who only want to walk the dog during the week in the middle of the day, sitters who want to bring the dog to their home because "it's easier that way", sitters who don't do cats, just dogs, etc. But New York is full of possibilities and full of pet sitters. We kept going.

Find the right fit for your pets, not going for a walk is way less harmful than going for a walk with the wrong walker. Prepare and think ahead of the pet sitter interview. Collect references and check them. Work up a set of questions and a list of what you are looking for on how to best decide on bringing in the safest and best dog walker into your home to work with the cats or dog or both.

Did they close the door after themselves when they came into the home? Take a glance behind them as they did, to make sure no animal was behind them? Leaving a door ajar or being unaware of who might be hovering at the door is what you do NOT want to see. Walking in and out of any pet owning home requires keeping your eyes fixed on the door, space around it, and making sure no animals are leaving or entering unless they are on leash with you.

Make sure to tell every sitter and walker, that you are not asking them to train your cat or dog. This is for relief walks for dogs, water changes, feedings, litter scooping and cleaning of pet messes. Just that. Double emphasis here, especially to say, that you *only use positive, force free methods and require them to do the same.*

Pay extra attention to questions the prospective pet sitter asks about your pets first off. Note, whether they ask about your dog and cat to learn more about them as individuals, as in inquiring as to character, what they like, and don't, and how you provide for your pet's preferences.

Do they know it takes time for your pets to like them and seem willing to take the time and put in the work to make that happen? Watch for the person telling you what an expert they are with animal relationships, even though they have never met your pets. Relationships are created with our pets through time, experience and trust and are not instant.

Ask open ended questions on how they feel about working with pets. The sort of questions that need more than a "yes" or "no" answer. Wait a minute or two for a response, avoid filling in the silence. That pause after a question asked does not need to be broken by more words on your part or your follow up question. Listen for their response instead. You want to know their answer, it's important.

Make sure those questions include what they do to correct behavior and what behavior do they think needs correcting? What happens when things go wrong? What do they do when a dog does not listen? When a cat hides or does not want to be petted? How would they handle that? Concretized. Let them tell you what that looks like.

Listen but don't fill in the blanks. Wait. Resist filling the silence. Breathe and allow for answers. What are they saying in response? Move forward? Pull? Lecture? Redirect with neutral or positive energy? Now think. Do you want those answers applied to your pet?

Ask again about corrections and redirections and what those looks like for them when they use them. Red flags for answers that include scolding or what a pet might perceive as punishment or mentions of "alpha," "results," or "balance." You do not want to hire any walker or sitter that advocates force or outdated dominance theories regarding companion animals, nor do you want such ideas applied where your pet is concerned for ethical and humane reasons.

Ask how they feel about dogs and sniffing? And wait to let them tell you first before you say anything. A leisurely sniff around the block relaxes and satisfies the dog brain far more than any timed walk not including a "sniffari" can ever do.

Be very wary of the group walk. Dogs go at different speeds. Necessary relief walks require a good sniff around and elimination, this is easily compromised on lead with a pack of other dogs moving forward. Individual dog walks also can make sure that your dog benefits from the full attention of the walker and prevents unexpected or unwelcome complications from another dog that is not familiar to your pet.

Not to mention, those occasions where the walk you think you are paying for, is your dog parked on a sidewalk instead.

What about cats? There are cat people and dog people and there are cat and dog people. If you have both, you need someone who is both as well. Someone who appreciates the differences in their care, and not just who has the right answers.

Watch them in action are they good with cats and dogs? Speak gently? Avoid direct eye contact initially? Use a "jolly camper" voice for a dog and an "elevator" voice for a cat? Not loom over your pet? Approach from the side? Worth every word of repeating: Immediate, large, and glaring red flags for corrections, including scolding or punishment.

How do your pets respond to them? You know what it looks like when your animals likes somebody or not, so watch closely to see how they feel about the meeting.

- Did they come bearing treats? Or ask if you had your pet's favorite treats around? Did they know how to approach, pet, and play with a cat? Did they ask you first, what the cat liked best? Know that "let them come to you and make them want to" has cat written all over it.

- Is there phone out of sight? Ear buds off? Do they stay off their phone throughout this and all visits? How can anyone watch your pet when they are looking at their phone? Or listening to pod casts or tunes? Non-negotiable.

   Are they able to commit to a dependable and routine schedule for visits and walks? Have an emergency contact? Who is their back up in case they cannot make a visit? How good are their references? Make sure you meet them too.

- Can they let you know in real time how a visit went? Send a photo?

Can you monitor the situation behind the scenes to make sure everything is as is should be? With pets home alone and sitters and walkers coming and going, keeping an eye on what is happening at home while you are not there is more than a good idea. Digital cameras are a necessary component of making sure your pet is cared for properly.

Remotes provide an essential view on how things are going and allow you to make sure the schedules your pet expect and you have asked for, are maintained. They also provide an up-close view to make sure nothing but positive handling and love is happening in your home in your absence. Watch the remote feed after each visit.

For the cats, morning visits are a must for meal times and we ask for that. We also specify a time, as in between 6 am and 9 am and ask that the same time be kept for the duration of the visits. Cats seem the most eager for the morning meal, so we respect that. Additionally, we also know that regular routines including feeding times can offset stress, schedules are known to offer a sense of control to companion animals who are subject to most things not necessarily being in their control, as in when they eat or walk or the litter is changed, etc.

Necessary parts of any cat visit are: Doing a "cat check" - making sure they are all safely present and accounted for, changing water, cleaning litter and fresh food. Any uneaten wet food needs to be discarded (we once had a cat sitter who would add new pate to the old so it "wouldn't go to waste") and to clean plates in between feedings.

As highly professed with cat knowledge a sitter may have, a good one knows, the basics may be all some cats want with a new person around. For the cat that is unsure and keeping out of sight, let them be, as long as they can be visually located (under the bed or sofa counts).

Over the course of days, such cats may be more forward. Leave it up to the cats. For the cats that do easily approach people, show how these cats like to play, where the toys are and where to keep them out of sight.

- Remind your sitter that *cats, especially yours, are only to be pet in a cat friendly manner – confining touch to the head, as in behind the ears and along the muzzle. No matter what their cat at home likes, touching in any other manner is too much, too soon, for a new acquaintance.* The right sitter should focus on how your individual cat responds, not anyone else's cat. Accentuate this.

- With owners out of sight and new people caring for them, one of the most important things for cats, dogs, or any companion animals, is not to overly stress them. Not forcing contact can also offset stress. This means that if the cat does not want to interact initially, allow for this. If the dog wants to go back home after elimination, let them. If the cat is hiding from you, they are no doubt doing this because they are frightened of a new presence. Likewise, if the dog is hesitant with petting and play, tone it down.

When working with a new pet sitter, please remember to give your pets enough time to get to know the new person who will be in your home before you leave. As many walks or visits as possible, before you leave them alone with the sitter, are necessary to begin acclimating your animals to a novel presence in your home.

Schedule these for initially when you are around and then when you are not at home for maximum benefit. Again, again, and again, emphasize there is to be no "training", no force, no corrections.

I arranged for our new pet sitter to walk Daisy three times over the next three days. On each day, I am part of that process, both as social support for my dog as well as to physically show how we do dog things. From approach -always from the side, no looming over (yes, the walker should have passed knowing this when they got the gig but bears repeating), putting on her halter, leash, holding the leash and walking pace, route, and style. I talk about which are the preferred walks, where the shady side of the street is for summer days, and scaffolding stretches are found for when it rains.

I accompany both Daisy and the walker on the first walk. I start holding the leash and after a block (half a block, if Daisy is really relaxing), hand the leash to the walker with the dog between us and continue for another a block in this fashion. During the walk, I tell Daisy what a wonderful dog she is just for walking with us.

The next step, with a comfortable dog, when a block from home, is to allow the walker to step around so they are between you and the dog and walk in this fashion for half a block, then return the dog to between the two of you for the last half of a block home. Take your cues from the dog on this last block, if they are unsure, do steps instead. A few steps at comfort level, are way more effective than half a block under pressure.

It is important to keep these introductory walks short. As soon as Daisy relieves herself and turns for home, we follow. This is not about timed walks; this is about allowing the dog to be comfortable with a new and strange person walking them. And why a short positive walk is the best walk to desensitize and counter condition.

When your first introductory walk has gone well - the handling is good, the praise has flowed, no signs of minor stress such as yawning or lip licking, etc., or major signs such as trembling or slinking (stop and start again with a slower first walk if you see this), try for a second walk. Go through your leashing process together but allow the walker to put on the leash and hold it this time. This is another short and positive walk. The third walk is when you can ask the walker to walk your dog alone for a very brief walk, while you wait at home for them to return. Make sure and confirm that just a pee on this third solo walk is fine, a poo is great if it happens but **the dog must be allowed to return home when they want** (be very suspicious of a walk that is not short here), where you calmly greet them with love.

Daisy appears unsure initially on that first walk, not as scared on the second and somewhat resigned by the third. On the solo third walk, the sitter reports that Daisy just wants to take care of business and return. Bingo. That is just fine. And that quick return is a good sign that the walker has heard what we asked for in letting the dog set the pace of these walks.

No prolonged time for sniffing on these walks unless the dog decides so. And because these new walks can be stressful initially, I have asked only for "relief" walks for these walks to begin with. While you may value a longer walk for your dog, recognize that with a new person walking your dog, shorter walks are less stressful in the beginning of the relationship.

Sniffing at will needs always to be allowed for, but distance takes a back seat with a new handler. Weather, always an important component can also determine walking time. Same for cats, the cats are only used to you and the people you know in your home. A new presence in the home is stressful for most cats. Allow plenty of time to go by before asking the sitter to spend extra time with them.

Maintain as much of the usual structure your animals expect, keeping to the predictability of schedules to lessen stress for your pets of when meals are fed and walks are given.

Email, text, and leave a paper printout, in plain sight of a detailed, written list of contact information, how you do things and where to find supplies. Stack cans of wet food, clean plates, and paper towels next to the list, paper bags of kibble need to be behind cupboard doors so nobody helps themselves.

Leaving your radio turned to a classical music (studies show this type of music has been shown to lessen stress related behaviors) station which the sitter can turn on and off on alternate days, can alleviate some of the angst over your absence. Plug-in's such as *Pet Remedy* may also help.

We find the right one (fingers and paws crossed). As our new sitter gets to know our pet family, we are struck again by the time it takes to create a relationship. No matter how much of an animal person one is, there is still time to be taken in building a bond with a new animal comfort and trust need to be established. Above all else, do not expect an instant bond, steps can help. Your companion animal has a relationship with you built on time, history, and trust, with the right sitter, only time and positive experience can allow for the new relationship to develop.

References

- https://nyc-business.nyc.gov/nycbusiness/description/animal-care-and-handling-course#:~:text=The%20NYC%20Health%20Code%20requires%20that%20all%20pet,up%20to%20twice%20a%20year%20depending%20on%20demand. Retrieved February 7, 2024

- Shelley-Grielen, F. (2018) Behind the Scenes. *Barks from the Guild* (32). 44-47.

- Shelley-Grielen, F. (2018) Understanding Animals. *Barks from the Guild* (33). 50-51.

- Shelley-Grielen, F. (2019) Room for Improvement. *Barks from the Guild* (34). 40-43.

- Shelley-Grielen, F. (2019) Maximizing Welfare. *Barks from the Guild* (35). 22-24.

MOVING WITH YOUR PET

I am not a fan of moving, who is? Putting down roots, staying in one place has its appeal. Cats and I have this in common and sure we are in good company. Even so, there was a period of transitions when we moved five times in six years. I truly liked the fifth place for multiple reasons, but I was more set on staying there for the next several years just not to pack up house and move again. No one had to tell me (or the cats or the dog), that moving is one of the top three stressful experiences for humans. And when the move includes our pets, the stress can affect everyone in the family. However, humans do get to offset some anxiety through exercising some choice and control over what the move might bring while our animals simply get none.

Planning and executing a move for us, includes all sorts of think ahead moments from scheduling, timing, and anticipating potential issues, along with figuring how best to address them, to what to look to forward to in your new home. Each new experience, from acquiring and packing boxes, to moving trucks, to rearranging furniture, to finding your next favorite neighborhood restaurant and shops, are events that your cat or dog or both, get to show up to without knowing ahead of time they are coming.

All the plans and expectations you have for your new place, your pet simply cannot know or share them. This new space, is all foreign, uncharted territory outside of the realm of the routine world your cat or dog has come to know and depend on. Cats, perhaps even more than dogs, are attached to the permanence and familiarity of home territory and will find a new home more traumatic than dogs, who while still stressed, are more assured in a new place by the very presence of their caretakers. Knowing this, how can you make that move less stressful for the companion animals in your household?

Before you go: Micro chipped or not, make sure all pets have collars with ID tags attached that include your phone number and new address. Pets are more easily returned if lost, with easily accessible tags.

Pack for your pet too- Toys, beds, scratch posts, food, water bowls, puzzle feeders, litter pans and other pet paraphernalia need to be as recognizable as possible to provide familiar smells and known object recognition (think "their stuff"). You know how you hear that packing a suitcase for humans is a good idea for that initial settling in period of your new place? Your pet needs one also.

Moving day: is sure to be chaotic with everyone coming, going and doors left ajar, open, slammed, not to mention all sorts of strangers to your pets in the form of moving men, etc. Know where your pets are at all times. Keep it safe the day of for cats, by making sure to confine carrier kept pets to the last room anyone will access with a securely closed door on which you have posted a very large sign barring entry. Place cats the morning of moving and packing days, after supervised food and litter breaks, into cat carriers with one of your worn t shirts to provide scent and comfort. Position the carriers next to each other and against a wall - not in the center of the room for additional security. Partially cover the tops of the carrier with a pillow case to block some visual exposure making sure there is sufficient airflow. A spritz of either *Pet Remedy* in the air or *Rescue Remedy* on the side of the muzzle and/or paws (avoid eyes) is helpful for both cats and dogs. Never add products to water, it can put your pet off drinking and there is no guarantee of getting the product into the animal.

Larger dogs should be kept with you on lead and smaller dogs can also be confined to carriers if they are accustomed to carriers. If not, keep them with you on lead as well. Make sure and bring your animals with you and **never** on the moving van. Check frequently on them throughout the day to monitor stress.

Your new place: Allow for less is more with cats initially. While you may think having an option to explore the new digs is the way to go for felines, it is not. All that new, strange, and unfamiliar space is overwhelming and scary. Keep cats in one room for the first week or two to allow them to gradually acquaint themselves to new surroundings in a more manageable space first. Your bedroom is the ideal place, especially since you will be spending all those nocturnal hours together and is full of your scent, when you are in it and when you are not. All of which is incredibly reassuring in unchartered territory.

Double check to make sure doors, windows, window screens and ceiling tiles are secure and escape proof in this new space. Install their familiar furniture to find refuge in, cat beds with at least three raised sides (don't forget the value of good carboard box), include cat houses or beds on a raised surface or cat towers and sprinkle opened capsules of Valerian Root on these surfaces to add soothing properties of this cat attractant. Allow for cats going incognito briefly when they arrive at the destination, site cat furniture with hiding spot potential in mind. Add toys, litter box, and a water and feeding station in the room.

Remember to greet pets by name on each entrance and exit to and from the room and home for additional reassurance. Leaving a radio set to a classical music station at a soft volume can also help reduce the stress of this new environment for everyone.

Explore your new surrounding with your dog. Take the time to allow the dog to sniff around while on a leash, explore the block, yard, and perimeter of your home. This walk around should also include the indoor areas of your home. Engage with your dog during this process by talking to them in a happy and calm voice, explain what is going on and where you are now even if it is to describe each new room. While the dog may not understand every word you are saying, your tone of voice and accompanying body language will reassure them that they are in a welcome and safe place. Keeping the dog on leash with you keeps the dog physically connected to you, helping to foster both emotional and physical security for the both of you.

Keep the routine: Maintain the predictability of routines and its comforts in schedules and consistency. This means playtimes, feeding and litter box cleaning at the same times as before, the same for walk schedules and other outings. Your pets do not get to control when things happen but do anticipate events like these and keeping on schedule will give much needed structure and reassurance to getting everyone familiar with a new environment.

Share what is good about your move: You know the things you can look forward to about your new neighborhood or home, your pet does not.

Consider the pet perspective in cat tower placement –in the great corner spot near that sunny window or other pet perfect spots for beds and cat houses. Remember to start small, all that great newness can be overwhelming so keep that familiar cat environment confined to one room for the first ten days or two weeks before allowing the cat access to the whole house.

For dogs, keep walks short for the first week or so, as in a good sniff around the block or up to the corner. Do save dog parks for later in approaching weeks when you are both settled and keep those visits on the shorter side initially even then. Map out the future routes to some great walks with lots of sniff potential. Find out where a good trail or dog park is for upcoming visits.

Allow for excitement/reaction: Your pets will respond to changes with a level of excitement or inhibition. Those responses will change over time as they acclimate. Be the social support that your pet needs by providing interactions, familiar furnishings, routine and calm energy. Keep talking to your pets, keep the environment as familiar as possible, keep schedules, keep playing classical music at home, add a plug in such as Pet Remedy or dust valerian root on surfaces on the first days, all soothing.

After the first week introduce interactive play time with cats and a five-minute morning and evening training session for dogs (sit, down, look, touch, etc. or any variation of three requests not more than three times in a row, all well praised and rewarded) to provide engagement, comfort, and structure with all that excitement.

Allow for mistakes: Even with the best house training your home is new to your dog or cat, there are no familiar smells in the walls, or safe and comfortable corners to be in, yet. Should there be a break in house-training, know that this loss of control may be stress related. Avoid any reprimands as they will only serve to create more stress.

Moving is a stressful and exciting experience. Where we can, here's to making the best of it from the canine, feline and human point of view.

WHAT TO KNOW ABOUT TIMING AND TRAINING

Timing is hard to come by, difficulties present from not being species specific to inescapable personal bias, it confounds. I have seen this over and over in others, and have experienced it myself, the challenge of first applying and then synchronizing timing in training and behavior modification. We, humans tend to watch what is happening in front of us first, to observe a response or behavior and wait/think about it to inform our own reaction and/or response. What can happen in this time lapse, is losing the association in associative learning. Mostly too much time has elapsed for that link or meaningful relationship you are looking to create to connect stimulus to response to marker to reinforcer. But, when your response is almost concurrent, within micro seconds of the event you have asked for, you have linked the behavior and response (the dog sits) with the marker ("Good Sit!") and you are able to more effectively, reward/reinforce with a treat or more verbal praise when treats are out of reach.

There is an almost magical moment of synergy when you find that rhythm, when you can see the behavior happening and respond to it in that exact moment, whether it is to remark on it, or change your own behavior, or cause the environment to change. That responsiveness, is where the maxim comes from, that training is a mechanical skill. When your reaction becomes an almost machine-like response in timing, there is a flow to the dialogue of learning that becomes intrinsically reinforcing.

To understand and work with other species you must start with the natural history of the species and understand how they communicate including their body language. And how they learn (a change in behavior based on experience). All animals learn, whether by trial and error (figuring it out), through association (what happens before, during and after an event), observation (watching), socially (watching others, especially significant others) and it can be argued, by insight (the aha or eureka moments). These concepts and how they work are part of learning theory, they inform much of how to train and modify behavior. Knowing the process allows us to gain awareness of what we are seeing before or after behaviors and how to shape and tailor the interpositions of cues or requests for behaviors, to acknowledge those behaviors have occurred and to reward them. Foundational theory and concepts are one thing on paper and intellectually and another when it comes to applying them. Much of the training and behavior modification we use, depends on associative learning and perfecting, most importantly the skill of timing.

This precision required in the skill of timing with training, means that your very own mastery of the moment in the Request, Response, Reward, Reinforce sequence is what makes the teaching and learning happen in association. Without it, both dog and human are out of sync and confused. As much as your dog desires to please you, and most do, they cannot understand English. As well programmed as our dogs might be in reading our body language to determine our mood and doing their best to appease us, they do not understand specific requests unless they have been taught them. They, unlike us, have not been to school and learned to understand the many words we used. Dogs can and do learn, socially and through requests and cues, if we break them down clearly and mark responses plainly, and this is where the timing comes in.

From basic requests, like "sit," "off," and "come," your pet, cat, or dog, needs to understand what you are asking them to do, know when they have done it, and be reinforced for it. Behaviors that are reinforced or rewarded, whatever those rewards might be, tend to repeat. Sounds simple so far, right? It is. Again, you need to have the all-important detail of timing down to the nanosecond.

Let's look at how we can request "sit": we can either lure a dog or cat into position by closely holding a treat directly in front of their nose and pulling it slowly back over the head so the haunches lower to the ground (lifting the treat away from the animal at that point is luring a jump, a common trainer error to watch for). We can also ask for the behavior if the pet already knows the request. Or we can capture/label/mark the behavior as it occurs, an often overlooked and extremely effective training technique.

The second the haunches hit the ground as the animal sits, is when we need to mark/label the behavior by saying "Good Sit!" followed by a reward – treats, pets, praise, toys, whatever is reinforcing for the individual. If we pause to watch the sit unfolding to completion first, looking without marking as it is completed, the moment passes and we lose the timing opportunity for coaching or teaching what that word/label "sit" means. Marking the behavior more than 2 seconds after it happens loses the immediate association and is less effective. Animals are able to learn "sit" means "put haunches on the floor" or "haunches are on the floor," only because of the immediate association of marking/acknowledging the behavior as it happens in real time with the label 'Good Sit," followed by the reinforcer of the treat following.

What, how, and when, we reinforce matters. Marking a behavior with "Good" in front of the label assigned is a nice way to add additional reinforcement to that label. The positive energy and the compliment, surely does not go unnoticed by the animal. Using "thank you" as a marker instead (of in this case "Good Sit"), can also help with timing. Think about it. We are more conditioned in our own interactions with each other to acknowledge a welcome behavior as it happens in real time without delay when saying "thank you;" when someone holds open a door, moves a bag off a seat to make space, passes a plate, etc. That sort of timing, is exactly the kind we are looking for when marking/acknowledging requested behaviors from our companion animals.

Beware of that other common trainer error, repeating the request, here "sit" repeatedly, even as the cat or dog is doing it or has done it. Putting the "Good (name of cued behavior)!" in front of the request once the action is underway or done, makes all the difference in asking for something, rewarding, and marking what the pet is doing in response.

Timing also plays into feedback and redirection. Using time-outs is a familiar concept, to be most effective, definitively identify context first. What has happened before (antecedent), event (behavior), likely motivation and the result (consequence). Knowing what happened before and what is happening now leads to more on why. Being well acquainted with individuals, personalities, understanding an animal's body language, history of a specific situation, including context in current and past events can inform not just understanding but your own response.

Not every scenario calls for an intervention. All animals work to avoid conflict with distance increasing behaviors, including our own species. What can limit the effectiveness of these communications is constrained spaces – lack of exits or egress, ineffective signaling – either not received or prevented (often by humans). Assessing when intercession is warranted, and when it is not, is important. Time outs, in those qualified scenarios with high arousal and high physical conflict potential, can help to prevent dangerous situations from getting more so.

Here it is again, when working with behavior -timing is integral to the process working. Timing in time outs refers not just to introducing the time out in synchronized concert with the behavior you are seeking to discourage; it also refers to an appropriate duration of time out. For this to be an effective tool or strategy, the time-out must follow the event within mere seconds and the length of the time out must be short enough (yes short enough, as in less than two to three minutes) to be associated with the immediate antecedent behavior -what just happened that you want the cat or the dog to have in the now of their short-term memory. For high arousal events, decompression times also play into the mix.

As much as you would like to, you can never, ever simply say to a cat or a dog, "Do you know why I put you in time out?" and get an answer, much less the answer you would from a human. If you do ask this, and people do all the time, despite the incomprehension of the pet, the "answer" and the appeasement behavior you do get from a dog – that looking away, lip licking, yawning, etc., is not an admission of guilt, more a request for you to stop the scolding. This type of "conversation" with a cat will most probably get you whiskers back, a head pulled further back on the body and depending on your delivery, flattened ears. This will also create a fearful cat who would like the scolding to stop as well.

In none of the scenarios will you be able to articulate the why for your pet with language nor will your pet ever be able to understand the lecture. We are limited to positive or negative associations here of the events in question, and for those associations to work, the briefer the period in between, the better. This may fly in the face of conventional practice, most probably because time outs are being used as if cats and dogs were children, who do understand lectures, and can use the same words we do, to let us know exactly what they do or don't understand. Pets just cannot use their words instead.

Duration for time outs, will differ for individuals and for species. Generally, a cat will typically take longer to settle from a highly aroused state than a dog and can benefit from a five to ten-minute interval whereas with dogs, less than two minutes, or three minutes for a more highly aroused dog, are more appropriate. And this varies for individuals, dependent on history, context, and personality.

Be very careful not to increase the behavior you are seeking to discourage with the use of time outs. When seeking time-out as a redirection, it is always helpful in how you use it. With a redirection we ask immediately for a different behavior. We also acknowledge the cessation of the behavior we are looking to stop with a "thank you," useful also for your own energy along with acknowledgement, followed by praise for the beginning of the redirection, as in "Good Quiet!"

When our timing in response to a behavior is off, we lose any meaningful association or link with the behavior and the time out. If we are separating animals, keeping a time out going for an excessive amount of time (from the pet perspective, not ours) also adds negative associations to the experience itself including environment, handlers etc. With dogs, we frequently see a time out after the fact, when the dog has been redirected, has responded to the redirection, or after the dog has done what has been asked. In this instance, praise or a "thank you" for listening is called for, instead of the time out. Instead, what often happens is, when the dog is lunging, barking, mounting, etc., and has been asked to stop, and has complied, they are given the time out - when they are basically punished for listening. Punished, because, after the dog has stopped whatever, it was that they were doing, as asked, is usually when they are scolded, grabbed, and rushed into a time out that exceeds the recommended three minutes.

Using timeouts incorrectly signals to a dog that just being around other dogs or situations they find objectionable is linked to punishment, creating fear and displacement behaviors. This also happens when frequently aversive or negative corrections precede or follow the time out. Doing things like scolding or shaking or scruffing, all, unnecessary rough handling, causing pain or discomfort, work against any positive benefit.

A much more effective way to change unwanted behavior is to reward the dog's redirected behavior and move forward. If the dog has done what you have asked, let the dog know that this is what you wanted by reinforcing the behavior you have requested. Immediately offering a treat to a dog that has stopped barking rewards the quiet and not the barking, especially when the treat is preceded by a "quiet" or "shush" or "good quiet" or "good shush," etc.

Allow for decompression. Your timeout can also simply mean relax. Training stationing before it is needed to create positive associations with the request and the place, for either cat and dogs can be helpful here.

Understanding how to use the skill of timing in training is work. It is even more work in applying it live on the ground. This, as with any skill, takes time to develop through repetition and focus. The trainer needs to learn how to be squarely in that moment to mark the exact instant of a response. Powers of observation and paying close attention to intention movements are vital to develop. With the sit, for instance, you will see a slight deliberation in the slowing of movement and haunches to begin lowering into the posture. As the haunches are touching the ground the words or vocal marker needs to come out of your mouth. The final moment is rewarded.

Knowing when the behavior is underway can also call for reinforcing a behavior you are asking for, with a recall, or asking a pet to come to you, the first step or steps forward in the movement you are asking for are the most important steps to mark and reinforce and continue to mark and reinforce to arrival. With a behavior like a sit or a down, you need to mark and reinforce the completion of the action sequence.

An excellent exercise to study and break down timing, and one when done correctly, to stop pulling, is "Red Light. Green Light." This is highly effective executed properly and highly confusing done incorrectly. Worth learning, this is an exercise all about exquisite timing in association. A dog pulls and the handler stops movement –

the "Red Light." The pulling relaxes and the handler moves forward – the "Green Light." Sounds straight forward, right? Sure. Try it. No lingering, no pausing, no watching if the dog is pulling or has stopped pulling, no thinking. Lock steps are required in Pull/Stop. Stop Pulling/Go Forward. Each aspect needs to happen simultaneously. Concurrent movement is required for the dog to make the association that pulling equals stopping and not pulling equals going forward. Micro seconds count. With a dog that pulls, this will look more like shuffling down the street. An almost sequence of baby steps when done in sync advancing over a half a block at a time at top speed to advance to longer intervals of not pulling. Red Light, Green Light is an excellent training exercise if you can master it.

There are no short cuts with timing. It comes with experience and application. Dexterity is acquired that way. Whatever aids we rely on also require expertise. Clicker training seeks to assist in marking behaviors at their exact moment only when the trainer engages the clicker in time. I use voice instead, it is always there, starts up in symphony with seeing the behavior and does not require another device to manipulate or delay response.

The very best way to develop the skill of timing in training is to begin practicing with the focus being on your timing in concert with what the dog is doing in the Request, Response and Reward sequence. Think of the work being on the trainer and not the dog, as improving on your own response and timing to what you are asking the dog to do and how quickly you can mark it and reinforce it. There is a definite learning curve with this, keep in mind, the more you practice, the better you get. Remember, they can't learn if we can't teach.

ADJUSTING PET FEEDING TIMES, THE LOW STRESS APPROACH

I was late to a meeting once because I had forgotten to reset the clock when the time changed the Sunday before. I had no idea I was an hour behind everyone else until I arrived. As far as I was concerned, I was "on time" including when I had fed the cats that morning, the last morning on time for a while as far as they were concerned.

In most of the United States, a Sunday, in the fall or spring, is usually the beginning or end of daylight savings time. This mechanism for bringing more light to farmers at work and children waiting at rural bus stops, can leave city dwellers off to work and back home in the dark. When daylight savings time makes our fall days start even later, that's the time of year, I hear more often from cat owners that kitty complaining has reached new heights, for food that is. People tell me those plaintive meows are making them crazy. Well, how do you think your cat feels about all of this? Or pleading puppy eyes from your dog? Food bowls can get rocked around and that crying, zooming, bouncing things off the dresser might just be hungry meets time-to-eat-and-find-something-to-do-about-it.

This time change means scheduled activities are set back one full hour in autumn or forward one full hour in spring, including the set times we may feed our pets. Such change is doubtless upsetting to companion animals that are aware only of a sudden shift in feeding times, especially when they are made to wait for that unexpected extra hour. No one set their clock back. Schedules and the dependable routine they afford to animals, who have limited, if any, ability to control events in their lives, are a way to have control over an environment they have little actual control over.

There are several things to pay attention to here, with the number one being the issue of choice, agency, and control. As much as we love our companion animals, we most definitely deprive them of much of the choice and control over the resources in their lives. We are mostly in charge of; when to eat, what to eat, when and where to eliminate, in what, when it's cleaned, when to go out, when not to, what to do, what not to do and with what, what to play with and who, where to sleep and on what and who, the list goes on and on. Being deprived of choice and control is inherently stressful for all animals and with a resource so integral to survival such as food, the stress is greatly amplified.

An animal's welfare is often directly and forcefully impacted by routine and environmental events. A pivotal study done in 2011, found that disruption to routine, resulted in sickness behaviors in healthy cats and that providing an enriched environment to sick cats, resulted in a significant decrease in the number of sickness behaviors and/or symptoms exhibited. A good number of important factors played into the findings. The study found that keeping the time the same every single day for each feeding was paramount to stress reduction. Other notable factors were: providing for the same caregiver, playing classical music, offering interactive playtime, keeping litter boxes clean and in the same locations, and avoiding manual restraint.

Anticipation or looking forward to an event is another factor influencing behavior. Depending on the starting emotional state, how long the wait is, or how desirable the event is, behavior is affected. Anticipation, if it is met with satisfaction, can border on the pleasant. When observing stereotypic behavior in captive tigers (an indicator of poor welfare) pacing prior to meal times is not classified as being stereotypic rather as "anticipatory." This behavior is also apparent in the intertwining anticipatory dance your own cat may do while you open food cans or fill kibble bowls and when your puppy may nudge or carry that food bowl towards you.

When an event is delayed, feelings can border on frustration. If there is a cat or dog analog for that human low blood sugar, "hangry" feeling before a delayed meal, it might be seen with pronouncements, reminders, and gauging human responses to the intrusion of wet noses, presentations of empty food bowls, knocking objects from heights, patting eyes, cheeks, zoomies, or even short tempers around other household pets.

With a resource so integral to survival such as food, anticipatory stress can give way to greater anxiety when meals are delayed and the stress is greatly amplified. The cat or dog cannot fulfill the desire to eat, the need to eat, or a method to procure their own nourishment and is entirely dependent on the "when", "what", "where," and "how" of the owner. Foraging and/or hunting account for a significant portion of how wild animals spend their time, leaving our pets with a whole lot of free time with little to fill it save for what we provide. And then there's time and who's keeping it.

When it comes to timing for life events, all animals have internal or biological clocks and are subject to circadian rhythms which relate to light and dark cycles in their environment and impact behavioral, cognitive, and physical changes in the animal. These cycles are not dependent on clocks or calendars, nature keeps them. And a day, according to nature, is not how long or short as we might think. Circadian rhythms are not a strict 24-hour time span. This rhythm ranges with species and individuals from 23.5 hours to 24.5 hours more or less. Therefore, mechanical time clocks which measure out an exact span of hours, may not be keeping time with an actual day set by circadian and biological time keeping. Along with the anticipation of looking forward to the event, this time discrepancy may help us in understanding why some of our pets may always be on the "earlier" side when it comes to reminding us of mealtimes. It certainly goes to explaining why daylight savings time is a huge interruption in schedules.

Humans are diurnal, most active during daylight hours, unlike cats and dogs who are crepuscular animals, which means they are naturally most active during twilight or dawn (more insight into early morning food requests). Domesticated animals being dependent upon us for food, become accustomed and conditioned to our diurnal routines. We feed according to our own patterns of when breakfast and dinner are for us. Dogs, whether more obliging, stoic, or longer suffering, usually appear less put off by timing changes. While the change may be vexing for them, accommodation, for their humans, is usually more apparent. Why felines are more affected by waiting for a meal may be due to several factors. During the history of our domestication of the cat, we have depended on this animal to often procure their own food whether for own utility or their nourishment or sport. Cats are focused predators with superior hunting skills who mostly target small mammals and reptiles, this hunt for mice has helped us store foodstuffs without a rodent "problem," while supplying the cat, food to catch and the sheer fun of pursuit of prey.

Our pets do adapt to our waking and sleeping patterns, if we maintain them in our environments. A 2013 study which compared nighttime behaviors of cats housed indoors, with cats let out for the evening (nine pm - eight am), found that the indoor cats had established activity patterns of rest and sleep which were in concert with their humans. The outdoor cats were, you guessed it—mainly active at night. Foraging and/or hunting account for a significant portion of how wild animals spend their time. Not having to forage or hunt, leaves our pets with a whole lot of free time with little to fill it save for what we provide.

To offset boredom and provide for the opportunity for your cat or dog to indulge those natural foraging or hunting behaviors, feeding either the morning, evening meal or both, with a puzzle feeder can be intrinsically satisfying in time spent and the opportunity to indulge prolonged natural behaviors cats and dogs need to do.

No more dry food in bowls. I suggest this to promote welfare and provide enrichment to clients as a matter of course. Clients who love their pets and doubt the cat or the dog will figure it out sufficiently and go hungry, may question this is a good idea. It is. They will figure it out, especially if you are using the most appropriate puzzle feeder for their species to channel and satisfy natural behaviors. Puzzle feeders are best used to allow for natural behaviors to express, as in the kibble standing in for the mouse they would paw at or bat around before dining for cats or the bone they would worry or morsels to sniff out for dogs.

For felines, rolling feeders are the most interactive while large stationery tray puzzle feeders are the easiest to learn and use, will not roll under furniture, can be positioned to be shared by more than one cat, and can be placed on an elevated surface away from dogs who may co-opt rolling feeders.

For canines – puzzle feeders to chew, gnaw or sniff food out of. No "slow feeders" which frustrate as solely designed to make food more difficult to get to or not intended for the species (as in silicone mats to lick food from, not a good choice unless the pet is geriatric or has dental issues). Free feeding (food available at their disposal) has multiple benefits linked with reducing stress and giving your pet that something to do that solves problems, provides control over objects in their environment and adds to your pet's overall well-being.

Encourage puzzle feeder use to newbies by demonstrating how they work and filling them with extra special finds. Cats and dogs, like many animals, are social learners, so showing the pet how to use a puzzle feeder is a good idea. Make sure to make the puzzle simple to solve at first. Place food at the edges of tray grids so they are readily scooped or food stuffed close to openings on chew out feeders. Set the openings on the rolling feeders wide initially, so food and treats fall out with minimal effort. Mingle high value treats in both feeders, along with dry food. For chew puzzle feeders for dogs, add wet food, moistened kibble and/or whatever extras the dog can have to heighten value as in plain Greek yogurt, apple sauce, peanut butter, etc. Replenish each day's dry supply daily in the puzzle feeder, making sure to check the supply before bedtime and leave where it is most available to encourage your pet's "at will" engagement.

Keep routine times to feed pets, especially when those times are complimentary with what the animals might choose for themselves. Feed early in the morning and early in the evening to translate to the best feeding times for feline and canine natures - that crepuscular timing, being most active before dusk and dawn (probably when the best hunting is) means early morning and late afternoon meals are best times to feed.

Knowing their natural diets can help to know how to feed them. Dogs are omnivores, like us, cats are "obligate carnivores," needing amino acids and other nutrients only available to them in meat. Feed the best possible meat-based diet, no vegetarian formulas. Offer the best possible food for satisfaction, if your pet won't eat their food, chances are they don't like the taste. Find other choices they do like, rather than waiting them out until hunger forces them to eat. More frustration.

Contentment can help to lessen other stresses surrounding feeding times. Food needs to be more than nutritionally complete to satiate, it must taste good to fully satisfy. Finding the most nutritionally complete and best tasting food can take effort, especially in a self-regulated pet food industry. Keep trying for the pet's sake, they cannot just scrounge or catch a breakfast or dinner they would like to eat for themselves.

For the guardian, already leaving dry food out for their cat or dog at all times ("ad libitum" or free), such pets and owners, can be blissfully oblivious to the whole spring forward fall back routine when it comes to kibble. For those free feeders, do raise the enrichment factor and give your pet more to do by feeding with a puzzle feeder (for all meals and not just treats). For the rest of us it will soon become apparent that your cat and or dog just did not get the memo about daylight savings time. Depending on whatever hour kitty expects breakfast, expect a reminder at the pre-daylight savings time hour, you know the reminders: the pat on the cheek, the plaintive cry, the books toppling off the bookshelf. A short bark, food bowl nudge, paw, wet nose on the cheek are the more typical dog reminders. All maybe what you would not like with an available extra hour of sleep but when was the last time you gave them kitchen privileges? Those reminders are all they have to get fed already.

Disruption in routines can also affect feeding schedules, work from home, days off, trips away, etc., can all impact on when pets end up being fed. Even sleeping in on a weekend. Using puzzle feeders takes some of the pressure off. Keep those puzzle feeders restocked when empty for at will feeding. And do make sure to specify times to feed with friends and pet sitters you are paying. I ask for before nine am with cats. Using and viewing a remote in-home camera, can ensure the time you have contracted for is kept for everyone's sake.

Offset the stresses of schedule changes and add to your human-cat and/or human-dog relationship by providing for interactive play time in the evenings. A minimum of two to three minutes of play with a fishing wand toy (experiment to determine preference and remember to draw objects away from and across your cat's line of vision) will give your cat some predatory fun time. Tug for dogs and a walk before bed is a good way to end the evening and discharge energy. And just as with feeding times, keeping activities on schedule, and keeping routines consistent, can help to give your cat and dog that elusive sense of control over meaningful life events pets can benefit from.

Reset incompatible feeding expectations by keeping mealtimes the same time each day and for each feeding. This is key to stress reduction. Remember, research and practice shows companion animals thrive on schedules.

Begin now to alter changed or erratic schedules through a span of 15-minute adjustment periods over the course of a week. For example, with autumn daylight savings change, all activities were set back one hour including pet feeding times which meant that a cat fed at seven A.M. was now being fed a full hour later. Allow for adapting to the change by feeding 45 minutes earlier for the first two days, then 30 minutes earlier for the third and fourth day and 15 minutes earlier for the fifth and sixth day to recalibrate timing. If your cat petitions just as vociferously with 15-minute delay shift to 10 minutes. You can change the increments to shorter time spans, just allow for a longer span of days to fit the adjustment period into.

Sounds overwhelming? Getting up earlier, feeding pets and then climbing back into bed is always a thing. Using an automated feeder to help you with the time transition is also an option, but do phase it out for the much-preferred puzzle feeder for needed enrichment and maximum pet satisfaction during feedings. Flexibility on your part for adjusting pet feeding times when you are serving and scheduling mealtimes during this transition can only help to make the change less nerve-racking for your pets and in turn, less demanding for you.

References

- Piccione, G., Marafioti, S., Giannetto, Panzera, M., Fazio, F. (2013) Daily rhythm of total activity patterns in domestic cats (Felis silvestris catus) maintained in two different housing conditions. *Journal of Veterinary Behavior: Clinical Applications and Research*. Published online January 7, 2013.

- Stella, J.L., Lord, L.K., Buffington, C.A.T. (2011). Sickness behaviors in response to unusual external events in healthy cats and cats with feline interstitial cystitis. *Journal of the American Veterinary Medical Association*, 238, 1, 67-73

CHOOSING THE RIGHT VET FOR YOU AND YOUR PET

"She's still sulking" front desk veterinary staff told me at six-month old Katie's discharge from spay surgery. The kitten was huddled tightly with face turned away in the back of the carrier. Sulking? A few hours post invasive surgery with a six-inch incision? Traumatized, scared, in pain, not considered? I could only hope that the entire of Katie's experience at the clinic had not been along these lines. When working with animals and seeing stress, discomfort and pain they experience, we can either dismiss or trivialize it, or use what we do know to mitigate it. We cannot make it all go away, no matter how uncomfortable or responsible we may feel.

In the study of behavior, it is emphasized that one cannot, with certainty, ascribe a motivation or thought to another and even less so to a disparate species. Vast research focuses on which behaviors of certain species indicate certain emotions by observation of body language and context. Such a basic understanding of which emotions the animal might be experiencing at certain times are necessary to determine feelings to get closer to discerning animal welfare. Whatever the margin of error, inference of cognitive processing or affect, influence our interactions and interventions with animals. "Critical anthropomorphism" calls for expert observation informed by species specific behavior and empathy. Veterinary and behavior professionals routinely rely on reading emotion in animals to indicate pain, discomfort and more. We need to know what we are looking at, we need to rule out medical concerns before behavioral or training presentations. "Sulking" is not on any official list; such an observation was off on all counts, save for being an indication of negative bias. Invasive procedures, surgeries, restraint, and confinement are overwhelming, traumatic, and distressing situations for animals and why Sophia Yin called her pioneering programs" Low Stress" and why there is no such thing, ever, as "Fear Free".

I took Katie home along with her littermate brother, Camden, who had been neutered that same day. They were both wobbly and disoriented from the sedation and procedures. It seemed safer to leave them in their carriers for another hour or two rather than giving them access to the whole apartment. Katie showed no objection to this arrangement and, foregoing all other known resting spaces, chose the carrier to retreat to for the next two days (the door of which was now removed). That association of carrier as safe haven, indicates much of the degree of trauma experienced. (To this day, Katie reacts to being restrained by a stranger with a hiss. She remembers.) Seeing this and other indications of discomfort in her body posturing, on the next day's follow up call to the vet practice, I asked for more pain medication than what had been dispensed and expressed concern for the behavior I was seeing. The office countered that such behavior is typical and not to be concerned about. I could not help, for Katie and any other cat, to say, it was of definite concern for what this behavior is telling us, exactly that they are not OK and we, who can help here, need to step up that game and not rely on endurance or learned helplessness.

Katie's spay surgery was not the first such surgery for cats I have lived with or have known. It was the first surgery where the incision was over six inches long. And the first where the animal's welfare was easily and matter of factly, dismissed. Incisions of shorter length benefit with lessened trauma and shorter healing time. Her discharge notes cautioned against activity post-surgery, a heightened caution, and no doubt a concern because of her many sutures. Camden wore a padded Elizabethan collar, while Katie wore a softer donut version. As she felt safer at home and more comfortable, Katie ran, jumped and a stitch came undone and then two. The vet's office then recommended cage confinement. To prevent added stress I didn't cage her, instead relying on a solo restriction to one room and more barrier protection to prevent access to her incision.

I had zero observations of Katie investigating her incision (Camden was a different story, it's all he wanted to do, find out what happened down there), still, out of an abundance of caution, she wore another added layer along with her donut collar. A newborn baby onesie, found in a local store, seemed suitable, the armholes were off but it worked. Worked along with the best medicine of all, one we have in infinite supply, love. Lots of it, and time given to sleep and sit with them quietly, pet softly and speak kindly.

But, here's the other thing, this barrier protection with cats, dogs too, is plain awful. Cats, who groom daily, on multiple occasions, for extended time periods, are prevented from the behavior, with good reason but they don't like it. The collar/cone varieties routinely provided, massively interfere with not just grooming but whiskers and proprioception. The first several days of navigating an environment are full of walking into walls, and just try and eat and drink water with one of those on. Those cute surgery recovery suits (four different ones were ordered online, none of which fit) are a better idea, if you can find one that will stay on or works on a cat's body. Communicating this, asking the practice, the place where the surgeries are performed, if they might have better options for cones or suits or after care, I heard there is not enough call for it. Perhaps because clients are unaware of options.

I first thought about becoming a veterinarian before deciding on being an animal behaviorist. I was not convinced I had the nerve for veterinary work. I knew I could give a cat a pill, even if they did not want one, catch the unsuspecting feral cat, walk shelter dogs, care for caged rehabbing wildlife, including TNR patients and train captive exotics. Much of this was at their expense, such condition not to be taken lightly. Add to that, the full-fledged immobilization, steady hands, and focus required to place catheters, draw blood, or fine needle aspirations? All those other seemingly impossible and delicate tasks to complete with sheer determination and little to no patient acquiescence? How did they do it? How did they manage knowing the mostly inevitable patient refusal responses to come? As far I could see, with most animals, they seldom agreed that a visit to the vet was what they wanted. They fought and hid and resisted most, if not all, of it. And the veterinarians and technicians that had the touch? The way to handle an animal with skillful hands and soothing voice? What made them different? An innate sense? A learned ability?

While I searched for and patronized (save for emergencies where choice is not an option), the most compassionate of practitioners for my pets, it appeared that in some practices, helping them medically, included hurting them in the prodding, poking, pinning, restraining, scuffing and even more unwelcome handling and holds. The good thing was, and is, there were and are, clinicians who listen to pets, who care what the animals say in response. There are those providers with empathetic skill and scholarship to practice veterinary medicine and do it with thoughtful consideration for patient and client. There are veterinarians, you can feel lucky to have found and happy to keep coming back to, vets to return to over and over.

Whether your own pet or someone else's, first steps in behavior and training work is ascertaining wellness and ruling out the medical. Signs of pain and illness can present in "problem behaviors." A veterinary work up is invaluable in assessing the health of the pet and in providing any necessary supports. You also need to have vets you can rely on and communicate with. It is worth it for, clients and pets, to find the right practice to feel good about patronizing and about the services and care being provided:

Ask other likeminded pet owners which vets they prefer and which they do not and why. Personal referrals count, drawn from our own experiences we are often keen to share the good interactions and the bad when it comes to our pets.

The Animal Humane Society and PetFinder.com suggest asking around for recommendations and opinions on local vets. Every vet has a different approach as well as different strengths and weaknesses. Identify what you need in a veterinarian: do they need to treat a variety of companion animals or be a dog or cat specialist? Are they force free? Is guardian being present with pet through the exam supported? Is being open to and able to practice complimentary medicine important? What size facility do you need? Small and personable or up to the minute with the latest technology?

Patronize practices that advertise their *Low-Stress Handling, Cat Friendly* or *Fear-Free* certifications. Advances in veterinary medicine now include growing recognition that trauma experienced by an animal during a visit can be balanced by more humane and thoughtful pacing and handling in response to the pet's body language during the visit itself. A less distressing experience reduces adverse expectations to future visits for both patient and client and increases the likelihood of return visits. The late veterinarian turned Animal Behaviorist, Sophia Yin pioneered a revolution in low stress handling and restraint and literally wrote the book on how veterinarians and technicians need to practice promoting welfare along with wellness in animal care. Commenting on current handling practices in 2009, Yin said:

> "If you went to a physical therapist and they just grabbed you and shoved you in a chair, you wouldn't trust them. And yet, that's what we do with dogs and cats all the time. We just move them. Nobody ever taught us the right way to do it." (Yin, 2009)

Sophia Yin's behavior book, did exactly that, taught the right way to do it. Animal expert, Steve Dale knew Yin personally, writing on the monumental significance of her contribution to veterinary medicine and animal welfare after her passing, he noted:

> "Yin conceded that even professionals continue to use forceful techniques. In 2005 she told me, "I want to change the way some veterinarians, technicians, dog trainers and groomers deal with their clients – it's animal abuse and it's wrong."

> In 2009 Yin released a 470-page book (it includes 1,600 color images and an instructional DVD), *"Low Stress Handling, Restraint and Behavior Modification of Dogs & Cats"* (Cattle Dog Publishing, Davis, CA; $118). By offering gentle handing alternatives to forceful methods, countless veterinarians, veterinary technicians, groomers and dog trainers no longer scruff cats by the back of their necks or forcibly turn dogs on their backs to simply clip their nails." (Dale, nd)

Sophia Yin's untimely death was instrumental in multiple ways, her legacy raised awareness of the need for highlighting understanding and interventions to prevent the growing rate of suicide in the profession. Her dedication and life work gifted the animal services industry the vital, necessary importance of humane, skilled handling and welfare in leaving behind the essential foundational manual. In addition to being generously illustrated, the book comes with a DVD with live demonstrations to model from, one I watched countless times, incorporating it into classroom lectures and presentations and learning from it at each turn. Her book remains required reading for any professional or handler who interacts with animals.

Yin's seminars and presentations included certification programs for clinicians to be versed in low stress handling and restraint practices. Of immense benefit to Low Stress Restraint and Handling, is the recognition, starting with the name itself, that there is no way to completely remove the stress or fear of any unsolicited or intrusive procedure, handling, or practice and for the impetus to be on the clinicians to learn a better way than outdated and traditional methods. Other programs promote welfare focused clinical interactions. Among these are, the certification and program of The American Association of Feline Practitioners' *Cat Friendly Practice* and the more popular *Fear-Free Pets*. International Cat Care, out of the United Kingdom, also has a program offering insightful, valuable, and countless free online webinars and videos to animal professionals in the veterinary services and others who might be interested.

While Low Stress Restraint and Handling was welcome and accepted by a segment of the veterinary community, and the popularity of Fear Free practices has increased, the impact is not sizeable, based on reports and field observations. Formal statistics are not available for the number of practices using low stress methods versus older, more traditional methods and behavior courses are not a required part of a veterinary education. Several vet schools are now introducing short segments in behavior, but concern remains as to what happens in the real world when applying the material. New veterinarians joining practices may have the best intentions, have learned the best procedures, but without the current culture supporting their use, they may not be able to implement them. Such a fact is not lost on the educators at Lincoln Memorial University College of Veterinary Medicine writing on instituting a new educational requirement in their curriculum:

"If we taught students to use gentle handling techniques as part of the formal curriculum, only for them to observe clinicians performing conflicting methods later as part of the informal or hidden curriculum, students would soon learn that veterinarians treat animals differently than they have been taught." (Johnson, Hunt, et al., 2021)

It is time consuming to adjust clinical approach in pace with monitoring of body language and tailoring handling to respond to and temper an animal's reactions. Desensitization and counter-conditioning takes time. Protocols and methods that depart from the typical routine can add minutes which add to hours in a day or call for an exam to stop when the animal has passed a level of the Fear Anxiety and Stress scale of body language relied on for handling finesse in Fear Free.

Going over time allotted to a visit with a full schedule and a waiting room full of patients and clients can be problematic. It is quicker and often easier to use the older more traditional and oftentimes forceful methods. Clinicians know this too. During my back-office visits to groomers, doggy day cares, shelters, and at the lunches and dinners at veterinary and other behavior conferences I would hear one phrase repeated often - "I don't have time for behavior."

That giving time to behavior during a visit, is in fact more efficient, not just for the ethical approach, but for future visits where the animal is less likely to resist handling, is not considered sufficiently. Those future visits and owner buy-ins are serious considerations along with current research on how to mitigate stress before, during, after exams and visits. Value of owner presence, low stress handling and behavior knowledge, are highlighted in the workshop and presentation I give to veterinary and animal professional called *"Making Time For Behavior."*

It would be remiss not to say again and again, that there are those practices already putting low stress handling and restraint into place, and practices that have never deviated from using any other method without having done the certifications and the workshops. We need to keep these practices going with our patronage. Find them.

Are the offices clean? Is the front office staff helpful or off-putting and intimidating? There is an emotional dynamic to consider with pets and clients. Research shows stress can start for both even before leaving home. You already know your pet knows when it's time to go to the vet and that is before the cat carrier even comes out. Even how close your car can be parked in the parking lot matters, whether you have to wait outside matters, if the waiting room has separate waiting areas for dogs and cats matters (In spaces where this is not the case, elevating cat carriers off the floor and away from dogs can help).

Staff adds to the energy. Establishing a caring relationship for your pet starts with everyone you or your pet interacts with being both respectful and compassionate towards the both of you, on the phone, in email and in the clinic. Staff needs to care that you care. PetFinder.com offers additional questions on the office and staff:

> "Do they acknowledge you when you walk in or are you ignored? What is the overall appearance of the clinic? Is it clean? Odor free? What is the attitude of the staff toward the other clients who may be present? How about to those on the other end of the phone line? You can learn a lot by just observing."

Ask for a tour of the entire facility, the whole space, including the back. PetFinder.com has additional tips:

> "It is also legitimate to request a tour at a time that is mutually convenient. There may be times of the day when a tour is not advisable but your request should be granted at some point."

On your tour, look around and make sure to pay extra attention to the condition of the areas not normally in public view, these should stand up to the same scrutiny the front office does. Cleanliness and cage arrangements count equally. Cats and dogs should be separated and cat cages elevated.

If your pet needs to be hospitalized, does the practice permit visits? The College of Veterinary Medicine at Ohio State University advocates visiting as often as the clinic allows. Make sure the clinic does. Being hospitalized is extremely stressful for your animal, visiting will offer the comfort so necessary in supporting your pet during recovery time and healing.

Is appropriate time dedicated to each visit? Do technicians and doctors take the time to begin a relationship with you as the authority on your pet before commencing an exam? An initial conversation should be with you from both the technician and veterinarian on your concerns for your pet and inquiry made into your pet's temperament, handling preferences and prior vet experiences. You are the expert on your pet. Remember, you are the advocate for your animal. Your pet does not converse in English -you really are their voice.

Are the veterinarians (and the technicians) good listeners and are they willing to take the time to answer questions about your visit and pet? CompendiumVet.com urges vets to utilize the ask-tell-ask technique:

> "This approach is based on the notion that client education requires identifying what the client already knows and building on that knowledge, it shows that you are willing to listen to and negotiate the client's agenda".

Time is short in any visit, keep the focus on your end on the animal, situation, and reason for your visit. Tangents on off topics are easy to go on but add little to the consultation.

As with any relationship, what comes first is respect. This goes both ways. Clients must be as respectful with veterinarians and technicians as they would like them to be with them and their animals. Vets and techs are there to help pets. Respect that and go from there. Take deep breaths. Establish a relationship where you are the person someone wants to speak to.

Any professional, approaching your animal should first stop, address the pet by name and offer a soft touch on consent before anything else. A pause on entering the room and announcing a new presence by greeting the animal allows the pet to alert to their presence first before any additional movement forward. Cats and dogs differ in how they react to human voices. Dogs are mostly good with the jolly camper voice, while cats prefer our softer, elevator voices.

Keeping surroundings as familiar as possible can also ease anxiety for your pet, a worn article of your clothing placed in your pet carrier can help to ease fretfulness.

Where exactly are patients being examined and how? Exams done with the least disturbance and to the pet's greatest comfort can lessen stress. Dogs rarely want to be on the exam table and cats rarely want to leave a carrier unless it is to go home in one. Why increase the stress when we can easily adapt where we examine them? This is further emphasized in the following study:

> "Examine patients where they are most comfortable; cats and small dogs may feel more comfortable being examined on the veterinarian's lap. Cats often prefer being examined in a structure with sides (e.g., on weighing scales or in the bottom half of a carrier) and can be partially hidden under a blanket. Many dogs are more comfortable being examined on the floor, rather than on the exam table; or in some difficult cases, outdoors in a secure environment" (Lloyd, 2017)

Large dogs should be examined on the floor and for cats. Your vet can and should be performing most of the exam with the cat in the carrier. The American Association of Feline Practitioners suggests:

> "With respectful handling, even fearful cats are often calmer and easier to work with if at least part of the examination is done within the bottom half of the carrier" (the one the cat came to the practice in).

Make sure the carrier you bring your pet in can be easily taken apart either by unzipping the top half or removing it. Practice this at home so you can help.

Avoids taking your pet to "the back" -Are you present for routine procedures: vaccines, blood draws, etc., or are these done in the back room? Be extremely wary of the practice keen to whisk your pet away for routine procedures. Moving an animal, especially a cat, to yet another strange and unfamiliar room adds more discomfort to the visit and prolongs absence from an owner. While this may be thought to be easier for personnel working with your pet, remember that pets are often fearful of procedures and past experiences. As the client, your insistence on staying in the room can be the determining factor. Make this a non-negotiable for your pet.

The American Association of Feline Practitioners further advises vets that performing these procedures in the exam room:

> "can comfort the client and remove the fear of the mysterious "back room." When you remove the cat from the examination room, your client wonders, "What is being done to poor Fluffy that couldn't be done here?" This anxiety worsens if they can hear "yowls" from their cat or even sounds from dogs and other cats that are also in the "back." Always offer to have the client leave the room if they seem uneasy or uncomfortable, as many are happy to do so. It is better for the client to leave the room than to increase the cat's stress moving them to the "back," as it can take 10 minutes for a cat to acclimate to a new environment."

The exam, conversations, handling, and most procedures, should always be done with your assistance or with you present to offer additional assurance to your animal. Your presence and oversight is necessary for the welfare of your animal, staying with them enables this. Find the practitioner as committed to that welfare as you are. Extensive research continues to support that your presence and support is comforting in a stressful situation. Be that support. Animals associate most vet visits with intrusive pokes, prods, and painful injections from strangers - all in the name of health but still distressing and uncomfortable. The literature shows fear is reduced by owner assistance with handling and presence in multiple studies:

> "Various strategies are recommended for reducing fear in dogs during veterinary care, and we found that previous exposure to some of these strategies (i.e., providing treats, acclimating to the exam room, allowing owner assistance with handling) was associated with lower dog fear scores...Similar to the current results, previous studies have found that dog fear levels are reduced by owner presence (Stellato et al., 2020) and

interaction with the dog (Csoltova et al., 2017) in the examination room. " (Stellato, Flint, et al., 2021).

You would not send a more dependent family member in your care off for a vaccination without your hand to hold, or your presence in the room. You have the right to ask these procedures be done in front of you. Competent, caring professionals will be willing to work with you and your pet. Be mindful of your own behavior here, if you cannot stand the sight of blood or faint at the sight of a needle, please look away and do not act as if the sky is falling because it is not.

Study after study has corroborated multiple findings that low stress handling and owner presence was beneficial to welfare in the clinic environment. Specifically, studies show the social support of the owner can offset the stress of the visit. One 2017 study found that signs of stress dogs exhibited at the vet office included increased heart rate and lip licking, these signs were reduced significantly when their owners were present, petting and talking to them. In addition to lowering heart rates, owner presence lessened the number of attempts to jump off the exam table. According to the study:

> "owner—dog interactions improve the well-being of dogs during a veterinary examination."

> "Owner separation coupled with physical examination location can result in clinically significant increases in perceived stress in cats, and compromise vital sign assessments. Whenever possible, physical examinations and procedures should take place with the owner present with separation from unfamiliar dogs and cats." (Griffin, Mandese, et. al., 2021)

There may be times when the emotion of worrying over how handling is being conducted with your pet overcomes you, be aware of the emotional contagion and know that you have a decision to make. Either ask for the exam to stop or step away from the room at that point. Research shows where owner nerves can make things worse in this scenario:

"Only owner-reported nervousness during different aspects of the veterinary visit was associated with aggression scores. If the owner reported not being nervous during nail trims by the veterinarian at the veterinary clinic, aggression scores were lower, whereas if the owner reported being nervous when the veterinarian handles their dog, aggression scores were higher. Owner nervousness in response to the following situations were not significantly associated with aggression scores at the clinic: putting the dog in the car, transporting the dog to the clinic, the dog staying overnight at the clinic, the dog being weighed, and the dog receiving a vaccination." (Stellato, Flint, et. al., 2021)

It is not just that we want to be with them throughout the visit, it is that being with them makes the visit less traumatic and a healthier one for them. Stay there.

Owner presence is beneficial unless owners and staff correct a dog for responding fearfully or reacting to aversive situations. Aggression is mostly fear based, especially in veterinary settings where agency and consent is curtailed. Do not punish it. Listen and work on changing the setting to lessen the fear. Stopping or giving distance when asked for can make all the difference. A study on predictors of fear and aggression-based behavior found that a history of negative experiences at the veterinary clinic set a stage for fear for future visits and aggressive behavior going forward, especially when exams fail to consider time considerations and handling:

"Primary predictors of aggression were previous exposure to particular examination strategies in the clinic (e.g., towel restraint and shortened examinations), and the use of particular strategies for managing both unwanted behaviours generally (i.e., ignoring unwanted behaviour and using a spray bottle for correction) and fear or aggression within the clinic (i.e., ignoring aggression and forcing interactions)." Stellato, Flint, et al. 2021

How is handling and restraint being done? Over restraint is highly problematic, as is scruffing. None of us are mother cats carrying kittens around. At home or in the clinic, no matter what you hear, please know, scruffing is an outdated and cruel way to restrain any animal. Make sure clinicians and or the practice avoids this.

The American Association of Animal Hospitals and International Cat Care are among the organizations that condemn the use of scruffing as inhumane. Alternate methods of gentler restraint are advocated and instruction on how to implement them are freely available online from multiple sources, including the respected ASPCApro and International Cat Care.

When your concerns are voiced on scruffing, they may be dismissed as it being a necessary practice and one cats are used too. It is not. Mistakes, it has been said, serve us best, if as we make them, we learn so as never to repeat each and if they open a door to insights, to new ways of seeing, that might be the best part of them.

Again, none of us is a mother cat moving a very small kitten. Taylor, in writing targeted for veterinary practices, but informative for all, outlines in summary the dangers and disadvantages of the practice of scruffing:

> "• Discomfort/pain caused by the handler during the scruffing process
>
> • Scruffing is interpreted negatively by the cat as it is used during mating and fighting as a means of control
>
> • It is only used by a queen to move her kittens to a place of safety – not as a form of control. This is usually in the kittens first few weeks of life, so they weigh much less than an adult cat and have a more flexible neck
>
> • Scruffing does not allow the cat to feel in control, causing further distress in an environment where the cat may already feel vulnerable
>
> • Increases the likelihood of injury to staff
>
> • Exacerbates feelings of anxiety, fear, and frustration, which may influence how the cat behaves at the next visit

• Breakdown in the relationship between staff and the patient, making examinations and treatment difficult

• Other methods such as a towel wrap are recommended as a less stressful and more effective means of restraint for cats

• Scruffed cats are more likely to demonstrate negative behavioural and physiological responses compared to cats that are restrained more passively" (Taylor, 2020)

Practices and handler may resort to more forceful methods in the interest of "effectiveness" and "efficiency." And while overpowering an animal may work for that visit, that animal has learned that the next visit is to be feared and resisted to an even greater extent. In researching the effects of over restraint as being more of a time saver compared to gentler methods of restraint, a separate study by Moody, Picketts et al. showed that more force, is in the end, less effective and stressful.

A note on towel wraps – not every animal, especially cats, like them, often because a considered and low stress approach is not put into practice when using them.

The use of force at home or at the vet's office can make a current and future vet visit worse not better. Respect what your pet is telling you at the vet's office and make sure the staff does too. An owner or clinician, punishing or ignoring a distance increasing behavior can establish a history for the pet where the vet visit becomes even scarier over time. Do not permit or perform scolding, manhandling or admonishing in the clinic:

"In relation to dog behaviour at the veterinary clinic, aggression scores were higher if owners reported that they either ignore aggression displayed towards the veterinary staff, or force the dog to interact if their dog shows fear towards the veterinary staff."

"Use of particular training techniques by owners was associated with reported aggression scores at the veterinary clinic. If owners reported responding to unwanted behaviours in general contexts by using a spray bottle, aggression scores in the veterinary context were higher; however, if they reported ignoring unwanted behaviours in these contexts, aggression scores were lower." (Stellato, Flint, et al. 2021)

No one, including dogs, want or need to be reprimanded for defensive behaviors or emotions. Instead, focus on reinforcing for the behaviors you do want rather than punishing the behaviors you do not. Any appeasement behavior is just that, an effort to stop correction and not an admission or recognition of guilt.

Does the vet explain procedures, medications, vaccinations, etc. and get your permission before commencing treatment? Make sure you are clear on and have agreed to what the plan of care is before it is underway or has happened. Ask about the side effects of the medications in addition to their purpose. Where side effects can be deleterious (for instance, affecting balance or increasing aggression) ask about substitutes. Check which vaccinations can be less taxing when spread over a course of visits. Costs can add up, find out what is a priority, what payment plans are offered and if any procedures can wait 'til the next visit if your budget is too tight.

Home veterinary visits are another option, if available in your area, and with the right vet can be an excellent choice. This is true especially for cats, who are highly reactive to changes in environment and most comfortable in their home settings. Containing your cat in the carrier shortly before the vet arrives is a good idea to prevent that scramble and stress when they do arrive. A 2017 study on home visits, highlighted benefits, finding:

> "A significant difference was found in blood pressure, heart rate, and respiratory rate between the home and veterinary hospital environments" (Quimby, Smith, et. al, 2017)

Not every practice will be a match. Do vote with your feet and make sure a complete set of your pet's records go with you. If things are not right, first, always try and communicate this clearly and respectfully to your vet and staff. Stay on topic substantively and avoid personal attacks. Sometimes differences cannot be resolved but often simple, clear communication is key, the Humane Society writes:

> "If you feel that your veterinarian isn't meeting your needs as a client or the needs of your pet as a patient, it may be time to find a new one. But sometimes simple misunderstandings cause conflicts, which you and your vet can resolve by talking things out and looking for solutions."

Communication requires two parties to work. Make sure the vet you choose is both willing and able to establish rapport and respect with you and your pets. CompendiumVet.com advises vets:

> "A great deal of communication in small animal practice involves providing information, although this does not mean that communication should be largely one-way."

Caretakers are charged with being informed in knowing what to look for when locating the best veterinarian, practice and keeping present to advocate throughout the visit with our animals. With time, some groundwork, holding fast to what you know is best for your animal and a bit of luck you are can find the best partner in caring for a pets' health. And do not forget - according to the Humane Society:

> "You're doing more than searching for a medical expert. You're looking for someone to meet your needs and those of your pet, a doctor who has people as well as animal skills".

References

- Humane Handling of Cats https://www.aspcapro.org/catholds

- Humane Restraint of Animals https://www.aaha.org/about-aaha/aaha-position-statements/humane-restraint-of-animals/

- Csoltova. E., Martineau, M., Boissy, A., Gilbert, C. (2017). Behavioral and physiological reactions in dogs to a veterinary examination: Owner-dog interactions improve canine well-being. *Physiology & Behavior*, 1(177) 270-281.

- Dale, S. https://positively.com/contributors/one-month-two-great-losses-in-veterinary-behavior/ retrieved 2/20/2023

- Griffin, F. C., Mandese, W. W., Reynolds, P. S., Deriberprey, A. S., & Blew, A. C. (2021). Evaluation of clinical examination location on stress in cats: a randomized crossover trial. *Journal of Feline Medicine and Surgery*, *23*(4), 364–369. https://doi.org/10.1177/1098612X20959046

- Pledge to Go Scruff Free https://icatcare.org/our-campaigns/pledge-to-go-scruff-free/

- Johnson, J.T., Hunt J., Perkins, J., Nibbe A.N. Teaching Gentle Canine and Feline Handling as a Veterinary Clinical Skill [version 1]. *MedEdPublish*, 10:158 (https://doi.org/10.15694/mep.2021.000158.1)

- Lloyd J. (2017). Minimising Stress for Patients in the Veterinary Hospital: Why It Is Important and What Can Be Done about It. *Veterinary sciences*, *4*(2), 22. https://doi.org/10.3390/vetsci4020022

- Moody, C., Picketts, V., Mason, G., Dewey, C., Niel, L. (2018). Can you handle it? Validating negative responses to restraint in cats. *Applied Animal Behaviour Science*. 204. 10.1016/j.applanim.2018.04.012.

- Quimby, J.M., Smith, M.L., Lunn, K.F. (2017) Evaluation of the Effects of Hospital Visit Stress on Physiologic Parameters in the Cat. *Journal of Feline Medicine and Surgery*. 2011;13(10):733-737. doi:10.1016/j.jfms.2011.07.003

- Stellato, A.C., Flint, H.E., Dewey, C.E., Widowski, T.M., Niel, L. (2021). Risk-factors associated with veterinary-related fear and aggression in owned domestic dogs, *Applied Animal Behaviour Science*, Volume 241

- Taylor, A. F. (2020). Literature review on the handling and restraint of cats in practice and its effect on patient welfare. *Veterinary Nursing Journal*, 35(6), 162–166. doi:10.1080/17415349.2020.1754986

- Rodriguez, P. (2009) Problem Solver: Sophia Yin, DVM, Veterinary Practice News
https://www.veterinarypracticenews.com/problem-solver-sophia-yin-dvm/ retrieved online 2/21/2023

ACKNOWLEDGEMENTS

This is the part of the book where the author gets to acknowledge the parties who have supported and contributed to the work. For my students, clients, readers, for the rescuers, researchers, scientists, trainers, and behaviorists, for the behavior nerds - friends, and fellow learners, where the conversations never end, all - who have taken the time to do the considering, thinking, and wondering about the welfare of the animals, about how we share and shape the world for them, and how their world matters just as much for them. For the questions asked and possible explanations that give rise to behavior matters, the ones interested, dedicated, motivated to look for a deeper understanding for the companion animals we share our lives and homes with. And to make the changes for the both of you. I am indebted to you for the animals and I thank you. And to my own animals, from as far back as I can remember, for forever being fascinated with how you see the world and what you had to say about it. All the words and thinking begin with you. And the rest of my family, I love you, you already know how much I owe you.

ABOUT THE AUTHOR

Frania Shelley-Grielen is an animal behaviorist, trainer, author, speaker, and educator who holds a Masters Degree in Animal Behavior and Conservation from Hunter College and a Masters Degree in Urban Planning from New York University. In addition to working with cats, dogs, birds and their people, she writes extensively on animal behavior, training, and humane management for urban wildlife. Her first book on behavior was: "Cats and Dogs Living with and Looking at Companion Animals from their Point of View." Frania founded AnimalBehaviorist.us in 2009, to share her work on how welfare based, force free, science focused strategies from the canine and feline point of view can be more effective and make everyone happier, including the humans.

Frania is licensed by the NY State Education Department to teach the Pet Care Technician program in vocational schools, a registered therapy dog handler, certified *Doggone Safe* Bite Safety Instructor, and professional member of the Pet Professional Guild and International Society for Applied Ethology.

Frania's career in animal behavior, includes five years developing and teaching full-time, New York City's only Pet Care Technician program at a post-secondary vocational school for individuals with disabilities. Her work in the classroom and behind the scenes in pet services gave Frania unique insight into this industry and the need for higher standards for both worker education and services. This experience led her to create PetCenterEd, Inc., an innovative not for profit, science and welfare-based learning and services center for people and pets. Frania also taught the ASPCA's Fundamentals of Dog Care course for the Houlton Institute.

Frania has presented on parrot behavior at the Long Island Parrot Society, on scent and walking the urban dog at the Harlem Pet Services Fair, lectured on animal communication for the undergraduate course in Animal Behavior at Hunter College, offered webinars for The Pet Professional Guild including, "Animal Welfare, From the Five Freedoms to the Five Domains and Beyond" and "To click or not to click the how to train question". Her webinar on aggression to affiliation in multi cat households was part of Pet Professional Guild's 2020 Geek Week. She presented further on this topic at the 2021 International Society for Applied Ethology conference. She presented in October 2021 on "Making time for behavior, bite and scratch prevention for veterinary professionals" at a private veterinary practice, and "Feline Aggression" at the University of Pennsylvania School of Veterinary Medicine.

She lives and works in New York City.

www.ingramcontent.com/pod-product-compliance
Lightning Source LLC
Chambersburg PA
CBHW071300140726
47996CB00005B/1571